Springer Tracts in Civil Engineering (STCE) publishes the latest developments in Civil Engineering - quickly, informally and in top quality. The series scope includes monographs, professional books, graduate textbooks and edited volumes, as well as outstanding PhD theses. Its goal is to cover all the main branches of civil engineering, both theoretical and applied, including:

- Construction and Structural Mechanics
- Building Materials
- Concrete, Steel and Timber Structures
- Geotechnical Engineering
- Earthquake Engineering
- Coastal Engineering; Ocean and Offshore Engineering
- Hydraulics, Hydrology and Water Resources Engineering
- Environmental Engineering and Sustainability
- Structural Health and Monitoring
- Surveying and Geographical Information Systems
- Heating, Ventilation and Air Conditioning (HVAC)
- Transportation and Traffic
- Risk Analysis
- Safety and Security

Indexed by Scopus

To submit a proposal or request further information, please contact:
Pierpaolo Riva at Pierpaolo.Riva@springer.com (Europe and Americas)
Wayne Hu at wayne.hu@springer.com (China)

Springer Tracts in Civil Engineering

Nuh Bilgin · Cemal Balci

Critical Issues in Selecting Conventional and Mechanized Tunnelling Methods

Nuh Bilgin
Department of Mining Engineering
Istanbul Technical University
Maslak, Türkiye

Cemal Balci
Department of Mining Engineering
Istanbul Technical University
Maslak, Türkiye

ISSN 2366-259X ISSN 2366-2603 (electronic)
Springer Tracts in Civil Engineering
ISBN 978-3-031-89113-7 ISBN 978-3-031-89114-4 (eBook)
https://doi.org/10.1007/978-3-031-89114-4

This Springer imprint is published by the registered company Springer Nature Switzerland AG
The registered company address is: Gewerbestrasse 11, 6330 Cham, Switzerland

This book is dedicated to our lovely wives
Ayfer Bilgin and Nurgül Balci
and our beloved children
Damlanur Bilgin and Cem Eren Balci

Preface

ITA-AITES' (International Tunneling Association) survey shows that the global output for tunnels and underground space construction in 2019 was €125 bn, up from €86 bn in 2016, representing growth of between 7 and 9% each year. In summary, in the future, we will need to drive more and more tunnels and underground facilities efficiently and economically. Unfortunately, some projects don't end on scheduled time and with a planned budget due to difficult geologies and needing the profit from past experiences. There are excellent books published on tunneling. However, it is scarce to find books on the shelf of a bookstore on the comparative studies between conventional and mechanized tunneling. This book, which is intended to fill the gap in this respect, is a result of several years of experience of the authors and an intensive literature survey. The main topics covering the book are the geological and geotechnical parameters affecting tunneling methods, summarizing conventional tunneling and mechanized tunneling methods, the factors affecting the choice of tunneling methods such as the cost of the initial investment, the length of the tunnel, project scheduling, time for mobilization, emerging new technologies. Some examples of changing the tunneling method from conventional to mechanized tunneling or vice versa during the same ongoing project and hybrid tunneling methods in the same project are also given. The last chapter resumes the innovations made for the tunneling industry, summarizing advancements in safety, non-circular TBMs, robotics, new instrumentation, new materials, and methodologies to decrease carbon footprint.

This book is the fourth in this series. The first was *Mechanical Excavation in Mining and Civil Industries*, published in 2014; the second was *TBM Excavation in Difficult Ground Conditions, Case Studies from Turkey*, published in 2016; and the third was *Practical Management of Tunneling with Tunnel Boring Machines*, published in 2024. We believe that the fourth book will be a supplementary contribution to the success of tunneling.

Maslak, Türkiye

Nuh Bilgin
Cemal Balci

About the Authors

Nuh Bilgin graduated from the Mining Engineering Department of Istanbul Technical University. He obtained his Ph.D. at Newcastle upon Tyne University, UK, in 1977 on "Rock cutting mechanics of high-strength rocks." He was appointed a full-time professor at İstanbul Technical University (ITU) in 1989. He worked one year at Colorado School of Mines, USA, in 1994 as a visiting professor and Fulbright scholar and one year at the University of Witwatersrand, South Africa, in 1989 as a visiting professor.

He was formerly the head of the Mining Engineering Department at Istanbul Technical University and a Senator representing the school in the university's Senate. However, he currently works as a Mining and Tunneling Company consultant and is the Chairman of the Turkish Tunneling Society. He has published over 150 papers on Tunneling and Mining, including one book in Turkish and five books in English.

Cemal Balci graduated from the Mining Engineering Department of Istanbul Technical University (ITU). He obtained his Ph.D. at ITU Mining Engineering Department, Mine Mechanization and Technology Division in 2004 on "Comparison of Small and Full Scale Rock Cutting Test to Select Mechanized Excavation machines".

He joined the research group at Earth Mechanics Institute, Colorado School of Mines, USA between 2001 and 2004 as a researcher. He has authored over 100 publications including three books in English with more than 25 of these appearing in SCI-indexed journals. His works have been cited over 2150 times, and he holds an h-index of 19 in the mining engineering discipline and supervised the thesis of 2 Ph.D. and 5 M.Sc. students. He has been carrying out many national and international implementation and research projects funded by national organizations including The Scientific and Technological Research Council of Türkiye (TUBITAK) and doing consultancy. He is currently working as a full professor and teaches mechanization in mining and tunneling excavations at ITU and board member of the Turkish Tunneling Society.

Chapter 1
Critical Issues in Selecting Conventional and Mechanized Tunneling Methods

Abstract This book is written to discuss the critical issues in selecting conventional and mechanized tunneling methods and lessons learned from the past. The book covers the following main topics: geological and geotechnical parameters affecting tunneling methods, summarizing conventional tunneling and mechanized tunneling methods, the factors affecting the choice of tunneling methods such as the cost of the initial investment, the length of the tunnel, project scheduling, time for mobilization, emerging new technologies. Some examples of changing the tunneling method from conventional to mechanized tunneling or vice versa during the same ongoing project and hybrid tunneling methods in the same project are also given. The book is unique in the subject, aiming at civil, geology, and mining engineering graduate and undergraduate students, researchers, and those dealing with the tunneling business.

1.1 Introduction

The population is increasing rapidly in urban areas. By 2050, it is predicted that 70% of all people will live in cities. Parallel to this, the need for a better life will necessitate tunnels for rapid mobility, tunnels to solve energy, water, and sanitation problems, underground construction for protection against disasters, to satisfy the need for infrastructure for food, etc. Broere (2016). This is reflected in the current worldwide tunneling market, which is growing fast. ITA-AITES' (International Tunneling Association) survey shows that the global output for tunnels and underground space construction in 2019 was €125 bn, up from €86 bn in 2016, representing growth of between 7 and 9% each year, ITA (2019), Thomas (2020). In summary, in the future, we will need to drive more and more tunnels and underground facilities efficiently and economically. In light of these, this book is prepared to summarize the basic requirements of conventional and mechanized tunneling comparatively and to give several examples from the past to make it clear to understand the problems encountered while applying these two tunneling methods.

There are two tunneling methods: conventional and mechanized tunneling. *Conventional tunneling can be defined as the construction of underground openings*

N. Bilgin and C. Balci, *Critical Issues in Selecting Conventional and Mechanized Tunnelling Methods*, Springer Tracts in Civil Engineering,
https://doi.org/10.1007/978-3-031-89114-4_1

Table 1.1 Basic comparison of conventional and mechanized tunneling methods

Parameter	Drill and blast	Roadheader	TBM
Geometry	Any shape	Any shape	Circular, recently any shape
Safety	Low	High	High
Storage and handling explosives	Terrorism, and accident risk	Avoided	Avoided
Advance rate	Low	Depending on the rock mass properties	Highly depending on the rock mass
Tunnel length	Any	Any	A limited length
Flexibility	High	High	Not
Operation	Cyclical	Cyclical	Continious
Initial investment	Low	Low	High
Total cost	Depending on the tunnel length and the geology	Depending on the tunnel length and the geology	Depending on the tunnel length and the geology
Overbreak	High	Low	Low
Noise and vibrations	High	Low	Low

of any shape with a cyclic construction process composed of excavation by using the drill and blast method or very basic mechanical excavators, mucking, placement of the primary support elements such as steel ribs or lattice girders, soil nails or rock bolts, sprayed or cast in situ *concrete. The second tunneling method is generally named mechanized tunneling or TBM tunneling,* ITA (2009, 2016). However, driving tunnels with roadheaders may also be accepted as a mechanized tunneling method, Bilgin et al. (2004, 2005). The basic comparison between both approaches is made by different authors, Tarkoy and Byram (1991), Girmscheid and Schexnayder (2002), Macias and Bruland (2014), and it is summarized in Table 1.1.

Although understanding the basic advantages and disadvantages of both methods is not complicated, the choice between them may be critical. The following sections of the book will discuss the critical points.

1.2 Some Critical Issues in Choosing Tunneling Methods. How Important Are the Lessons Learned from the Past?

Although it is unusual to give a long introduction in a book, we prefer to prolong it here to make it clear and show the readers how important it is to understand the following items.

1.2.1 Squeezing and High Convergence Characteristics of the Formations

The incorrect choice between the two methods may lead to tunnel construction lasting several years more than expected, and the cost maybe 3 or 4 times higher than originally estimated, as happened in the Yacambu-Quibor Water Tunnel in Venezuela, Ramonet (2008) and the T26 high-speed tunnel in Turkey, Bilgin (2016).

Yacambu-Quibor tunnel:

The project began in 1974, but it took 34 years to complete for technical, political, and financial reasons. Over those 34 years, the project was stopped and started again eight times using TBMs, roadheaders, and conventional tunneling methods. Instead of costing $173 million as initially expected, the tunnel ended up costing more than $800 million in current dollars, Ramonet (2008). The geology consists of sandstone, phyllite, and limestone. The graphitic phyllite has a squeezing characteristic and compressive strength change between 15 and 50 MPa, and the estimated GSI value is about 24, https://tunnelbuilder.com32-33/08. Graphitic phyllite is a very poor quality rock that causes serious squeezing problems, which, without adequate support, results in the closure of the tunnel at an important level, as seen in Fig. 1.1, Jimenez and Senent (2012). Tectonically deformed graphitic phyllite on the tunnel face is seen in Fig. 1.2 Hoek and Guevara (2009).

T26 High-Speed Tunnel/Turkey:

The excavation started with the conventional tunneling method in 2010. The tunnel is in the North-East of Turkey. Due to geological problems, shear zones, fault zones, and low RQD values, the daily advance rates were very slow. The tunnel excavation method was changed to mechanical excavation using a 13.7 m diameter single shield TBM at the end of 2011. Karakaya is the primary formation in the tunneling area between the Pazarcık melange. It contains fault zones in several places. This formation has similarities with the Karakaya Formation found in the Uluabat Energy Tunnel, which will be discussed in the following paragraph. T26 tunnel is planned to pass through the predominated graphitic schist, which frequently includes chlorite schist with the size of some tens of meters and rarely marble sheer bodies with a dimension of a few tens of meters, Bilgin (2016). Graphitic schist is moderately weathered to fresh, weak to weak, and highly sheared along the foliation surface, which causes extreme loads to the precast segments. Finally, segments started failing on 16th November 2012, and the tunnel collapsed. TBM was buried. After tremendous efforts and remedial work, tunnel excavation was carried out using drill and blast. Figure 1.3 shows the failure of the segment, and Fig. 1.4 at the top is the core samples taken from the tunnel wall at Ring 484 (9 o'clock direction) and a thin section from these samples. Please note the similarity of the rock formations from the T26 tunnel and the Yacambú-Quíbor tunnel.

Uluabat power tunnel/Turkey:

Fig. 1.1 Large convergences at Yacambú-Quíbor tunnel, Jimenez and Senent (2012)

Tunnel excavation commenced in 2002 from the portal next to Uluabat Lake, using conventional excavation methods with roof bolts, shotcrete, wire mesh, and steel arches as the primary tunnel support. Tunneling operations were halted in November 2003 due to extreme roof deformations and floor heaves. Figure 1.5 shows the tunnel's cross-section closure after a specific time for different stations. After the tunneling operations had been stopped for about two and a half years, it was decided that the project should be continued with an EPB-TBM. During tunnel excavation, the TBM was stuck a total of 18 times. 192 days were spent on TBM rescue operations at different tunnel locations. A general view of the squeezing ground around the TBM shield in the Uluabat Tunnel is shown in Fig. 1.6. The excavation terminated in March 2010 after a series of mitigation factors, such as injecting bentonite through the shield. It is reported that although the squeezing characteristics of the formations depend on several factors, it is closely related also to Q values, Barton and Bilgin (2016).

The three examples given above show how important it is to understand the formations' squeezing characteristics and the mitigation factors to overcome the problems arising. This topic will be treated separately within the book since it dictates the time scheduling of a tunneling project, the selection of the correct tunneling methods, and the mitigation factors to overcome.

Fig. 1.2 Graphitic schist at the Yacambú-Quíbor tunnel, Hoek and Guevara (2009)

1.2.2 Using TBM and Conventional Tunneling Methods in the Same Project–The Hybrid Solution

As Barton mentions, "*There are significant numbers of TBM projects that end up with difficult decisions to be made, namely to complete the projects by drill-and-blast from the other end of the tunnel. The deliberate selection of TBM and drill-and-blast may often be a simple matter of common sense, giving schedule advantages and cost savings. This is the preliminary level of hybrid tunneling. Very often, time is lost while waiting for TBM delivery and assembly, and great advantages could be gained by selective use of drill-and blast for more than just the standard TBM assembly chamber and starter tunnel*", Barton (2002a, 2002b).

A typical example of a hybrid solution is the Kargı power tunnel in Turkey. The 7.8 km of the total length of 11.8 km was excavated with a double shield TBM of 9.84 m diameter, and 4 km of the tunnel from the inlet part was opened with NATM. TBM was launched in April 2012. After only boring of 80 m, TBM became stuck in a section of the collapsed mixed ground face of hard rock and running ground. In the first 2 km of the tunnel, the TBM jammed 7 times due to face collapses, and bypass tunnels were required to free the TBM. Continuous probe drilling was carried out for 5 km along the tunnel, and an umbrella arch was applied in 9 different areas to stop face collapses. Excavation of the tunnel ended with success. Drilling and blasting at

Fig. 1.3 Failure of the segment from T26 tunnel. Photograph from Bilgin, the author

one end of the tunnel and TBM at the other was a good alternative for decreasing the job termination time, Bilgin (2016). Figure 1.7 compares advance rates of TBM and blast tunneling through the Beynamaz Volcanites. It can be seen that the drill and blast operations took twenty-three months to complete 4000 m of tunnel whereas it took less than 8 months for the TBM operations to complete the same length of the tunnel. Based on these production rates it would have taken approximately 32 months for the drill and blast operations to achieve the same production as the TBM achieved in 8 months, Clark (2015).

In most cases, conventional tunneling is an inevitable part of metro projects planned to be driven by TBMs. Table 1.2 gives the length of tunnels opened with conventional tunneling and with TBMs mutually in 5 recently terminated Metro projects in Istanbul. As seen from this table, 17% of the total length of the tunnels is excavated with conventional tunneling methods and 83% by TBMs.

1.2.3 Extreme Cases, the Story of One of the Shortest Tunnels Opened by a TBM

The north entrance tunnel of Sirkeci Station is a part of the Marmaray Project/ Istanbul, as outlined in Fig. 1.8. This section has a length of 120 m, and it was

Fig. 1.4 At the top are core samples of graphitic schist taken from ring 484, and below is the thin section representing core samples. Photograph from Bilgin, the author

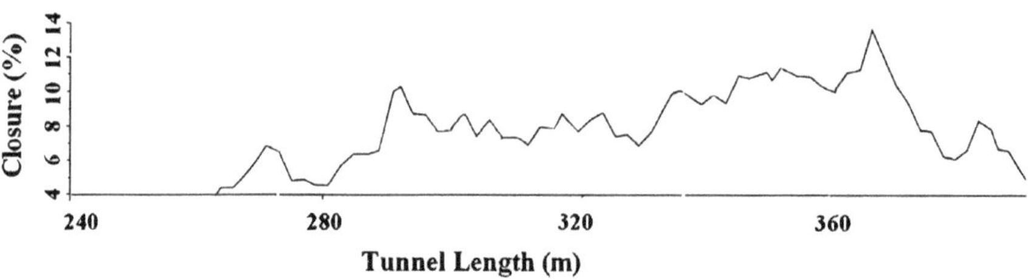

Fig. 1.5 The closure in percentage in cross-section area in the Uluabat tunnel after 6 months of excavation, Bilgin and Algan (2012)

initially designed to be opened using the conventional tunneling method. Detailed geological investigation showed that the conventional tunneling method was risky due to bad ground conditions and a low overburden of 20 m. It was later decided that the tunnel should be driven with an EPB-TBM of 6.2 in diameter, which was the 6th TBM used in the project's first phase, where 5 TBMs were slurry type. The launch EPB-TBM in the north entrance tunnel is shown in Fig. 1.9. The excavation of the

Fig. 1.6 General view of the squeezing ground around the TBM shield in Uluabat tunnel, Turkey, Bilgin (2016)

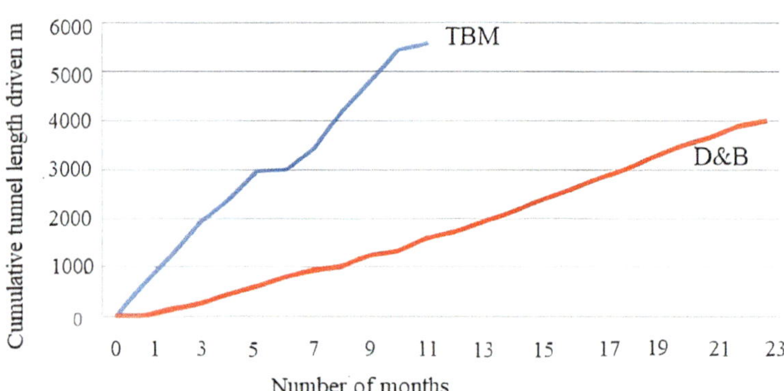

Fig. 1.7 The comparison of advance rates of TBM and drill and blast tunneling in Kargı Project, Turkey, credit must go to Clark (2015)

Table 1.2 The length of tunnels opened with conventional tunneling and with TBMs in 5 recently terminated Metro projects in Istanbul, from the archive of Bilgin, the author

Metro project tunnels	Cut and cover, m	Tunnels opened with conventional methods, m	TBM tunnels, m	Total, m
Gayrettepe-New Airport	–	8368	68,282	76,650
Halkalı-New Airport	–	6961	55,617	62,578
Bakırköy-Kirazlı	2700	2120	12,100	16,920
Ataköy-İkitelli	1640	8560	16,200	26,400
Başakşehir-Kayaşehir	720	6925	3928	11,573
Total	5060	32,934	156,127	194,121

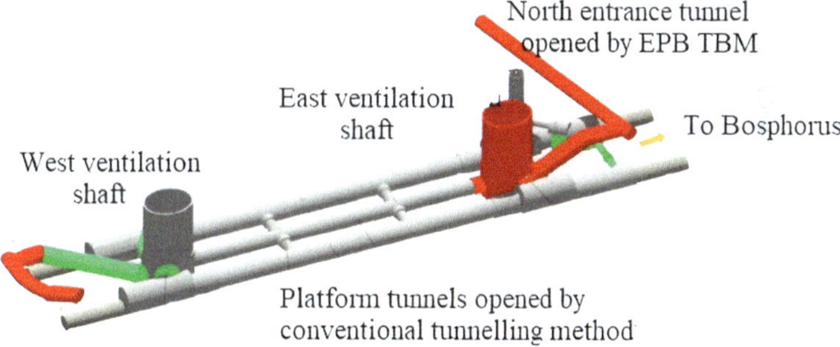

Fig. 1.8 Marmaray project, Sirkeci underground station, the figure is kindly provided by Şennazlı

tunnel lasted one month without severe surface deformations. The final profile of the north entrance is shown in Fig. 1.10. The readers are referred to consult Fig. 1.11 to understand the construction concept of the Marmaray project.

1.3 How Will the Critical Issues in Selecting the Proper Tunneling Methods Be Handled in This Book?

This book aims to explain understandably the geological and geotechnical factors affecting the success of tunneling methods by analyzing past experiences and aiming to transfer lessons learned to the readers. A quick look at the conventional and mechanized methods, with machines used, performance prediction of the machines, project planning and scheduling, the cost of the initial investment, and emerging new technologies will be significant topics that will be discussed within the book.

Fig. 1.9 The launch of EPB-TBM in the North entrance tunnel. The Photograph is kindly provided by Şennazlı

Fig. 1.10 The north entrance tunnel is in the finalized profile. The Photograph is kindly provided by Şennazlı

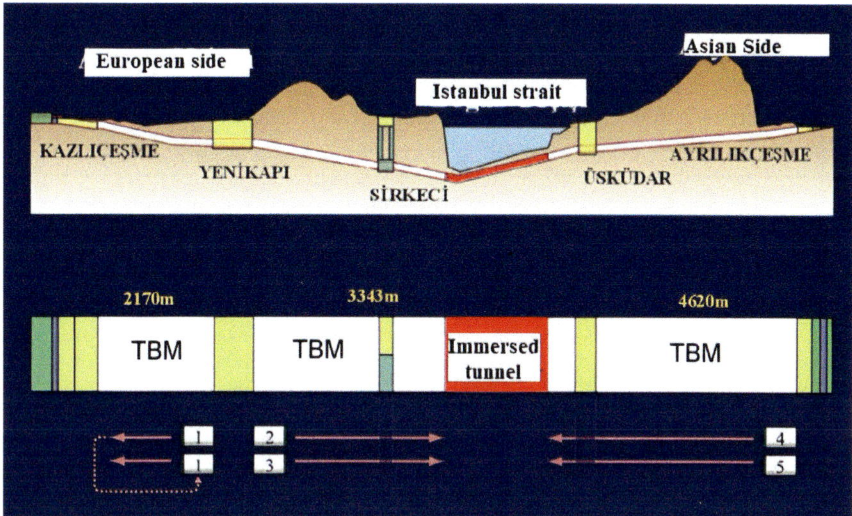

Fig. 1.11 The general construction concept of the Marmaray project, from Bilgin, the author

1.4 Emerging New Technologies

New developments in both tunneling methods considerably increase efficiency. Double-density TBMs, crossover TBMs, TBMs developed to drive square-shaped tunnels, development in data acquisition systems, new technologies in geophysical instruments for probing the formations in front of the tunnel face, and development in environmentally friendly chemicals will be major topics that will be explained within the Chap. 13.

1.5 Concluding Remarks

There are two tunneling methods: conventional and mechanized tunneling. The choice between them is critical. The incorrect choice may lead to tunnel construction lasting several years more than expected, and the cost is 3 or 4 times higher than initially estimated. This book is written to discuss the critical issues of selecting conventional tunneling against mechanized tunneling and lessons learned from the past. The book covers the main topics: geological and geotechnical parameters affecting tunneling, a quick look at conventional tunneling, a quick look at mechanized tunneling, the factors affecting the choice of tunneling methods, such as the cost of the initial investment, the length of the tunnel, project scheduling, time for mobilization, emerging new technologies. Some examples of changing the tunneling method from conventional to mechanized tunneling or vice versa during the same ongoing project and hybrid tunneling methods in the same project are also given.

Several examples are given of choosing conventional and mechanized tunneling methods against each other.

The book is unique in its subject, aiming at civil, geology, and mining engineering graduate and undergraduate students, researchers, and those working in the tunneling business.

References

Barton N (2002a) Reducing risk in long deep tunnels by using TBM and drill-and-blast methods in the same project–the hybrid solution. J Rock Mech Geotech Eng 4(2):115–126

Barton N (2002b) Hybrid TBM and drill-and-blast from the start. Tunneling J 22–32

Barton N, Bilgin N (2016) Fast or slow progress with TBM in ideal or faulted conditions. In: Ulusay et al (eds) Rock mechanics and rock engineering: from the past to the future. Taylor & Francis Group, London. ISBN 978-1-138-03265-1

Bilgin N (2016) An appraisal of TBM performances in Turkey in difficult ground conditions and some recommendations. Tunneling Undergr Space Technol 57:265–276

Bilgin N, Algan M (2012) The performance of a TBM in a squeezing ground at Uluabat, Turkey. Tunneling Undergr Space Technol 32:58–65

Bilgin N, Dincer D, Copur H, Erdoğan M (2004) Some geolological and geotechnical factors affecting the performance of a rodheader in an inclined tunnel. Tunn Undergr Space Technol 19(6):629–636

Bilgin N, Tumaç D, Feridunoğlu C, Karakaş AR, Akgül, M (2005) The performance of a roadheader in high strength rock formations in Küçüksu tunnel. In: 31th ITA-AITES world tunneling congress, 6p

Broere W (2016) Urban underground space: solving the problems of today's cities. Tunneling Undergr Space Technol 55:245–248

Clark J (2015) Extreme excavation in fault zones and squeezing ground at the Kargi HEPP in Turkey. In: "SEE tunnel: promoting tunneling in SEE region" ITAWTC 2015 congress and 41st general assembly May 22–28, Dubrovnik, Croatia

Girmscheid G, Schexnayder C (2002) Drill and blast tunneling practices. Pract Periodical Struct Des Constr 125–134

Hoek E, Guevara, R (2009) Overcoming squeezing in the Yacambú-Quibor tunnel, Venezuela. Rock Mech Rock Eng 42(2):389–418. https://tunnelbuilder.com/News/Yacambu-Quibor-Water-Tunnel-Breaks-Through.aspx. Uploded at 14th April 2023

ITA (2009) Report n°002, general report on conventional tunneling method. ISBN: 978-2-9700624-1-7, April

ITA (2016) Report n°17, recommendations on the development process for Mined Tunnels. ISBN: 978-2-9701013-6-9, April

ITA (2019) Tunnel market survey, p26

Jimenez R, Senent S (2012) Teaching the importance of engineering geology using case histories. In: McCabe, Pantazidou & Phillips (eds) Shaking the foundations of geo-engineering education. Taylor & Francis Group, London, ISBN 978-0-415-62127-4

Macias FC, Bruland A (2014) D&B versus TBM: review of the parameters for a right choice of the excavation method. In: Alejano L, Perucho A, Olalla C, Jiménez R (eds) Rock engineering and rock mechanics: structures in rock masses. Taylor & Francis Group, London, p 823–828. 978-1-138-00149-7

Ramonet D (2008) Venezuela completes Yacambú-Quíbor water transfer tunnel. EIR 35(32):32

Tarkoy P, Byram JE (1991) The advantages of tunnel boring: a qualitative/quantitative comparison of D&B and TBM excavation. Hong Kong Eng 7

Thomas T (2020) Tunneling market outlook: 1385 billion reasons to be cheerful. Tunneling J News

Chapter 2
Geological and Geotechnical Parameters Affecting Tunnelling Methods

Abstract Geology is essential for understanding the behavior of rock, soil, and other parameters such as rock mass, lithology, groundwater and hydrology, seismicity, faults, and climate for tunneling projects. These parameters help engineers design and select the methodology for tunnel excavation. Understanding the mechanical and physical behavior of soils, rocks, and rock mass together is critical for selecting conventional and mechanical excavation tunneling methods. This chapter briefly describes factors affecting tunneling performance, rock and soil properties, rock mass properties, major engineering rock mass classification systems, gas emission, underground water, testing methods and standards for using the mechanical miners, and understanding the rock mass squeezing characteristics.

2.1 Introduction

This chapter describes, in general, the physical and mechanical properties of rocks, soils, and rock mass properties for excavation method and selection, performance prediction of mechanical miners such as tunnel boring machines (TBM), roadheaders and solving numerous other engineering problems like blasting patterns, subsidence, dewatering, etc. for tunneling excavations. It is essential to carry out tests applying regular standards. Sample preparation and testing are necessary for obtaining meaningful data. The most widely used standards are the American Society for Testing Materials, ASTM (2013) and International Society for Rock Mechanics ISRM (2007) standards, providing guidance for performing laboratory and in situ tests and also non-standardized recommended tests such as full and small scale linear or rotary rock cutting, Bilgin et al. (2014).

Site investigations for tunneling are defined as the characterization of the rock mass conditions in which owners, planners, engineers, and contractors use geological, geotechnical, and other information about the area. In other words, site investigations are done for mining and tunneling projects in order: (revised by Nilsen and Ozdemir 1999a).

- To define the physical and mechanical properties of the rock formations,

N. Bilgin and C. Balci, *Critical Issues in Selecting Conventional and Mechanized Tunnelling Methods*, Springer Tracts in Civil Engineering,
https://doi.org/10.1007/978-3-031-89114-4_2

- To provide rock or soil design parameters, stability, and support requirements,
- To help define the overall planning, including siting and design optimization,
- To remove uncertainties of formation conditions for the bidder,
- To estimate cost, schedule, and help to prepare tender documents,
- To improve the safety of the work,
- To provide experience working with the material, which, in turn, will improve the quality of field decisions made during construction,
- To decide on excavation methods, selection, and performance prediction of the mechanical miners.

The general question for many owners is how much money should be spent on on-site investigations to provide sufficient information for tunneling activities. The question is difficult to answer since there is no single set price, and geological conditions are naturally variable. A famous saying is, "You pay for a site investigation whether you have one or not." As a general rule of thumb and from previous experience, the cost of the site investigations generally ranges from 1 to 8% of the project's total cost. If the project is less complicated, the ratio may go as low as 1%; for complicated projects, the percentage could go as high as 8%. The price also changes country by country because of the geological complexity.

The US National Committee on Tunneling Technology USNC/TT (1984) recommends site exploration budgets averaging 3.0% of the estimated project cost, while in hard rock countries like Norway, site investigation costs typically have been 0.5–1% of the total construction cost (higher for recent urban projects, Nilsen and Ozdemir 1999a).

There has not been an international standard for site investigation procedures and practices, but work has been accelerating to prepare international standards, Norbury (2004). The British site investigation standard, The British Standard Code of Practice BS5930 (2010), generally deals with civil engineering and building works purposes. In the USA, geotechnical investigations by the US Army Corps of Engineers (EM 1110-1-1804 2001) typically involve military and civil construction works.

There are some risks in excavation methods, drilling and blasting, or mechanical excavators in tunneling projects due to unexpected geology, wrong machine selection and performance prediction, and insufficient site investigations. These parameters cause increasing costs of the project, delays, and claims. This chapter outlines rock and rock mass investigation procedures for tunneling projects. It does not give a comprehensive site investigation manual but essential information to show how important it is for contractors, owners, and engineers. The organization of a site investigation and the techniques used in the tree can be seen and summarized in Fig. 2.1 by Fookes (1967) and Bell (1975).

In any excavation project, the first step is to conduct field investigations, search, and review existing geological data, papers, and records related to the project area. Field investigation covers all geological and geophysical data. Geological data should include maps, cross-sections, borehole layouts, and hydrogeological conditions. Geophysical data usually consists of the seismicity of the project area. Geophysical data are used to define the soil and rock interface. Experienced geologists or

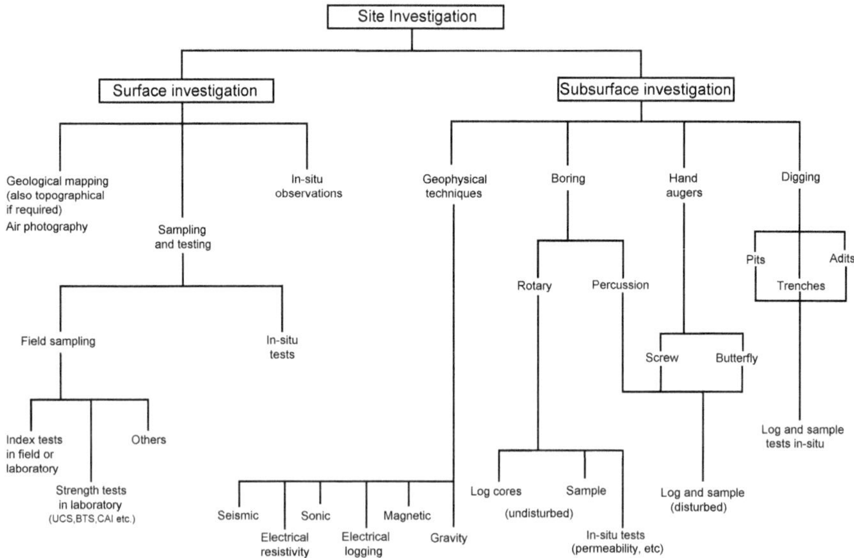

Fig. 2.1 Site investigation organization tree revised after Fookes (1967) and Bell (1975)

engineering geologists must perform geological mapping and data collecting during field investigations. During field works, rock types and geological conditions, representative sample collection for mechanical tests, and joint mapping are crucial for mechanical excavation projects. After geological mapping and preparing a geological cross-section, a general framework is planned in the detailed pitting, boring, trenches, and drilling program to get samples. Borehole numbers, layout, and depth are always questions for the site investigation. Boreholes will be drilled deeper than the tunnel line. No definite rules can be established for boring spacing and layout. They depend upon many factors, such as the nature and conditions of rock and soil and the complexity of the geological conditions. Core drilling (most commonly NX size) is used to obtain rock samples, auger-type drills for disturbed soil, and Shelby tubes for undisturbed soil sampling Bilgin et al. (2014).

Core drilling has some essential purposes: Nilsen and Ozdemir (1999a)

- To verify the geological interpretation,
- To obtain more information on rock type boundaries and degree of weathering,
- To supplement information on orientation and character of weakness zones,
- To provide samples for laboratory analyses,
- Hydrogeological and/or geophysical testing.

During field investigations, in addition to field mapping, field testing is an important parameter to obtain rock mass conditions. In-situ rock stress measurement, such as flat jack and hydraulic fracturing testing, is done to define the stress conditions of the rock mass. For groundwater characteristics of the rock mass, water injection or pumping tests are commonly used to determine the groundwater conditions of the project area.

Laboratory investigations should include standard rock and soil properties tests to prepare site investigation reports; standard laboratory tests are divided into three categories: physical, mechanical, and chemical. The essential role in laboratory studies is the representative sample supply and testing methods for selecting and designing mechanical excavators. If the sample does not represent the rock mass conditions and is not adequately tested in the laboratory, mechanical excavator selection and performance prediction cannot be reliable. The number of samples required will depend on several factors, such as the complexity of geology, length of the tunnel, intended test methods, type of project, and contract, Nilsen and Ozdemir (1999b).

2.2 Rock Properties

In geology, rocks occur in a solid aggregate of minerals connected by strong and permanent cohesive forces and classified according to their origin: igneous, sedimentary, and metamorphic. The three significant rocks are subdivided into many groups, and petrology is the scientific study. The rocks' minerals, grains and microscopic properties give some vital information about the rock properties such as strength, abrasivity, hardness, drillability, and cuttability. This section will explain these properties and the methodology of applying the tests without giving details.

Determination of rocks' physical and mechanical properties can be found directly or indirectly. For example, the uniaxial compressive strength (UCS) test can be measured directly from the UCS test or indirectly from point load strength tests. The rock laboratory testing must be based on standards and conducted according to the standards guidelines. Available standards, suggested methods, and descriptions for rock laboratory testing for mechanical miners are given in Table 2.1. These tests must be carried out for proper selection and performance prediction of mechanical miners.

2.2.1 Uniaxial Compressive Strength (UCS)

Uniaxial compressive strength (UCS) is one of the most essential rock strength parameters and the most common strength determination performed for mining and tunneling projects. It is generally measured by the procedures recommended in ASTM D2938 or ISRM, usually with NX-sized core samples (54 mm diameter). The samples are prepared to satisfy the requirements of ASTM D4543 or the methods suggested by ISRM. All core samples are cut and ground to a length-to-diameter ratio of 2.0–3.0. A minimum of three to five UCS determinations is recommended for statistical significance of the resulting average.

Compressive strength calculation is formulated below in Eq. (2.1).

Table 2.1 Rock testing methods and standards for mechanical miners based on, Nilsen and Ozdemir (1999b), Bamford (1986), Howarth (1987), Bilgin et al. (2014)

Test name		ISRM (2007) Suggested methods	ASTM standards	Other recommended methods
Mechanical strength	UCS	1979	D 2938	
	BTS	1978	D 3967	
	Static elastic constants	1979	D 3148	
	Dynamic elastic constants	1978	D 2845	
	Triaxial	1983	D 2664	
	Direct shear	1974	D 5607	
	Point load	1985	D 5731	
	Cone indenter			NCB (1964)
Hardness	Moh's			Nilsen and Ozdemir (1999b)
	Vickers			Nilsen and Ozdemir (1999b)
	Siever's J			Nilsen and Ozdemir (1999b)
	Shore	2006		
	Schmidt hammer	1978	D 5873	
Toughness/ Brittleness	Punch penetration test			Nilsen and Ozdemir (1999b)
	Fracture toughness	1988		Nilsen and Ozdemir (1999b)
	Brittleness value (S20)			Zare and Bruland (2013)
	Sievers' J-value (SJ)			Zare and Bruland (2013)
	Abrasion value (AV)			Zare and Bruland (2013)
	Abrasion value cutter steel (AVS)			Zare and Bruland (2013)
Abrasiveness	Cerchar		D 7625	CSM (1996)
	Schimazek			Schimazek and Knatz (1970)
	NTNU AVS			Zare and Bruland (2013)
	Taber			Tarkoy (1979)

(continued)

Table 2.1 (continued)

Test name		ISRM (2007) Suggested methods	ASTM standards	Other recommended methods
Rock cutting	Small scale linear rock cutting			Fowell and McFeat (1976), Balci (2004)
	Full scale linear rock cutting			CSM (1996), Eskikaya et al. (2000)
Other	Petrographic analysis	1978		Nilsen and Ozdemir (1999b)
	X-ray			Nilsen and Ozdemir (1999b)
	Sound velocity (P and S waves)	1978	D 2845	
	Density	1979		
	Porosity	1979		

$$\sigma_c = \frac{F_{Max}}{A_{Sec}} \tag{2.1}$$

where;

σ_c Uniaxial compressive strength, MPa
F_{Max} Maximum force on the sample before failure, kN
A_{Sec} Cross-sectional area of the sample before testing, mm^2.

Foliation and failure type are the most essential criteria for UCS testing and influencing mechanical miners' borability/cuttability parameters. Before and after UCS testing, great attention should be paid to how the sample fails. Joints, fractures, bedding, or foliation effects are classified as structural failure and do not represent the real rock strength.

2.2.2 *Indirect (Brazilian) Tensile Strength (BTS)*

Brazilian tensile strength (BTS) measures rock toughness and strength. This parameter is estimated using NX-sized core samples (54 mm in diameter) cut to a 0.5 length: diameter ratio and following the procedures of ASTM D3967 or ISRM suggested methods. The diameter must change to less than 0.5 mm over the length of the sample, and the core ends must be perpendicular to the core axis up to a precision depending on the standard applied. The Brazilian tensile strength is calculated by using Eq. (2.2).

$$\sigma_t = \frac{2P}{\pi LD} \tag{2.2}$$

where;

σ_t Brazilian tensile strength, MPa
D Diameter of the sample before testing, cm
P Maximum force on the sample before failure, kN
L Length of the sample before testing, cm.

The foliation and failure type are also the most essential criteria for BTS testing and should be noted before and after the testing on the report.

2.2.3 *Point Load Strength Index (PL)*

Point load strength is widely used as an accepted index test for strength classification and to measure the strength of a core or irregular piece of rock to determine UCS. The test can be performed with portable equipment and may be conducted in the field or the laboratory Broch and Franklin (1972), Brown (1981). The piece of rock is loaded to failure between two standard conical plates. The Point Load Strength index (Is) is calculated as the failure load and equivalent core diameter ratio. (Is) must be corrected to the standard equivalent diameter (De) of 50 mm and called Is (50) diameter of 50 mm. The procedure for size correction can be obtained graphically or mathematically as outlined by the standard applied procedures.

2.2.4 *Cerchar Abrasivity Index (CAI)*

The Cerchar Abrasivity Index (CAI) indicates the degree of rock abrasivity, for classifying and predicting cutter wear rate and cutter costs. The Cerchar test and associated CAI were developed at a time of more demand for the application of mechanical excavation machines at the Laboratoire du Center d' Études et Recherches des Charbonnages de France (CERCHAR) lately as standardized by "ASTM D7625-10 Standard Test Method for Laboratory Determination of Abrasiveness of Rock Using the CERCHAR Method". The tests are performed on freshly broken rock surfaces, free of weathering effects. The remnant pieces from indirect (Brazilian) tensile strength tests are usually used. A conical 90° hardened steel pin, fastened in a 7 kg head, is set carefully on the fresh surface and drawn 1 cm across it in 1 s. The test is repeated for five pins for each sample test. Minimum and maximum wear diameters are measured for each pin, and the shape of the wear is recorded. The tips of the pins are then examined under a reticular microscope, and two perpendicular diameters of the resulting wear flat are recorded for each pin. Coating the pin tips with a dye before testing makes the wear flat more visible. The Cerchar Abrasivity Index is then calculated by Eq. (2.3) below:

$$CAI = \frac{1}{10}\sum_{i=1}^{10} d_i\left[\text{unknown template}\right] \tag{2.3}$$

where di is pin wear diameter in (1/10 mm). A general Cutter Consumption Rate (CCR) for pick cutters is estimated by Eq. (2.4) below:

$$\mathrm{CCR}\left(\mathrm{cutter/m^3}\right) = 0.25\,\mathrm{CAI} \tag{2.4}$$

The lower the CAI, the softer and less abrasive the rock is for cutters. ASTM D7625 publishes a criterion for abrasiveness: A CAI of 1 is very smooth, while 6 is extremely abrasive.

2.2.5 *Schmidt Rebound Hardness*

The Schmidt hammer is a portable, small, cost-effective instrument capable of estimating the rock surface hardness as an index value used in the laboratory and situ. It was initially developed to assess the in situ strength of concrete. Since then, a lot of research has been carried out using Schmidt Hammer to estimate the intact and rock mass properties, characterize mine roof stability, estimate the performance of roadheaders, etc. Based on the statistical relationships, the Schmidt Hammer rebound value can be used as an index value or converted to an unconfined compressive strength value. The plasticity index can also be estimated using rebound values. The operation mechanism is simple: a plunger released by a spring impacts against the rock surface, and then the rebound distance of the plunger is read directly from the numerical scale, changing from 10 to 100. Schmidt hammers are available in several energy ranges, including Types L, N, and M, with 0.735 Nm, 2.207 Nm, and 29.43 Nm impact energies, respectively. The hammer should be calibrated before testing. A core sample, block sample larger than 20 × 20 × 20 cm, or an excavation face can be used to apply the Schmidt Hammer test. The rock surface should be flat and clean. The test is usually applied horizontally, and correction should be applied for angled or vertical rebound readings, Bilgin et al. (2002). ASTM D5873 and ISRM describe the procedure for testing rock.

2.2.6 *Shore Scleroscope Hardness*

Shore scleroscope hardness is one of the simplest methods, given the surface hardness of the tested material. It is determined from the rebound height of a diamond or tungsten-carbide-tipped hammer dropped onto a horizontal smooth surface. A diamond tip is dropped from a fixed height into the rock specimen in the Shore scleroscope test. The hammer then rebounds, but not to its original height, because some of the energy in the falling tip is dissipated in producing an indentation. The instrument used is supplied in models designated Model C and Model D. A diamond tip is dropped from a specified height and rebounds within the glass tube. According to the suggested methods published by the International Society for Rock Mechanics (ISRM 2007), a test specimen having a minimum surface area of 10 cm^2 and a minimum thickness of 1 cm is necessary. Measurement points should be at least 5 mm from each other, and only one test must be carried out at the same spot. The minimum number of tests for each rock is recommended to be 20 for statistical reliability (ISRM 2007), explained in detail in, Tumac et al. (2007) and Altindag and Guney (2006).

2.2.7 Density, Porosity and Water Content

Density, porosity, and water content are common physical properties of rocks. Density is a measure of mass per unit of volume and generally changes between 2.2 and 2.8 kg/cm^3 for most rock types. Porosity is a measure of the void spaces in a rock. It is the ratio of the non-solid volume to the total volume of material. The rock's water or moisture content indicates the amount of water the rock material contains. It is the water volume ratio to the rock material's bulk volume.

2.3 Soils

Soil is an aggregate of mineral grains that can be separated by such gentle means as agitation in water. Soils can be classified into two categories: cohesive (clays, silts, clayey silts, silty clays; particles bound together with clay minerals forces) and cohesionless soils (sands and gravel), loose particles without strong inter-particles. Grain shape, size, and clay minerals are essential, especially in designing and selecting mechanical miners. Grain size distribution or sieve analysis and the description of the clay minerals are the first soil properties that must be analyzed before choosing mechanical miners. In soft ground, the selection and design of mechanical excavators for tunneling and mining and the main parameters to be identified are given in Table 2.2. Those parameters are generally used for designing and prediction purposes on:

- surface settlement
- face stability
- earth pressure
- slurry system
- excavability

Table 2.2 Soil testing methods and standards for mechanical miners selection Bilgin et al. (2014)

Test name		AASHTO*	ASTM standards	Other recommended methods
Physical and mechanical properties	Cohesion	T 236	D 3080	
	Angle of internal friction	T 236	D 3080	
	Specific gravity	T 100	D 854	
	Water content	T 265	D 4959	
	Atterberg limits (liquid limit, plastic limit, and plasticity)	T 89	D 4318	
	Standard penetration test	T 206	D 1586	
	Unconfined compressive strength of cohesive	T 208	D 2166	
	Triaxial strength	T 296, 297	D 2850, 4767	
	Modulus of elasticity			USACE EM 1110-1-1904
	Direct shear strength	T 236	D 3080	
Petrografic and other properties	Grain size distribution (sieve and hydrometer analysis)	T 88	D 422, D 2487, D 1140	
	Mineral contents (quartz content)			X-ray diffraction
	Clay mineralogy			X-ray diffraction
	pH		D 4972	
	Groundwater condtions			
	Sulfate content	T 290	D 4230	
	Chloride content	T 291	D 512	
	Abrasiveness (grain shape and hardness)		ASTM G75	SAT; Nilsen et al. (2006)
	Swell potential of clays	T 256	D 4546	
	Collapse potential of clays		D 5333	
	Permeability for granular soil Permeability for all soil	T 215	D2434 D 5084	
	Possible existence of boulders and cobbles: type, amount, sizes, strength and abrasivity			

*American Association of State Highway and Transportation Officials

Table 2.3 Descriptive terms for soil fractions from U.S. Standard Sieve

Material		Upper limit 100% passing	Lower limit 100% retained
Boulders		–	300 mm
Cobbles		300 mm	75 mm
Gravel	Coarse	75 mm	19 mm
	Fine	19 mm	4.75 mm
Sand	Coarse	4.75	2 mm
	Medium	2 mm	425 μm
	Fine	425 μm	75 μm
Silt		75 μm low dry strength	
Clay		75 μm—moderate to high dry strength	

- separation plant
- sticky and swelling potential of clay minerals
- muck disposal and transportation.

2.3.1 Grain Size Distribution

Grain size distribution is widely used in the classification of soils and is defined by sieve analysis and hydrometer tests. The data obtained from grain size distribution curves is used to select tunnel boring machines and determine suitability for using different TBMs. The standard grain size analysis test determines the relative proportions of various grain sizes distributed among specific size ranges. A hydrometer test is used for sizes below 0.074 mm. Unified Soil Classification System (USCS) is the most common classification system to describe soil properties for selecting mechanical miners. The method is standardized in ASTM D 2487 and D422. The definition of the grain size of soil particles according to their texture (particle size, shape, and gradation) is given in Table 2.3

2.3.2 Clay Minerals

Tunnels are excavated in soil or rock formations, including clay minerals, which are essential for tunneling works. Clay minerals can cause swelling and squeezing ground conditions. The three main groups of clay minerals are kaolinite, illite, and montmorillonite. Water and moisture cause the swelling (expansion) of clay formations having high swelling capacities, such as montmorillonite. In addition to size distribution, clay minerals can be defined by X-ray diffraction and mineralogical analysis.

2.3.3 *Permeability*

Permeability is commonly measured in terms of the rate of water flow through the soil in a given period. The permeability coefficient is usually expressed as k in meters per second (m/s) or centimeters per second (cm/s). The coefficient of permeability for soils is an essential parameter in selecting mechanical miners, especially for earth pressure balance and slurry tunnel boring machines.

2.4 Rock Mass Properties

Rock mass classification for tunnel engineering problems has been improving and developing over the years, such as design criteria for supporting, quality and structure of the rock, faults, discontinuities, joints, weathering, and strength, etc., to define characteristics by quantifying and classifying the quality of rock masses. The science of rock mechanics has been developing since the '50s and has been used to solve mining problems. Some critical rock mass classification systems that are mainly used for tunneling projects are summarized in Table 2.4

Table 2.4 Major engineering rock mass classifications currently in use Bieniawski (1989), Coşar (2004)

Name of classification	Originator and date	Country of origin	Applications
1. Rock load	Terzaghi (1946)	USA	Tunnels with steel support
2. Stand-up time	Lauffer (1958)	Austria	Tunneling
3. NATM	Pacher et al. (1974)	Austria	Tunneling
4. Rock quality designation (RQD)	Deere et al. (1967)	USA	Core logging, tunneling
5. Rock structure rating, (RSR)	Wickham et al. (1972)	USA	Tunneling
6. RMR system	Bieniawski (1973) (last modified, 1989)	South Africa	Tunnels, mines, slopes
	Nakao et al. (1993)	Japan	Tunneling
7. Q-system	Barton et al. (1974), Barton (2002)	Norway	Tunnels, chambers
8. Strength-block size	Franklin (1975)	Canada	Tunneling
9. Rock mass index (RMi)	Palmström (1996)	Sweden	Tunneling
10. Geological strength index (GSI)	Hoek and Brown (1997)	Canada	All underground excavations

2.4.1 *Terzaghi's Rock Mass Classification*

The earliest reference to the use of rock mass classification for the design of tunnel support is in a paper by Terzaghi (1946) in which the rock loads carried by steel sets are estimated based on a descriptive classification. Terzaghi (1946) classified rock conditions into nine categories ranging from hard and intact rock to swelling rock. In practice, there are no sharp boundaries between these rock categories and the properties of the rocks. Table 2.5 summarizes rock condition and rock load factor classification with remarks by Terzaghi for steel arch-supported tunnels. The mechanics of the ground arc action given in Fig. 2.2 and the rock load Hp are determined by different equations explained by Terzaghi (1946).

In Fig. 2.2, the ground arch is seen by shaded area a c d b, and B1 is the ground arch width with tunnel height H.

Table 2.5 Rock conditions and rock load factor classification by Terzaghi (1946)

Rock condition	Rock load Hp in feet	Remarks
Hard and intact	Zero	Light lining required only if spalling or popping occurs
Hard stratified and schistose	0–0.5 B	Light support
Massive, moderately jointed	0–0.25 B	Load may change erratically from point to point
Moderately blocky and seamy	0.25 B to 0.35 (B + Ht)	No side pressure
Very blocky and seamy	(0.35–1.1) (B + Ht)	Little or no side pressure
Completely crushed but chemically intact	1.1 (B + Ht)	Considerable side pressure. Softening effects of seepage towards bottom of tunnel requires either continuous support for lower end ribs or circular ribs
Squeezing rock at moderate depth	(1.1–2.1) (B + Ht)	Heavy side pressure. Invert struts required. Circular ribs recommended
Squeezing rock at great depth	(2.1–4.5) (B + Ht)	Heavy side pressure. Invert struts required. Circular ribs recommended
Swelling rock	Up to 250 feet, irrespective of value of B and Ht	Circular ribs required. In extreme cases use yielding support

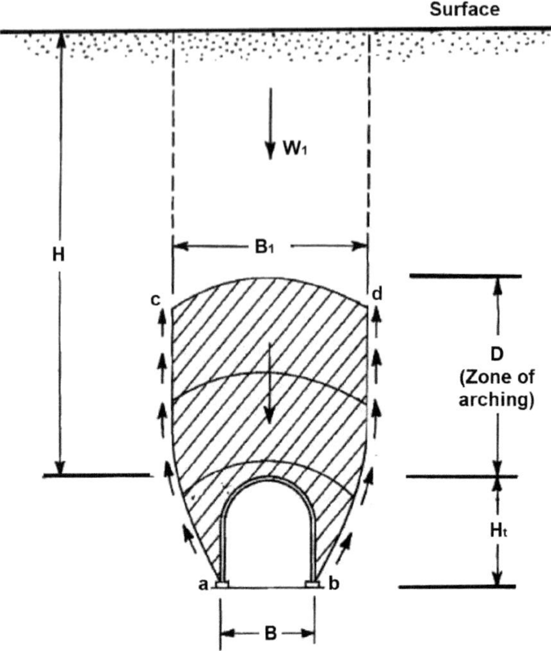

Fig. 2.2 The mechanics of tunnel ground arch Terzaghi (1946)

2.4.2 *Rock Quality Designation Index (RQD)*

A quantitative helpful index logging rock core is a rock quality designation (RQD) developed by Deere (1968). The RQD is a modified core recovery percentage in which all the pieces of sound core over 10 cm long are counted as recovery. The smaller pieces are considered to be due to close shearing, jointing, faulting, or weathering in the rock mass and are not counted. The RQD is a more general measure of the core quality than the fracture frequency. Core loss, weathered and soft zones, and fractures are accounted for in the determination. The RQD provides a preliminary estimate of the variation of the in situ rock mass properties from the properties of the "sound" portion of the rock core. Thus, a general estimate of the engineering behavior of the rock mass can be made. An RQD approaching 100% denotes an excellent-quality rock mass with properties similar to an intact specimen's. RQD values ranging from 0 to 50% indicate a poor-quality rock mass having a small fraction of the strength and stiffness measured for an intact specimen. An example of determining the RQD from a core run of 60 in. is given in Fig. 2.3. For this particular case, the core recovery was 50 in., and the modified core recovery was 34 in. This calculation yields an RQD of 57%, classifying the rock mass in the fair category Deere et al. (1967), Deere (1968).

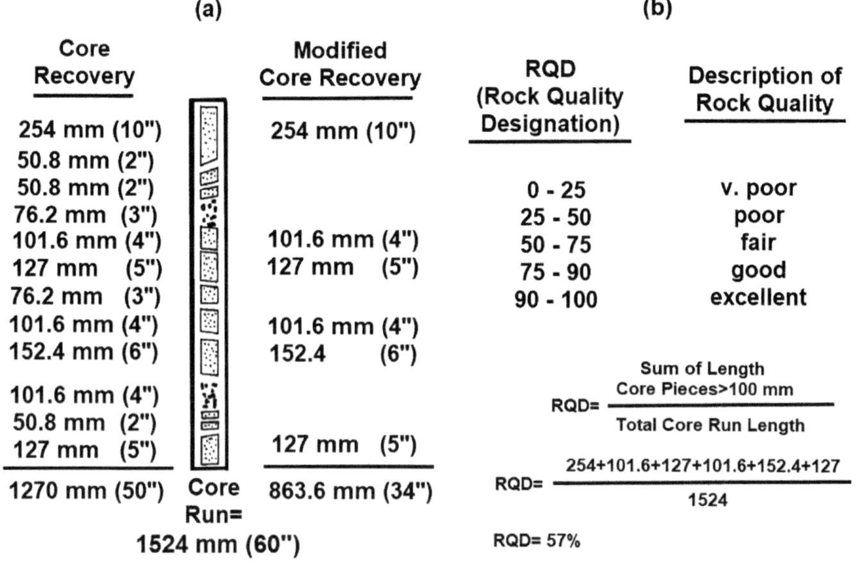

Fig. 2.3 Modified core recovery as an index of rock quality

2.4.3 *Rock Mass Rating (RMR) System*

Rock mass rating (RMR) was developed by Bieniawski in 1973 and published the details of a rock mass classification called the Geomechanics Classification or the Rock Mass Rating (RMR) using 49 case histories and then significant changes were carried out over the years to develop classification system adding new mining and tunneling case histories Bieniawski (1974, 1976, 1978 and 1989). RMR classification system includes the following six parameters measurable in the field and obtained from borehole data Bieniawski (1989):

Uniaxial compressive strength of intact rock material,
Rock quality designation (RQD),
Spacing of discontinuities,
Condition of discontinuities,
Groundwater conditions,
Orientation of discontinuities (Table 2.6).

2.4.4 *Q System*

The Q-system was developed by Barton et al. in 1974, analyzing 212 tunnel case histories from Norway to use tunnel support requirements. Barton presented the

Table 2.6 Guidelines for excavation and support of rock tunnels by the RMR system Bieniawski (1989)*

		Support		
Rock mass class	Excavation	Rock bolts (20-mm dia fully grouted)	Shotcrete	Steel sets
Very good rock I RMR:81–100	Full face 3-m advance	Generally, no support required except for occasional spot bolting		
Good rock II RMR:61–80	Full face 1.0–1.5 m advance Complete support 20 m from face	Locally, bolts in crown 3 m long, spaced 2.5 m, with occasional wire mesh	50 mm in crown where required	None
Fair rock III RMR:41–60	Top heading and bench 1.5–3 m advance in top heading Commence support after each blast Complete support 10 m from face	Systematic bolts 4 m long, spaced 1.5–2 m in crown and walls with wire mesh in crown	50–100 mm in crown and 30 mm in sides	None
Poor rock IV RMR: 21–40	Top heading and bench 1.0–1.5 m advance in top heading. Install support concurrently with excavation 10 m from face	Systematic bolts 4–5 m long, spaced 1–1.5 m in crown and wall with wire mesh	100–150 mm in crown and 100 mm in sides	Light to medium ribs spaced 1.5 m where required
Very poor rock V RMR: < 20	Multiple drifts 0.5–1.5 m advance in top heading. Install support concurrently with excavation. Shotcrete as soon as possible after blasting	Systematic bolts 5–6 m long, spaced 1–1.5 m in crown and walls with wire mesh. Bolt invert	150–200 mm in crown, 150 mm in sides, and 50 mm on face	Medium to heavy ribs spaced 0.75 m with steel lagging and forepoling if required. Close invert

*Shape: horseshoe; width: 10 m; vertical stress: < 25 MPa;._Construction: drilling, and blasting

Q-system classification system in (2002), adding more tunnel case histories to the database and distributing igneous, metamorphic, and sedimentary rock types.

The Q-system is a quantitative classification system that uses six different parameters as follows:

RQD
Number of joint sets
Roughness of the most unfavorable joint or discontinuity
Degree of alteration or filling along the weakest joint
Water inflow

Table 2.7 Q-system rock mass classification description

Class	Q-value	Definition Q rating
A	400–1000 100–400 40–100	Exceptionally good Extremely good Very good
B	10–40	Good
C	4–10	Fair
D	1–4	Poor
E	0.1–1	Very poor
F	0.01–01	Extremely poor
G	0.001–0.01	Exceptionally poor

Stress condition

These six parameters are grouped into three quotients to give the overall rock mass quality Q as follows:

$$Q = \frac{RQD}{J_n} \times \frac{J_r}{J_a} \times \frac{J_w}{SRF} \tag{2.5}$$

where

RQD Rock quality designation
Jn Joint set number
Jr Joint roughness number
Ja Joint alteration number
Jw Joint water reduction number
SRF Stress reduction factor.

The rock quality can range from Q = 0.001 to Q = 1000 on the logarithmic rock mass quality scale, and rock classes are divided into the following groups, as shown in Table 2.7.

The Q-system support chart is shown in Fig. 2.4, and the Q-value and the Equivalent dimensions are plotted along the horizontal axis and the Equivalent dimension along the vertical axis on the left-hand side Handbook by NGI (2022).

2.4.5 *Geological Strength Index (GSI System)*

The Geological Strength Index (GSI) has been used in many countries in recent years. It was proposed by Hoek (1994), Hoek et al. (1995), Hoek and Brown (1997), Hoek et al. (1998), and Hoek (1999) to estimate the strength and deformation modulus of the jointed rock mass using the Hoek–Brown criterion. The original Geological Strength Index (GSI) chart was developed on the assumption that observations of the

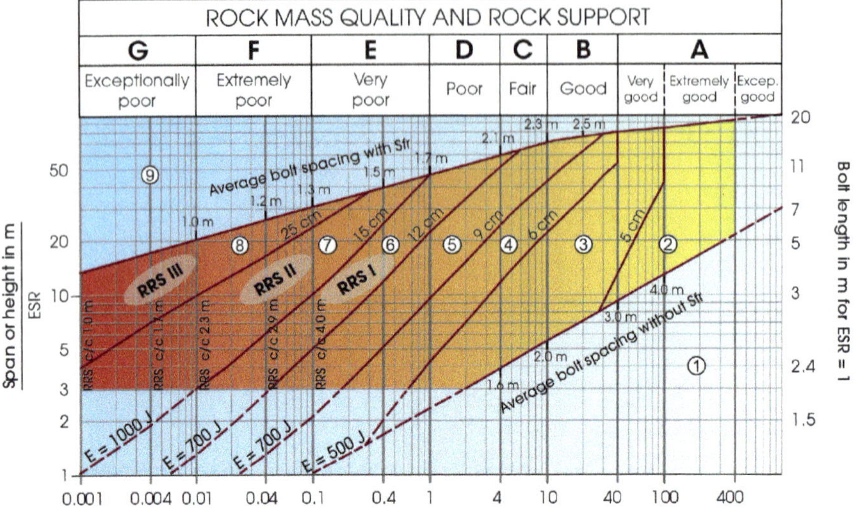

Fig. 2.4 The Q-system support chart handbook by NGI (2022)

rock mass would be made by qualified and experienced geologists or engineering geologists Hoek et al. (2013). GSI has been modified and improved to cover complex geological engineering problems such as sheared and folded series and heterogeneous rocks by many authors, Cai et al. (2004), Hoek and Marinos, (2000a, b), Hoek et al. (1998), Marinos and Hoek (2000), Sonmez and Ulusay (1999). The relationship between rock mass structure (blocky, very blocky, blocky/disturbed, disintegrated, foliated/laminated/sheared conditions) and rock discontinuity surface quality is used to estimate an average GSI value represented in the form of diagonal contours, as is seen in Fig. 2.5. It is recommended to use a range of values of GSI in preference to a single value Hoek et al. (1998).

2.4.6 NATM Classification

The New Austrian Tunnelling Method (NATM) was developed between 1957 and 1965 in Austria. It was given its name in Salzburg in 1962, and Rabcewicz explained it (1964). NATM is a scientific approach improved by Müller (1978), Golser and Mussger (1979), Brown (1981), Sauer and Gold (1989), and Kovari (1994). The basic principle of NATM is to reduce support and construction costs by using flexible support systems, allowing some ground deformation and, thus, providing ground arching after stress redistribution and some load-carrying capacity to the ground during construction. The ground deformations must be continuously measured, and information on ground behavior must be obtained during the arching period. Based on the behavior of the ground, the support elements can be reduced in the competent

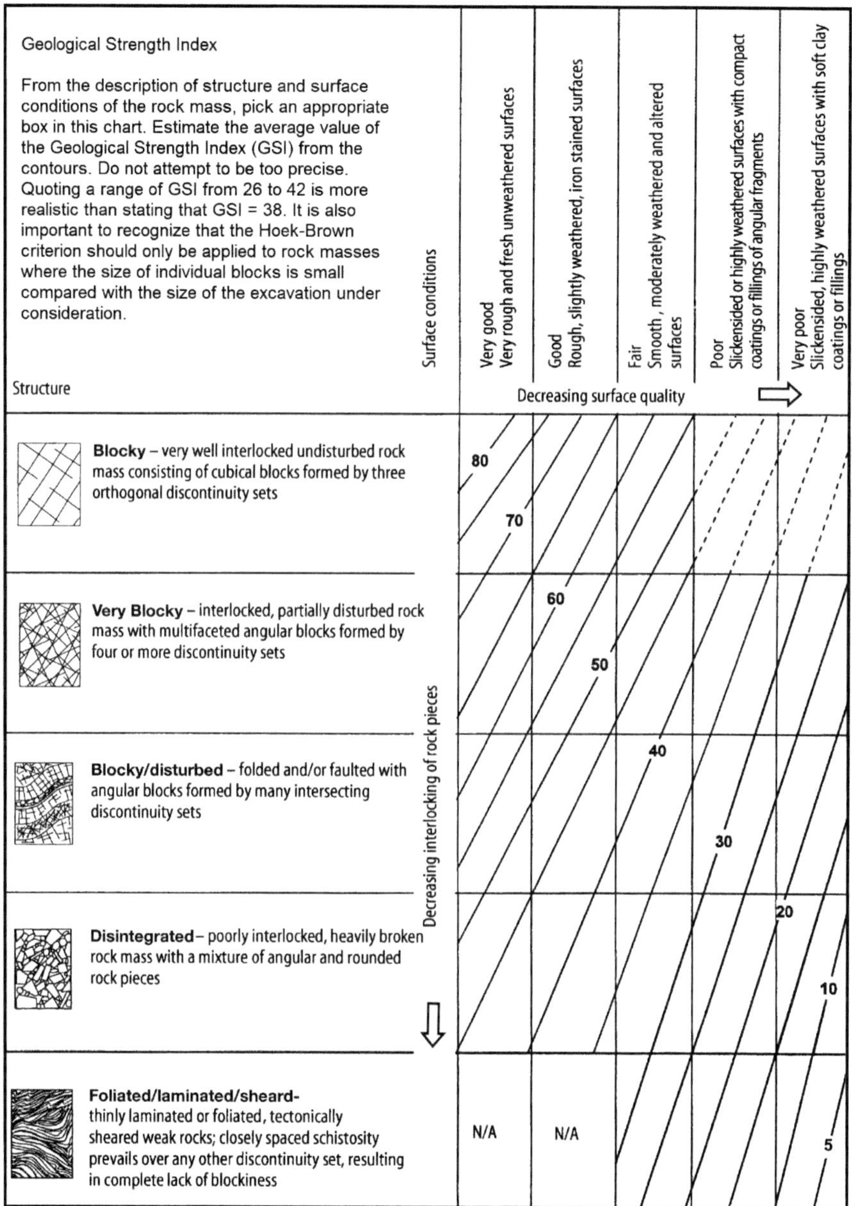

Fig. 2.5 Modified table for estimating the geological strength index Hoek et al. (1998)

ground and increased in weak ground conditions. Such flexibility in support design and application would reduce the costs in the competent ground and increase the safety in weak ground.

The main conditions that the New Austrian Tunneling Method (NATM) should/can be applied are summarized below (Copur et al. 2024):

Where large sections of underground caverns requiring sequential/staged/stepped/multi-pass excavation are needed,
Where large section, short, and variable cross-section tunnels are necessary, but using TBM (Tunnel Boring Machine) is not economical,
Where there are not enough site investigations/drillholes (may be due to high costs),

Basic application principles of the New Austrian Tunneling Method are summarized below in the application order (Copur et al. 2024):

Defining possible initial (primary) support elements (flexible, temporary) based on rock mass classification,
Application of staged/sequential excavation,
Observations/monitoring/instrumented measurements of ground deformations and stand-up time,
Allowing the occurrence of load carrying zone (ground arching by stress redistribution) by allowing some deformation during this period,
Initial (flexible) support installation,
Ring/invert closure,
Waterproofing and final (permanent, secondary) support installation.

Flexible support systems such as wire mesh, shotcrete, and rock anchors (rock bolts) are used in NATM as primary (temporary) support. Thus, the support requirements, such as wire mesh numbers and sizes, shotcrete thickness, and rock bolts' length, can be instantaneously determined based on the ground conditions identified by instrumented measurements. Flexible support systems move/deform just after installation together with the ground and allow for ground arching (plasticized load-carrying zone). Since the plasticized zone has a load-carrying capacity, the required support amount will be minimized. If the deformations are allowed more than the stand-up time of the ground, it might result in a collapse. Rigid (stiff, non-flexible) lattice girder-type support systems and umbrella arch-type pre-support systems in cohesionless and/or blocky grounds may be used/applied (Copur et al. 2024).

2.5 Gas

Tunnel air is a mixture of water vapor, dust, and gases that fills underground work spaces. The primary purpose of tunnel air is to provide clean and breathable air for workers. Gases in tunnels are classified as flammable, toxic, and suffocating. Flammable gases include methane (CH_4), carbon monoxide (CO), and hydrogen

(H_2), which are examples of carbon dioxide to suffocating gases (CO_2), nitrogen (N), and methane (CH_4) can be given as an example. Carbon monoxide (CO), all oxides of nitrogen (N), hydrogen sulfide (H_2S), sulfur dioxide (SO_2), etc. is composed of toxic gases. Dust particles suspended in the air are one of the main byproducts of tunnel construction sites, pose a serious threat to workers' health, and increase the risk of pneumoconiosis. Tunnel air is contaminated by the amount of gas and dust contained in the rock formation, the tendency of formation to combine with oxygen, the dimensions of the workspace, the excavation method applied, the amount of air coming into the work area, and the degree of mechanization and the type of machinery used depends.

Every country has set up legal rules defined by laws to manage the safety and health of workers at the workplace. For example, in Turkey, at underground mines, workers do not work in places with less than 19% oxygen, more than 2% methane, more than 0.5% carbon dioxide, more than 50 ppm (0.005%) carbon monoxide, and other dangerous gases. For 8 h of work, the maximum allowable hydrogen sulfide content is 20 ppm (0.002%). Air currents that are impaired or overheated due to a decrease in oxygen or the mixing of other flammable, combustible, and harmful gases are immediately and most shortly, without allowing them to pass through other working places, thrown out of the workplace. To protect workers from the adverse effects of deterioration of air properties, heating, and oxygen depletion, the working area and time shall be limited when work is mandatory. World Health Organization (WHO), The NIOSH, and The OSHA determine recommended and permissible exposure limits of gases. For example, the USA's lowest acceptable oxygen level is 19.5%, and the maximum permitted carbon monoxide is 50 ppm. Russia has the lowest acceptable oxygen level, 20%, and maximum permissible carbon monoxide at 50 ppm.

2.5.1 Oxygen

The air in Earth's atmosphere is made up of approx 21% oxygen. Oxygen needed for normal combustion, which is necessary for the sustainability of life, is odorless, colorless, and tasteless. A decrease in the amount of oxygen makes breathing difficult. Tunnels, oxidation of rocks-ores, decay of timber, the addition of other gases, fires, explosions and detonations, and breathing generally cause oxygen deficiency. Effects and symptoms of the amount of oxygen by volume on human health are given in Table 2.8.

2.5.2 Carbon Monoxide

Carbon Monoxide (chemical formula CO) is a highly toxic gas and is very dangerous to human health. It is slightly lighter than air, with a specific gravity of 0.96(g/cm^3).

Table 2.8 Effects and symptoms of the amount of oxygen on human being

Oxygen (% vol)	Effects and symptoms
21	Normal breathing and typical concentration in air
19	Almost normal breathing minimum safe level (19% is often the low level alarm of detectors)
15–19	Acceleration and difficulty in breathing, first sign of hypoxia. Early symptoms in person with coronary, pulmonary or circulatory problems
12–14	Respiration increases with exertion, pulse up, impaired muscular coordination
9	Mental failure, fainting or loss of consciousness, ashen face, nausea, vomiting
6–8	6 min 50% probability of death, 8 min 100% probability of death
4–6	Coma in 40 s, convulsions, respiration ceases, death

Table 2.9 Effects and symptoms related to the amount of carbon monoxide on human beings

CO (ppm)	Effects and symptoms of CO poisoning
5–10	None
50	Maximum allowable concentration for continuous exposure in any 8-h period
200	Headache, tiredness, dizziness and nausea after 2–3 h
400	Life threatening after 3 h
800	Dizziness, nausea and convulsions within 45 min. Death within 2–3 h
1600	Headache, dizziness and nausea within 20 min. Death within 1 h
3200	Headache, dizziness and nausea within 5–10 min. Death within 30 min
6400	Headache, dizziness and nausea within 1–2 min. Death within 10 min
12,800	Death within 1–3 min

Colorless, odorless, and tasteless gas is called the silent killer. Carbon monoxide binds to hemoglobin with a much greater affinity than oxygen to form carboxyhemoglobin (COHb), reducing oxygen-carrying capacity and oxygen utilization in the brain and other parts of the organs. It is produced by breathing and the combustion of any substance containing carbon (wood, coal, fuel, gas, oil, etc.) combustion. Explosives used in the tunnel, diesel working machines, wood decay, explosions, and fires are the familiar primary sources of carbon monoxide in tunneling works. Effects and symptoms related to the amount of carbon monoxide by volume on human health are given in Table 2.9.

2.5.3 *Methane*

Methane (CH_4) is a colorless, odorless, and non-noxious gas. Contrary to what is known, methane is not an explosive gas; it is a flammable gas. Since the density of methane (0.716 kg/m^3) is lower than the density of air (1.293 kg/m^3), it accumulates at the crown of a tunnel. When it mixes with enough air, it becomes flammable, and if the burning is in a closed environment, an explosion effect is seen with high pressure and temperature. When the methane concentration within the air is between 5% (lower limit) and 15% (upper limit), it becomes flammable and becomes explosive in a closed area due to sudden combustion. A temperature source of about 650–750 °C is required to create an explosion. It cannot be flammable but can cause asphyxiation due to a lack of oxygen at over 15% concentrations. It just flares or burns at under 5% concentration. Methane can be efficiently emitted through geological discontinuities, joints, and ground pores, as Doyle (2001) reported.

2.5.3.1 Some Methane and Gas-Related Tunnel Cases in Turkey

A methane flame occurred inside the excavation chamber of an EPB-TBM in Silivri-Istanbul on 20 May 2010 Copur et al. (2012). It is considered that the explosion due to accumulated methane in the pressure chamber of TBM was initiated by sparks created by friction between the screw conveyor and its casing. The explosion forced the muck in the chamber to be blown through the screw conveyor and out of the discharge door. The EPB excavation chamber is usually full of muck and foam. It is considered that the amount of air necessary for methane explosion came from the foam. Future research studies are strongly advised to focus on developing foam functioning without air on gassy grounds. The TBM used in this tunnel had no automatic gas measuring device inside the working chamber or critical parts of the TBM. This case study shows also that bentonite injection in the working chamber of EPB decreases methane emission, entering fracture zones.

Another gas flaming case occurred in the Silvan irrigation project, which constructed tunnels between the Silvan Dam and agricultural lands in Diyarbakir, southeast Turkey. A double-shield TBM of 7.8 m diameter was used for excavation. The 4668 m tunnel was excavated on 21 April 2015 when the gas flaming accident occurred at the chainage 18 + 447 km, Bilgin et al. (2016). After the gas flaming incident in the tunnel, it was discovered that the Turkish Petroleum Corporation (TPO) had already conducted site investigations along the tunnel and found a natural gas reservoir with a capacity of 24,000 m3/day around 300 m away from the accident area. However, neither the project owner nor the contractor knew about this risky situation. The findings of TPO indicate that the natural gas reservoir is of anticline type and had some tensile fractures due to the folding of the surrounding rock formations. These tensile fractures reached the tunnel area, Bilgin et al. (2016). 36 personnel worked in the tunnel during the gas flaming on 21 April 2015 at around 9:40 a.m. Thirteen of the personnel were burnt due to the gas fire, of whom five were seriously burnt. The fire/flame/flare engulfed the tunnel, starting from the face area and going through the crown of the tunnel. All of the equipment used in the tunnel was not ex-proof, and the gas flaming extensively damaged the TBM.

A ground contaminated with petroleum products was encountered in the Uskudar—Umraniye—Cekmekoy Metro Project Gundogdu et al. (2017). A sudden water inrush occurred along with fuel from an old/unused underground storage tank of a fuel station through one of the forepooling drills, and thus, construction of the double-tube NATM tunnels with a 30 m^2 cross-section was stopped on 12 June 2015. Metro line included Earth Pressure Balance Tunnel Boring Machines and classical New Austrian Tunneling Method excavated 16 station tunnels. The water-fuel mixture leakage area consisted mainly of fractured limestone and mudstone-claystone lithology (average overburden is around 30 m). The fuel leakage occurred during the excavation of the tunnel face, which involved drilling and blasting since the face was massive and hard. The amount of leakage of the water-fuel mixture was around 8 L/min at the first 90 min, and then, it dropped eventually to 1 L/min. The contractor cooperated with experts to analyze the situation and determine a plan for re-starting the construction activities Bilgin et al. (2015). It was realized that the water-fuel mixture leakage through the 12 m forepooling pipe was coming from a fuel station's old/unused (for several years) underground storage tanks. An interview with the administrator of the fuel station indicated that the new fuel tanks were tested recently, and there was no leakage in them. It is concluded that the leaking fuel was deposited within the fractures of limestone in the past. The fractured nature of the limestone and the dip and strike of the bedding planes of the mudstone-claystone indicated that the leakage would continue through the ~ 60 m along the tunnel excavation.

Another example of a petroleum product contaminated area is Bakirkoy–Kirazli Mero Line, Istanbul. During site investigations, it was discovered that a location close to the portal of the tubes was contaminated by a leakage of petroleum products coming from a petrol station, which is situated near the tunnels. Gasoline found in the ISK-3 borehole located next to the portal of the tubes, Bilgin et al. (2021). As the gas station authorities declared, fuel oil contamination was also discovered under another gas station approximately 60 m northeast of the present gas station. Within a 100 m radius of the working site are residential apartments, parks, four schools, and an enormous shopping mall. During the geotechnical studies, it has been observed that the soil around the metro line was permeable. Thus, it was concluded that the leakage around the gas station had contaminated and spread through the surrounding area.

2.5.4 *Carbon Dioxide*

Carbon dioxide (CO_2) is a colorless, odorless, noncombustible gas with an acidic taste. It is heavier than air (1.52 g/cm^3). Carbon dioxide is produced during combustion, and in brewing and other fermentation processes, wood decay, explosions, and fires are the primary sources of carbon dioxide. The lowest level at which CO_2 effects have been observed in human and animal studies is about 1000 ppm. CO_2 emissions in the tunnels generally come from the machines' energy used and fuel consumption during the construction, as well as concrete support, excavation, and ventilation.

2.5.5 Hydrogen Sulphur

Hydrogen sulfide, highly toxic, is a chemical compound with the formula H_2S and colorless chalcogen-hydride gas. It damages the sense of smell, is poisonous, corrosive, and flammable, and has an odor similar to rotten eggs. It has a specific gravity of 1.19 (g/cm^3) and is an explosive gas with a concentration of 4% and 44.5%. Hydrogen sulfide gas is released during the tunneling for drilling and combustion of black powder, the explosion of sulfurous ores, and the dewatering of flooded areas. 50 ppm H_2S causes eye and throat irritation, while 400 ppm causes unconsciousness and fainting. 900–2000 ppm H_2S causes severe poisoning and death in less than 1 min.

2.5.6 Sulphur Dioxide

Sulfur dioxide (SO_2) is toxic but not a flammable gas and a colorless, reactive air pollutant with a strong sulfurous odor. The specific gravity is 2.26(g/cm^3). The primary sources of sulfur dioxide are generally burnt iron pyrite; emissions are from fossil fuel combustion and the blasting of sulfurous ores. It irritates the eyes, nose, and throat. Large amounts of sulfur dioxide in the inhaled air damage the lungs. 20 ppm SO_2 threat to human health and animal health causes coughing, eye irritation, and throat irritation, while 400 ppm causes death in less than 1 min.

2.5.7 Nitrogen

Nitrogen appears as a colorless, odorless, noncombustible, and nontoxic gas. It makes up the central portion of the atmosphere about 80%. It is not an explosive and toxic gas. It does not contribute to the combustion reaction or breathing. The importance of nitrogen in the air is that it dilutes oxygen. If there is pure oxygen in the atmosphere, fires are uncontrollable. So, diluting oxygen by adding nitrogen and reducing the amount of oxygen can cause suffocation. The primary sources are the decay of organic matter, spreading through rock or coal cracks. Tunnel excavation with drilling and basting method, the explosion of 1 kg of nitroglycerine releases 135 L of nitrogen (N_2) in the tunnel workplace.

2.5.8 Nitric Oxide (NO)

Nitric oxide (NO) is a molecule composed of one nitrogen atom and one oxygen atom; nitrogen dioxide (NO_2), nitrogen trioxide (NO_2), nitrogen tetraoxide (NO_2), and nitrous oxide (N_2O_4) nitrogen oxides is a colorless gas with the formula NO. They occur in nitrogen explosions and the exhaust of diesel engines. Exposure to air containing 0.1% nitrous fumes for more than 30 min is dangerous. Within an 8-h working period, the concentration must not exceed 25 ppm. The smell of gunpowder can recognize nitrogen oxides after explosions. Excessive nitric oxide can cause a significant drop in blood pressure and cause headaches, tissue damage, and inflammation.

2.6 Underground Water

Water in tunneling brings significant problems and sometimes makes it impossible to open the tunnels. When the tunnel is excavated below the water table, unstable ground conditions, flood safety issues, etc., are always risks. Risk levels of water income can be classified as low, moderate, and high. Geological conditions such as fault zones, rock types, porosity, and permeability affect water risk levels. Water and water pressure increase the probability of collapse, destabilizing, and stability problems in the tunnel face, especially metro tunnels under the city, which may create subsidence to the surface of the Earth. It affects advance rates and project costs negatively. In dealing with the water problem, the necessary research should be done before the water enters the tunnel. The typical appearance of water income in the Cayirbasi highway tunnel in Turkey excavated using NATM is given in Figs. 2.6 and 2.7, respectively. This water adversely affected support and drilling time, causing delays. Another water problem was encountered in Turkey's Gerede Water Transmission Tunnel excavated using TBM. A high water inrush of 1500 L/sec flowed into the tunnel, causing the TBM to become stuck. A new design was made for precast segment transportation called custom flat cars, as seen in Fig. 2.8 by Harding and Alpagut (2020).

2.7 Geology, Squeezing of the Rock Mass

The squeezing ground or rock conditions are defined as time-dependent large displacement occurring around tunnels taking place when a particular combination of induced stresses around the tunnel beyond the limiting shear stress at which creep starts by Barla (1995). The following essential works generally deal with squeezing ground Bilgin and Algan (2012). According to the International Society for Rock Mechanics (ISRM), squeezing is a time-dependent large deformation around the tunnel associated with creep caused by exceeding a limiting shear stress Barla (2001). Aydan et al. (1993) considered the experience gained in Japan and suggested that squeezing occurs if the ratio of uniaxial compressive strength of intact rock to overburden pressure is less than 2. Aydan et al. (1993) proposed five different degrees of squeezing classification and explained the tunnel behavior in detail with some

Fig. 2.6 Typical appearance of water income from the tunnel side, Cayirbasi tunnel in Istanbul

comments. Hoek and Marinos (2000a, b) suggested that the rock mass uniaxial compressive strength ratio to the in situ stress may indicate tunnel squeezing characteristics. Kovari and Staus (1996) found that the low strength and the high deformability of the rock, as well as the presence of pore water pressure, facilitate squeezing. The following rock types are especially prone to developing considerable pressure and large deformations: altered gneiss, schist, phyllite, clay, shale, mudstone, and tuff, especially with high content of such minerals as micas, chlorite, serpentine, and clay. Panthi and Nilsen (2007) combined the semi-analytical method suggested by Hoek and Marinos (2000a, b) for Himalayan rock mass conditions. They concluded that squeezing characteristics of the ground may be predicted using some statistical analysis concerning engineering geological properties of rock formations along tunnel alignment Bilgin and Algan (2012).

The squeezing ground conditions often cause high costs and prolonged delays for the projects excavated by NATM or TBM methods. Squeezing of TBM or jamming the cutterhead is a nightmare for tunnel engineers since it affects machine utilization time and the realization of the project's scheduled time. The salvation (rescue) of a jammed cutterhead can considerably reduce the mean advance rate. This problem was studied for the Kargi, Uluabat, and Dogancay tunnels, where the causes and effects of TBM squeezing are discussed concerning remedial works needed for these three tunnels in Turkey Bilgin et al. (2016). Tectonic stresses squeezed the TBM several times, causing considerable delays in tunnel drivage in Turkey's TBM tunnels. The

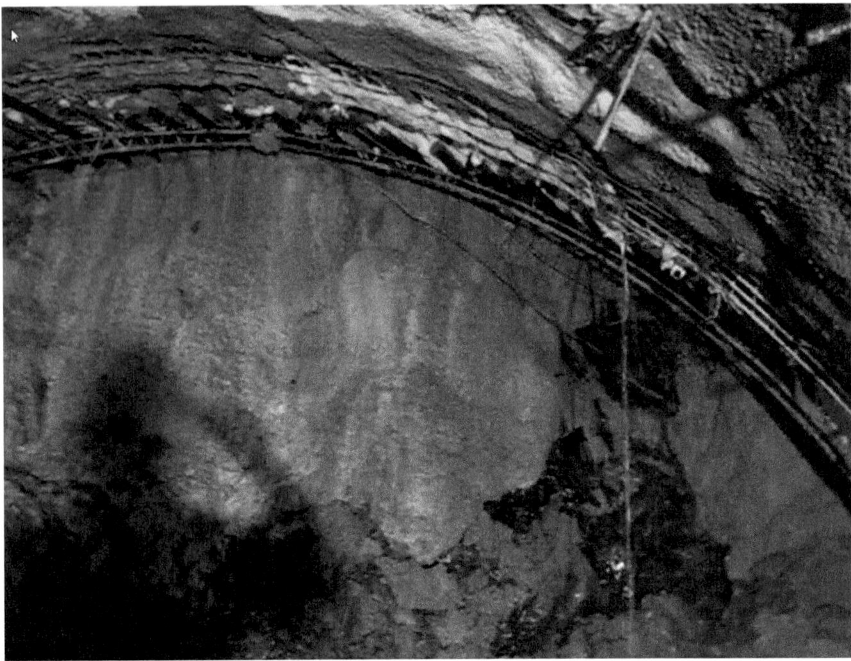

Fig. 2.7 Typical appearance of water income from the tunnel face and crown, Cayirbasi tunnel in Istanbul

effects of North and East Anatolian Faults on TBM performances in the Kargi energy tunnel, Dogancay energy tunnel, Nurdagi railway tunnel, and Uluabat energy tunnels are explained in detail, giving the causes, effects, and precautions to be taken to eliminate the problems created by two large sets of faults Bilgin et al. (2016).

2.8 Concluding Remarks

This chapter briefly summarizes the effects of geological and geotechnical properties of rock, soil, and rock mass on both NATM and TBM tunneling methods, but it does not go into detail. Understanding the geology is essential for underground excavations. Understanding the behavior of rock, soil, and other parameters ensures the successful opening of the tunneling project. These parameters help engineers understand that tunneling methods' selection, design, and methodology are critical for NATM and TBM tunneling methods. Geological, geotechnical, and other information for successful tunneling is obtained during the site investigation phases. Inadequate site investigation has potential consequences for the tunneling project. The question of how much money should be spent on adequate field and laboratory investigations during site investigations is essential. The budget to perform the geotechnical

Fig. 2.8 Custom flat cars designed for Turkey's Gerede water transmission tunnel Harding (2020)

investigations, evaluations, and report preparation should be sufficient to address the issues to an acceptable level, as explained for a medium-sized project (approximately $5,000,000). The inclusion of all the elements discussed in this chapter can typically run between 3 and 5% of the construction budget. The percentage could go as high as 8% for complicated small projects. It could go as low as 1% for a larger, less complex project. There is, therefore, no single set number that should be targeted. The budget and scope need to be established with an understanding of the complexity of the project and the owner's desire to obtain valuable information to the contractor.

References

Altindağ R, Güney A (2006) ISRM suggested method for determining the shore hardness value of rock. Int J Rock Mech Min Sci 43:19–22

American Society for Testing and Materials ASTM (2013) Standard test method for unconfined compressive strength of intact rock core specimens. In: Soil and rock, building stones. Annual book of ASTM Standards, vol 4.08 and 4.09.1828

Aydan O, Akagi T, Kawamoto T (1996) The squeezing potential of rock around tunnels: theory and prediction with examples taken from Japan. Rock Mech Rock Eng 29(3):125–143

Bamford WE (1986) Cuttability and drillability of rock. Civil College Technical Report Engineers Australia, July 11, 4

Balci C (2004) Comparison of small scale and full scale rock cutting tests to select mechanized excavation machines. Ph.D. Thesis, Istanbul Technical University, 269

Barla G (1995) Squeezing rocks in tunnels. ISRM News J 2(3&4):44–49

Barla G (2001) Tunnelling under squeezing rock conditions. eurosummer-school in tunnel mechanics, innsbruck. Logos Verlag, Berlin, pp. 169–268.

Barton N (2002) Some new Q-value correlations to assist in site characterisation and tunnel design. Int J Rock Mech Min Sci 39:185–216

Barton NR, Lien R, Lunde J (1974) Engineering classification of rock masses for the design of tunnel support. Rock Mech 6:189–239

Bell FG (1975) Introduction, site investigation. In: Bell FG (ed) Site investigations in areas of mining subsidence. Newnes Butterworth, London, pp 1–24

Bieniawski ZT (1973) Engineering classification of jointed rock masses: transaction of the South African. Inst Civ Eng 15:335–344

Bieniawski ZT (1974) Geomechanics classification of rock masses and its application in tunneling. In: Proceedings of the 3rd congress of international society for rock mechanics, vol 2, Denver, pp 27–32

Bieniawski ZT (1976) Rock mass classifications in rock engineering. In: Bieniawski ZT (ed) Proceedings symposium on exploration for rock engineering. A.A. Balkema, Rotterdam, pp 97–106

Bieniawski ZT (1978) Determining rock mass deformability experience from case histories. Int J Rock Mech Min Sci 15:237–247

Bieniawski ZT (1989) Engineering rock mass classifications, Wiley, Newyork, 251p, on pp 30–31

Bilgin N, Algan M (2012) The performance of a TBM in a squeezing ground at Uluabat, Turkey. Tunn Undergr Space Technol 32:58–65

Bilgin N, Balci C, Aslanbas A (2021) Case studies leading to the management of tunnel fire risks during TBM drives in an old coalfield Tunn Undergr Space Tech 112:103902

Bilgin N, Dincer T, Copur H (2002) The performance prediction of impact hammers from Schmidt hammer rebound values in Istanbul metro tunnel drivages. Tunn Undergr Space Technol 17:237–247

Bilgin N, Copur H, Balci C (2014) Mechanical excavation in mining and civil industries. CRC Press, Taylor and Francis Group, London

Bilgin N, Copur H, Fişne A (2015) Analysis of fuel leakage and precautions for re-starting the construction works. Report submitted to Doğuş Construction Inc., Istanbul Technical University Project (in Turkish)

Bilgin N, Copur H, Balci C (2016) TBM excavation in difficult ground conditions. Case studies from Turkey. Ernst & Sohn GmbH & Co. KG. Published by Ernst & Sohn GmbH & Co

Broch G, Franklin JA (1972) The point load strength test. Int J Rock Mech Miner Sci Geomech Abs 9:669–697

Brown ET (1981) Putting the NATM into perspective. Tunnels Tunn Int 13(10):137

BS 5930:1999+A2 (2010) Code of practice for site investigations. British Standard, 206

Cai M, Kaiser PK, Uno H, Tasaka Y, Minami M (2004) Estimation of rock mass strength and deformation modulus of jointed hard rock masses using the GSI system. Rock Mech Min Sci 41(1):3–19

CSM (1996) Test and model descriptions for performance and cost description of mechanical excavators for mining, underground construction and microtunnelling. EMI of Mining Engineering Department. 22.

Copur H, Cinar M, Okten G, Bilgin N (2012) A case study on the methane explosion in the excavation chamber of an EPB-TBM and lessons learnt including some recent accidents. Tunneling Undergr Space Technol 159Istanbul–167

Copur H, Balci C, Tumac D (2024) Tunnelling lecture notes. Istanbul Technical University

Coşar S (2004) Application of rock mass classification systems for future support design of the Dim Tunnel near Alanya. MSc thesis, Middle East Technical University, 238p

Deere DU (1968) Geological considerations. In: Stagg KG, Zienkiewicz OC (eds) Rock mechanics in engineering practice (Chp.1). Wiley, New York, pp 1–20

Deere DU, Hendron AJ, Patton FD, Cording EJ (1967) Design of surface and near surface construction in rock. In: 8th US symposium on rock mechanics: failure and breakage of rock. Society of Mining Engineers, American Institute of Mining, Metallurgical, and Petroleum Engineers, New York

Doyle BR (2001) Hazardous gases underground: applications to tunnel engineering. Marcel Dekker Inc., New York, USA

EM 1110-1-1804 (2001) Geotechnical investigations. US Army Corps of Engineers (USACE), Washington, DC, p 103

Eskikaya S, Bilgin N, Ozdemir L, et al. (2000) Development of rapid excavation technologies for the Turkish Mining and Tunneling Industries. NATO TU-Excavation SfS Programme Project Report. Istanbul Technical University, Mining Eng. Dept., Sept. 172

Fookes PG (1967) Planning and stages of site investigation. Eng Geol 2:81–106

Fowell RJ, Mcfeat-Smith I (1976) Factors influencing the cutting performance of a selective tunnelling machine. Int. Tunnelling '76 Symp., London. PP 301–318

Franklin JA (1975) Safety and economy in tunneling. In: Proceedings of the 10th Canadian rock mechanics symposium. Kingston, Canada, pp 27–53

Golser J, Mussger K (1979) New Austrian tunnelling method (NATM), contractual aspects. In: Tunnelling under difficult conditions, proceedings of the international tunnel symposium. Pergamon Press, Tokyo p 387–392

Gundogdu I, Kalayci G, Isler MK, Bilgin N, Copur H, Fisne A (2017) Safety precautions against fuel leakage into NATM tunnels of Uskudar—Umraniye—Cekmekoy Metro Project. In: ITA world tunnel congress (WTC 2017), surface challenges—underground solutions Bergen, Norway

Handbook NGI (2022) Using the Q-system rock mass classification and support design. Revised and new edition, Oslo, June 2022 56p

Harding D, Alpagut Y (2020) Tunneling through 48 fault zones and high water pressures on Turkey's Gerede water transmission tunnel. In: ITA-AITES world tunnel congress, WTC2020 and 46th general assembly Kuala Lumpur Convention Centre, Malaysia

Hoek E (1994) Strength of rock and rock masses. Int Soc Rock Mech New J 2:4–16

Hoek E, Brown ET (1997) Practical estimates of rock mass strength. Int J Rock Mech Min Sci 34(8):1165–1186

Hoek E, Marinos P (2000a) Predicting tunnel squeezing problems in weak heterogeneous rock masses. Tunnels Tunn Int Part 1(November):45–51

Hoek E, Marinos P (2000b) Predicting tunnel squeezing problems in weak heterogeneous rock masses. Tunnels Tunn Int Part 2(December):33–36

Hoek E, Kaiser PK, Bawden WF (1995) Support of underground excavations in hard rock. A.A. Balkema, Roterdam, Brookfield, pp 27–47

Hoek E, Marinos P, Benissi M (1998) Applicability of the Geological Strength Index (GSI) classification for very weak and sheared rock masses. The case of the Athens Schist formation. Bull Engg Geol Env 57(2):151–160

Hoek E (1999) Putting numbers to geology – an engineer's viewport. Quarter J Eng Geol 32:1–19

Hoek E, Carter TG, Diederichs MS (2013) Quantification of the geological strength index chart this paper was prepared for presentation at the 47th US rock mechanics. Geomechanics symposium held in San Francisco, CA, USA June 23–26, 2013

Howarth DF (1987) Mechanical rock excavation—assessment of cuttability and borability. In: Rapid excavation and tunneling conference proceedings, vol 1. Society for Mining, Metallurgy, and Exploration, Inc. New York, pp 145–164

ISRM (2007) The complete ISRM suggested methods for rock characterization, testing and monitoring 1974–2006. In: Ulusay R, Hudson JA (eds) Compilation arranged by the ISRM Turkish National Group, Ankara, 628

Kovari K (1994) Erroneous concepts behind the new Austrian tunnelling method. Tunnels Tunn 26(11):38e42

Kovari K, Staus J (1996) Basic considerations on tunnelling in squeezing ground. Rock Mech Rock Eng 29(4):203–210

Lauffer H (1958) Gebirgsklassifizierung für den Stollenbau. Geol Bauwesen 24:46–51

Marinos P, Hoek E (2000) GSI: A geologically friendly tool for rock mass strength estimation. In GeoEng2000: An International Conference on Geotechnical & Geological Engineering, International Society for Rock Mechanics, Melbourne, pp. 1422–1442

Müller L (1978) Removing misconceptions on the new Austrian tunnelling method. Tunnels Tunn Int 10(8):29e32

Nakao K, Iihoshi S, Koyama S (1993) Statistical reconsiderations on the parameters for the geomechanics classification. In: Proceedings of the 5th international congress on rock mechanics. ISRM, Melboume, pp B 13–B 16

NCB (1964) Methods of assessing rock cuttability. C.E.E. Report, No. 65(l)

NGI (2022) Handbook, using the Q system, rock mass classification and support design, revised version, Oslo, p 53

Nilsen B, Ozdemir L (1999a) Recent developments in site investigation and testing for hard rock TBM projects. In: Hilton DE, Samuelson K (ed) Rapid excavation and tunneling conference proceedings. Society for Mining, Metallurgy, and Exploration, Inc., Orlando, pp 715–731

Nilsen B, Ozdemir L (1999b) Recommended laboratory rock testing for TBM projects. Am Undergr Constr Assoc J 21–35

Nilsen B, Dahl F, Holzhäuser J, Raleigh P (2006) SAT: NTNU's new soil abrasion test. Tunnels & Tunneling International, May 2006, 43–45

Norbury D (2004) Current issues relating to the professional practice of engineering geology in Europe. In: Hack R, Azzam R, Charlier R (eds) Engineering geology for infrastructure planning in Europe. A European perspective. Springer, pp 15–30

Pacher F, Rabcewicz LV, Golser J (1974) Zum derzeitigen stand der gebirgsklassifizierung im stollen-und tunnelbau. In: Proceedings of the XXII geomechanics colloquium, Salzburg, pp 51–58

Palmstrom A (1996) RMi—a system for characterizing rock mass strength for use in rock engineering. J Rock Mech Tunne Technol 1:69–108

Panthi KK, Nilsen B (2007) Uncertainty analysis of tunnel squeezing for two tunnel cases from Nepal Himalaya. Int J Rock Mech Min Sci 44:67–76

Rabcewicz L (1964) The New Austrian tunnelling method, part one, water power, November 1964, 453–457, part two, water power, December 1964, 511–515

Sauer G, Gold H (1989) NATM ground support concepts and their effects on contracting practices. In: Proceedings of the 9th rapid excavation and tunneling conference (RETC), Los Angeles, 11–14 June 1989, pp 67–86

Schimazek J, Knatz H (1970) The influence of rock structures on the cutting speed and pick wear of heading machines (in German). Gluckauf, 106:275–278

Sonmez H, Ulusay R (1999) Modifications to the geological strength index (GSI) and their applicability to the stability of slopes. Int J Rock Mech Min Sci 36:743–760

Terzaghi K (1946) Rock defects and loads on tunnel supports. In: Proctor RV, White TL (eds) Rock tunneling with steel support, vol 1. Commercial Shearing and Stamping Company, Youngstown, Ohio, 268p

Tumac D, Bilgin N, Feridunoğlu C, Ergin H (2007) Estimation of rock cuttability from shore hardness and compressive strength properties. Rock Mech Rock Engng 40(5):477–490

US National Committee on Tunneling Technology (USNCTT) (1984) Geotechnical site investigations for underground projects, vol 1. National Academy Press, Washington, D.C., 182p

Wickham GE, Tiedemann HR, Skinner EH (1972) Support Determination Based on Geological Predictions, International Proc. North American Rapid Excavation Tunneling Conference, Chicago, K.S. Lane and L.A. Garfield (Eds), New York, PP 43–64

Zare S, Bruland A (2013) Applications of NTNU/SINTEF drillability indices in hard rock tunneling. Rock Mech Rock Eng 46:179–187

Chapter 3
A Review of the Conventional Tunneling Methods

Abstract This chapter first summarizes the equipment used in conventional tunneling and its role in high-speed drives and safe operations. The next topic is the drillability of the rock and its role in conventional tunneling. Within this topic, the drilling index developed in NTNU, the rock drillability index developed by Hoseinie et al., destruction work developed by Thuro, field drillability tests carried out by Bilgin, full-scale laboratory drilling tests and drilling specific energy, a comparative study between cutting and specific energy is discussed. Some examples of conventional tunneling methods and past experiences are also given. The chapter continues with a summary of blasting technology for tunneling, explosives, detonators, drilling patterns, blasting sequences, and types of cuts. In the section on vibration due to drill and blast and its effects on buildings and monuments, one of the authors summarizes his experience in the T1 Tunnel to protect the water source supplying Bejaie in Algeria, detailing the blasting operations and measurement of vibrations carried out in this area of the world.

3.1 Introduction

The conventional tunneling method, also known as the New Austrian Tunneling Method (NATM) or the Sequential Excavation Method (SEM), may be defined as the construction of underground openings with a cyclic construction process. It consists of the excavation using the drill and blast methods or the basic mechanical excavators, mucking, and placement of the primary support elements such as steel ribs or lattice girders or rock bolts, shotcrete, or cast in situ concrete. The basic philosophy behind NATM is monitoring the ground movement and seeking to maximize the ground's inherent resistance and support capacity. Although mechanized excavation by TBMs surpasses conventional tunneling in many cases, such as higher advice rates, safety in operation, less overbreak, etc., the conventional tunneling method is the world's most widely used underground construction method. It is preferred especially in difficult ground conditions through shear and fault zones and in squeezing ground conditions where tunneling by TBMs would be risky, and for the construction of subway station

N. Bilgin and C. Balci, *Critical Issues in Selecting Conventional and Mechanized Tunnelling Methods*, Springer Tracts in Civil Engineering,
https://doi.org/10.1007/978-3-031-89114-4_3

caverns, highway tunnels, and cross passages between TBM-driven tunnels, Gall et al. (2017). A recent study done for Mineta Transportation Institute covering 67 long tunnels greater than 4.5 km with a total length of 1469.3 km showed the inevitable contribution of conventional tunneling within the total amount of tunnels driven in 30 different countries. Within the total of 67 tunnels, 25.4% were driven with the conventional tunneling method, 33.7% were driven with TBMs, and 40.9% were driven with the hybrid tunneling method, i.e., conventional tunneling jointly with TBMs, Pyeon (2016).

3.2 The Equipment Used in Conventional Tunneling and Its Role in High-Speed Drives and Safe Operations

The conventional tunneling method allows very flexible and well-suited operations to situations requiring changes in the excavation process and support measures due to changed ground conditions. As Jennemyr mentions in his article published in 2020, drill and blast are often the only possible methods for short tunnels, large cross-sections, cavern construction, cross-overs, cross passages, shafts, etc. It also can be more flexible to adapt to varying profiles than a TBM tunnel that always gives a circular cross-section, especially for highway tunnels, resulting in much over-excavation about the actual cross-section needed. From 2015 to 2018, in Norway alone, 4.2 10.6 m^3 of underground rock excavation was realized by drill and blast. Intensive academic research work is increasing in this field. An analysis done on 144 peer-reviewed articles indexed in the Web of Science Core Collection from 2000 to 2023, year by year, showed the increasing demand for high speed and safety requirements of this tunneling technique, including blasting vibration, numerical simulation, rock damage, and overbreak, He et al. (2024). Bearing in mind that this cannot be realized unless the practicing tunnel engineers follow the latest developments in tunneling equipment, a summary in this respect will be given in this section.

3.2.1 *Development of Drilling Equipment*

Some examples of the commonly used tunneling equipment will be summarized in this section. Most recently, drilling jumbos now have computer controls that allow an entire tunnel round to be drilled without an operator. The drill rigs also contain much feedback information that allows for continuous monitoring of drill rig performance. Finally, the hole placement accuracy is vastly improved, allowing for precise perimeter control and significantly reducing overbreak. Recent data suggest that the time to drill the tunnel face has been reduced from 50% of the round length in 1960 to around 20% today, Brierely (2015). Epiroc is pioneering in this field; in several cases, the underground jumbo drill rigs from Epiroc are used for blast hole drilling in underground mining and tunneling. The rigs have up to four booms, covering cross sections from 6 to 206 m^2. Underground jumbo drill rigs are equipped with a Direct Control System or a computerized Rig Control System to which different levels of automation may be added. An extensive range of rock drills is available for impact power, from 16 to 40 kW. Several models of the face drill rigs are available with an optional zero-emission battery electric drive line. A typical Epiroc drill jumbo is seen in Fig. 3.1.

The Boomer XL3 D is a robust three-boom face drill rig providing an extra-large coverage of 178 m^2 with the boom BUT 35L. Designed to reduce labor costs while increasing productivity, the hard working Boomer XL3 D is an excellent recommended for for large scale underground operations. A general view of the Boomer XL3D is seen in Fig. 3.2.

A recent development uses multi-function jumbos suspended from the tunnel crown, allowing multiple functions to proceed simultaneously, such as drilling and mucking. The jumbo can also be used to install lattice girders and shotcrete. This approach overlaps sequential operations in tunneling, resulting in time savings on the schedule, Jennemyr (2020). Such a system is seen in Fig. 3.3.

Fig. 3.1 Boomer + XE3, Epiroc drill Jumbo, courtesy of Epiroc

Fig. 3.2 A general view of the Boomer XL3D, courtesy of Epiroc

Fig. 3.3 Multi-function jumbos suspended from the tunnel crown after Jennemyr (2020)

3.2.2 *Criticism of the Mucking Equipment Used in Drill and Blast Tunneling*

As Jennmyr mentioned (2020), wheel loaders and trucks are still the most common means of mucking, used in drill and blast for tunneling. In the case of access via shafts, the muck will be carried mainly through the wheel loader to the shaft, where it will be hoisted to the surface for further transport to the final disposal area. However, using a crusher at the tunnel face to break down the larger rock pieces to allow their transfer with a conveyor belt to bring the muck to the surface is another innovation developed recently, often for long tunnels. This method dramatically reduces the time for mucking, especially for long tunnels, and eliminates the trucks in the tunnel, improving the working environment and reducing the needed ventilation capacity. It also frees up the tunnel invert for concrete works. It has an additional advantage if the rock is of such quality that it can be used for aggregate production. In this case, the crushed rock can be minimally processed for other beneficial uses such as concrete aggregates, rail ballast, or pavement. To reduce the time from blasting to the shotcrete application, in cases where stand-up time can be an issue, the initial shotcrete layer can be applied to the roof before the mucking is done.

Side Dump wheel loaders are specially designed for tunnel conditions and can conduct loading, transporting, lifting, and other operations; a typical side dump wheel is seen in Fig. 3.4.

GHH Group (Komatsu) manufactures articulated dump trucks for mining and tunneling capacities ranging from 20 to 45 tons and bowl sizes ranging from 6 to 24 m^3. A typical example is seen in Fig. 3.5.

The space is limited for tunneling projects, so a compact excavator with a short boom and arm is needed to offer an acceptable underground performance. A Hitachi excavator ZX85US(B)-6, with a rated power of 42.4 kW, an operating weight of 8–9 t, and a backhoe bucket of 0.28 m^3, is seen in Fig. 3.6. The largest of these types of excavators is ZX350-7G, with engine-rated power of 200 kW, operating weight changing from 32 to 35 t, and a maximum backhoe bucket of 1.86 m^3.

Mucking out with a conveyor belt is typical for TBMs. However, as far as the knowledge of the authors of this book, the Tagami tunnel in Japan was the first

Fig. 3.4 A typical side dump wheel loader, with the courtesy of SEM, a Caterpillar Brand, courtesy of Caterpillar

Fig. 3.5 A typical dump truck used in a tunnel, with the courtesy of GHH Group (Komatsu)

Fig. 3.6 ZX350-7G excavator, courtesy of Hitachi

tunnel that utilized a conveyor belt as the mucking out technique in drill and blast tunneling, Yoshitomi (1999). It is reported that the following benefits were obtained using conveyor belts: (a) the reduction of dump trucks reduced the diesel exhaust gases, which improved the working environment, (b) the reduction of construction time due to increased possibilities to execute invert works in parallel with excavation at the tunnel face, (c) reduced restrictions for entering the tunnel during mucking out, facilitating different works executed in parallel with mucking out, (d) reduction of maintenance of the temporary road during construction due to fewer trucks and less wear of the road, (e) reduced dirt, dust, and mud around the tunnel portal and nearby areas. Since the Tagami tunnel, the use of belt conveyors in drill and blast tunneling increased considerably.

Fossati et al. (2017), in their work done for the Solbakk twin tube tunnel in Norway, compared the alternatives of (a) rubber wheel loaders and dumpers and (b) rubber wheel loaders, crushers, and conveyor belts. They concluded that the alternative with dumpers needs 24% more energy and produces 43% more total greenhouse gas emissions. A significant impact on the higher energy consumption of alternative dumpers is the high amount of fresh air needed. This energy corresponds to approximately 27% of the total energy required. Although the system flexibility and reliability of the first alternative are higher, the second alternative provided a better working environment.

Railway haulage is best suited for long-distance transportation of excavated materials with a gradient of 1 in 200–300. The system is flexible compared to a belt conveyor system. Good road, efficient maintenance, large output, and adequate ventilation are the basic requirements for the success of this system. Well-drained, properly graded, and minimum turning with smooth curves are necessary for the success of this system. Locomotives for underground use are either diesel or electric-power-driven. When the diesel locomotive is used, adequate ventilation is required, Zou (2017).

3.2.3 Shotcrete Equipment

Shotcrete is used to line tunnel walls, mines, subways, etc. Shotcrete or sprayed concrete is concrete or mortar conveyed through a hose and pneumatically projected onto a tunnel wall at high velocity. The concrete is typically reinforced by conventional steel rods, steel mesh, or fibers. The dry mix method involves placing the dry ingredients into a hopper and then conveying them pneumatically through a hose to the nozzle. The nozzle operator controls the addition of water at the nozzle. The water and the dry mixture are not completely mixed but are completed as the mixture hits the receiving surface. The dry mix process is useful in repair applications when it is necessary to stop frequently, as the dry material is easily discharged from the hose. Wet-mix shotcrete involves pumping previously prepared concrete, typically ready-mixed concrete, to the nozzle. Compressed air is introduced at the nozzle to impel the mixture onto the receiving surface. The wet-process procedure generally produces less rebound, waste (when material falls to the floor), and dust than the dry-mix process. The greatest advantage of the wet-mix process is that all the ingredients are mixed with the required water and additives, and larger volumes can be placed in less time than dry-process concrete, https://en.wikipedia.org/wiki/Shotcrete.

If correctly selected, shotcrete equipment facilitates tunnel support and increases tunneling efficiency. Figure 3.7 shows the Adroit 450G Shotcrete machine, which is fully equipped with a lifting, tilting, telescopic boom system, rotation group, compact modular system, water and admixture pumps and tanks, energy cable reel, compressor, and generator on the platform. It has a 30 m^3/h shotcrete pumping capacity, 13.6 m lateral, 15.8 m vertical reach, 9.2 m^3/min compressor capacity, and a 74.5 kW diesel engine.

Figure 3.8 shows the Adroit 016 DT crawler shotcrete machine, a shotcrete pump on a crawler chassis with a diesel engine. The machine has a concrete pumping capacity of 16 m^3/h. It has a cable remote for driving, concrete pump control, and boom movements. It is an ideal solution for narrow profile tunnels not accessible by truck-mounted pumps or 4 × 4 vehicles, and notably for open surface applications. The series is characterized by low rebound and low operating and maintenance costs.

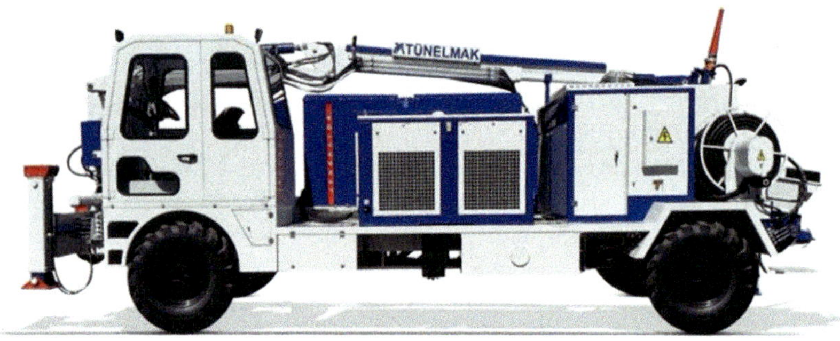

Fig. 3.7 Adroit 450G shotcrete machine, with the kind permission of Tünelmak

Fig. 3.8 Adroit 016DT crawler shotcrete machine, with the kind permission of Tünelmak

3.2.4 Explosive Emulsion Charging Track

Emulsions offer a significant safety advantage over traditional ANFO in blasting processes since they are inert until charged. This safety enhancement is crucial for the smooth operation of mining and tunneling activities. Normet-Charmec vehicle ensures a productive and safe underground charging process.

The use of bulk emulsion to charge the holes from a separate charging truck when the drill jumbo is being used for multiple headings or as a built-in feature to the drill jumbo when a single heading is being excavated is becoming more common unless there are local restrictions for this application. This method is commonly used in various areas around the world, with two or three holes being charged at the same time; the concentration of the emulsion can be adjusted depending on which holes are being charged. The cut and bottom holes are usually charged with 100% concentration, while contour holes have a much lighter concentration of about 25%. Bulk emulsion needs a booster in the form of a stick of packaged explosives (primer), which, together with the detonator, is inserted into the bottom of the holes and is needed to ignite the bulk emulsion pumped into the hole. Bulk emulsion reduces the overall charging time compared to traditional cartridges, where 80–100 holes/hr can be charged from a charging truck equipped with two charging pumps and one- or two-person baskets to reach the total cross-section, Fig. 3.9.

3.3 Drillability of the Rock and Its Role in Conventional Tunneling

The drill & blast method, still the most typical tunneling method for medium to hard rock conditions, requires careful planning and execution due to its cyclic nature. It can be applied to a wide range of rock conditions and features versatile equipment, fast start-up, and relatively low capital costs. However, the need for good work-site organization is important. Blast vibrations and noise also limit its use in urban areas. However, comprehensively and reliably assessing the rock drillability is a vital prerequisite for guiding the rapid construction of tunnels.

Fig. 3.9 Emulsion charging equipment, Jennemyr (2020)

The Drill and Blast technique includes the following steps: drilling, explosive charging, blasting, ventilation, mucking and scaling, support, shotcrete, bolting etc. Drilling impacts blast performance, efficiency, and tunneling operations cost. The drilling cycle includes positioning the jumbo, checking that the proper drilling pattern matches the position along the tunnel length, placing the drilling booms, and drilling the holes. As explained in Chap. 9, the percentage of the works done in Ovit Tunnel (one of Turkey's most successfully terminated tunnels), the time spent for drilling is 27.6% within the job distribution in all tunneling operations. The drillability of rocks depends on many factors: the equipment used, bit type and diameter, rotational speed, thrust, blow frequency, and flushing, which are the controllable parameters. On the other hand, uncontrollable parameters such as rock properties and geological conditions affect drilling rates and drilling-specific energy, so predicting these parameters will not only affect the correct scheduling of the job termination, it will also give rapid profiling of rock mass quality underneath the tunnel face, Wu et al. (2024); Zaho et al. (2024). This advantage of using drillability in predicting rock mass properties was used in the past by many researchers in classifying excavability of the rocks for different industrial applications, Bilgin (1983), Thuro (1977), Thuro and Plinninger (1999) Thuro and Plinninger (2003), Balcı et al. (2020). Within the scope and the importance of the subject, the drillability studies carried out in the past by different researchers will be discussed first. The industrial applications of these studies will be summarized later.

3.3.1 Drillability Studies Carried Out in the Past and Their Industrial Applications

In general, drillability is defined as the advance made in the unit time of the drill bit to overcome the rock strength. Rock drillability is one of the important properties for mining and tunneling operations. It can be determined by using several direct and indirect methods, as defined by Kahraman (1999) and Kahraman et al. (2003). However, there are some drillability prediction methods referred to by different researchers that one can not pass without emphasizing, which will be summarized below.

(a) Drilling Rate Index developed in NTNU

The drillability tests initiated at the Norwegian University of Science and Technology (NTNU) in the early 1960s have been continuously developed and formed the foundation of various hard rock equipment capacity and performance prediction models for mechanized and conventional tunneling. Due to the national and international success of the drillability tests, several other laboratories have also set up similar equipment to perform the same tests, Yaşar et al. (2015), Sakız et al. (2021). The Drilling Rate Index (DRI) developed in NTNU is assessed based on two laboratory-tested parameters: the Brittleness Value S20 and the Sievers' J-value (SJ), Figs. 3.10

and 3.11. The Brittleness Value S20 equals the percentage of material that passes an 11.2-mm mesh after the aggregate sample of the 11.2–16 mm fraction has been crushed by 20 impacts in a mortar. The Brittleness Value is the mean value of 3–5 parallel tests. The Sievers' Miniature Drill Test measures the rock's surface hardness (or resistance against tool indentation). The Sievers' value is the drill hole depth after 200 revolutions of a miniature drill bit, measured in 1/10 mm. The SJ-value is the mean value of 4–8 drill holes. Figure 3.3 assesses the DRI from the Brittleness Value S20 and the Sievers' J-value. The DRI may be described as the Brittleness Value corrected for the rock surface hardness.

Zare and Bruland (2013), in their paper titled "Applications of NTNU/SINTEF Drillability Indices," explain the practical application of these tests. They provide a case study of the San Francisco Tunnel in Ecuador, a water conveyance project where a TBM was employed. NTNU/SINTEF experts were involved in the project for laboratory tests, tunnel time, and cost prediction. They also present an example of a drill and blast method for a 40-m^2 D&B tunnel. In this project, the time and costs were less sensitive to the drillability parameters than for TBM since the drilling is only a part of the total drill and blast operations. By comparing the magnitude of the changes in the time and costs, the impact of the associated risk related to the rock drillability in D&B tunneling is one-fourth that in TBM tunneling documents.

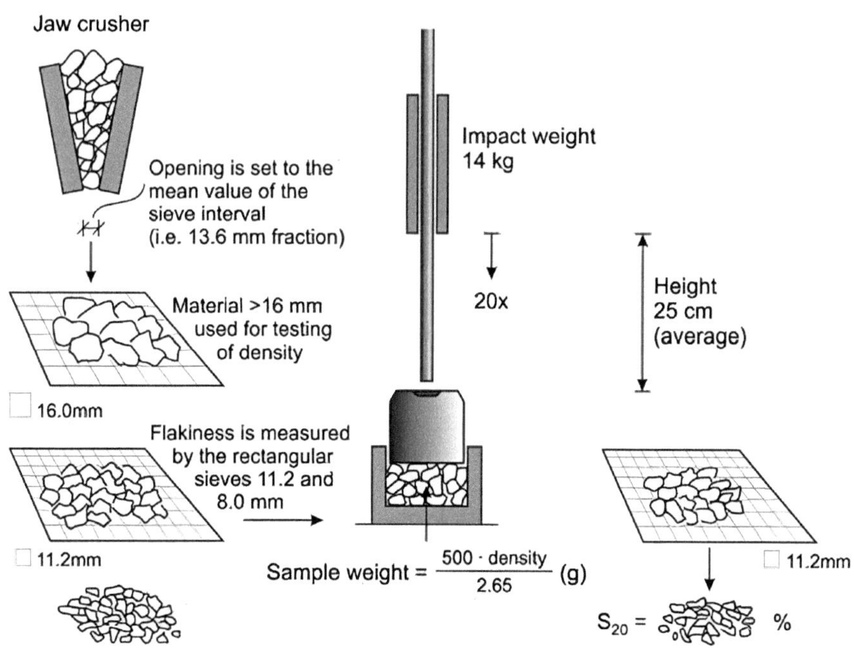

Fig. 3.10 The brittleness test, Zara and Bruland (2013)

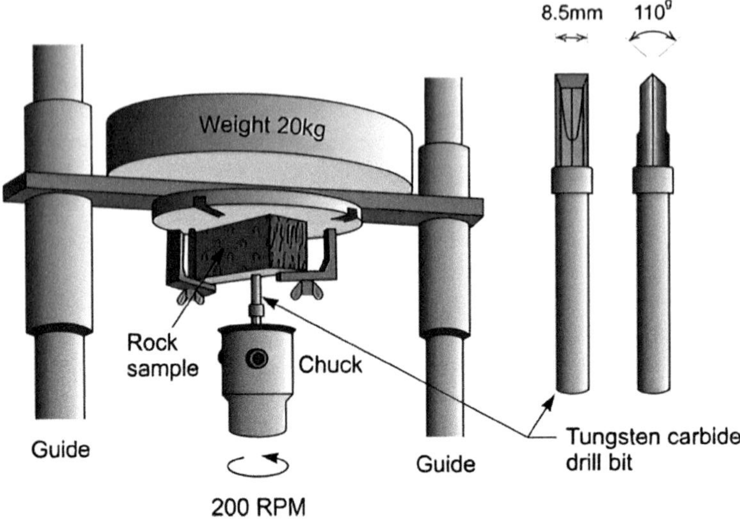

Fig. 3.11 The sievers' miniature drill test, Zara and Bruland (2013)

It is interesting to note that Beniawski and his co-authors (2008) developed a rating system to estimate rock excitability (RME), which included DRI as one of the parameters of the rating system.

(b) Rock Mass drillability index developed by Hoseinie et al. (2008)

Hoseiniea et al. (2008) developed a new classification system for defining the rock mass drillability index (RDi). For this purpose, six rock mass parameters, including texture and grain size, Mohs hardness, uniaxial compressive strength (UCS), joint spacing, joint filling (aperture), and joint dipping, have been used by physical modeling and rated. In particular, physical modeling has been used to investigate the effects of joint characteristics on the drilling rate. In the proposed RDi system, each rock mass is assigned a rating from 7 to 100, with a higher rating corresponding to faster drilling, as detailed in Table 3.1. Based on the RDi, the drilling rate is classified into five cases: slow, slow-medium, medium, and medium-fast. They found a significant relationship between RDi and the rate of DTH drilling in the Sungun mine from Iran.

Furthermore, Heydari et al. (2024a, b) used RDi to predict penetration rates of jumbo drills penetration rate in different hole diameters and concluded that RDi gives statistically significant estimates of penetration rates. However, after an intensive literature survey, they concluded that the penetration rate depends on several parameters, including rock mass properties (intact rock properties and structural parameters), machine specifications, and operational parameters, as defined in Fig. 3.13 They insisted that the proposed model is feasible and reliable in the prediction of the jumbo drill's penetration rate. However, they also concluded that this prediction is limited to the geological conditions and operations of the case studies in their

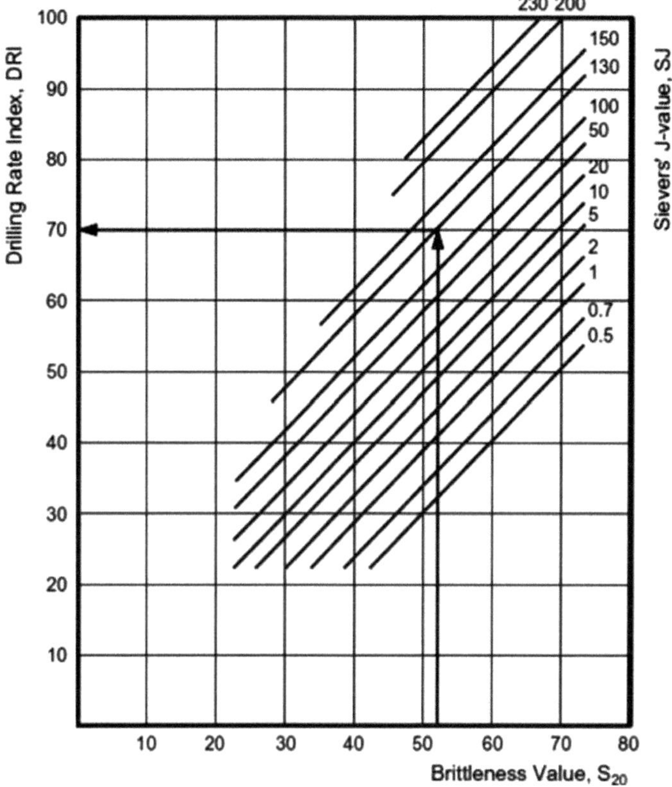

Fig. 3.12 Diagram for the assessment of DRI, Zara and Bruland (2013)

research study. The prediction models established in their research could be further refined, suggesting significant possibilities for drilling automation in underground mines.

(c) Destruction work/destruction energy developed by Thuro

Macias et al. (2017) reviewed the state-of-the-art and discussed the relevant parameters involving drillability assessments in hard rock conditions. They gave as an example the work done by Thuro (1977), emphasizing that according to Thuro, the principal parameters influencing drillability may be the jointing of rock mass, the orientation of schistosity (rock anisotropy), degree of interlocking of microstructures, porosity, and quality of cementation of clastic rock, degree of hydrothermal decomposition and weathering of a rock mass. Figure 3.14 illustrates the main factors, machine and geological parameters, influencing drilling in tunneling.

Thuro (1977) has introduced destruction Work/Energy as the best fitting parameter for drillability prediction in drill and blast tunneling, and the excavability of the rock by roadheaders. Destruction work is illustrated in Fig. 3.15.

Table 3.1 Rock mass drillability index (RDi) rating, Hoseinie et al. (2008)

Texture	Porous	Fragmental	Granitoid	Porphyritic	Dense
Grain size	–	> 5 mm	2–5 mm	0.05–1 and 2–5 mm	0.05–1 mm
Rating	15	10	7	4	1
Mohs hardness	1–3	3–4.5	4.5–6	6–7	47
Rating	18	13	9	4	1
UCS (MPa)	1–25	25–50	50–100	100–200	4200
Rating	22	16	11	6	
Joints spacing (m)	42	1–2	0.5–1	0.15–0.5	0–0.15
Rating	18	13	9	5	1
Joint aperture and filling (mm)	0–2	> 20	12–20	9–12	2–9
Rating	15	10	7	4	1
The angle between joint and borehole axis70^0–90^0		55°–70°	35°–55°	20°–35°	0–20°
Rating	12	8	6	3	1

Prediction of drilling rate Rdi 7–20 slow, 20–40 slow medium, 40–60 medium, 60–80 medium fast, 80–100 fast

(d) Field drillability tests

An approach to selecting roadheaders based on field drilling rates was developed due to a significant research project for the Turkish Scientific and Research Council. Detailed drillability studies were conducted in different roadways in Zonguldak coalfields, Turkey, Bilgin (1982), where drill and blast were used for roadway drives. The main objectives of this research program were pursued as given below:

The first objective was to investigate the relationships between rock properties and drilling rates for a given percussive drill rig and constant operational parameters. The specification of the percussive drill hammer used for in situ experiments was as follows:

Drill hammer weight: 25 kg
Operating pressure: 45 kPa
Blow frequency: 3.200 blows./min
Piston diameter: 80 mm
Nominal stroke: 40 mm
Bit type: Tungsten carbide, 110° edge angle.

The best correlations were obtained between drilling rate and shore stereoscope hardness (r = 0.976), tensile strength (r = 0.863), compressive strength (0.800), elastic modulus (0.794), and point load index (0.777).

The second objective was to investigate the relationships between drilling rates and cutting specific energy, leading to the classification of the rock formations in

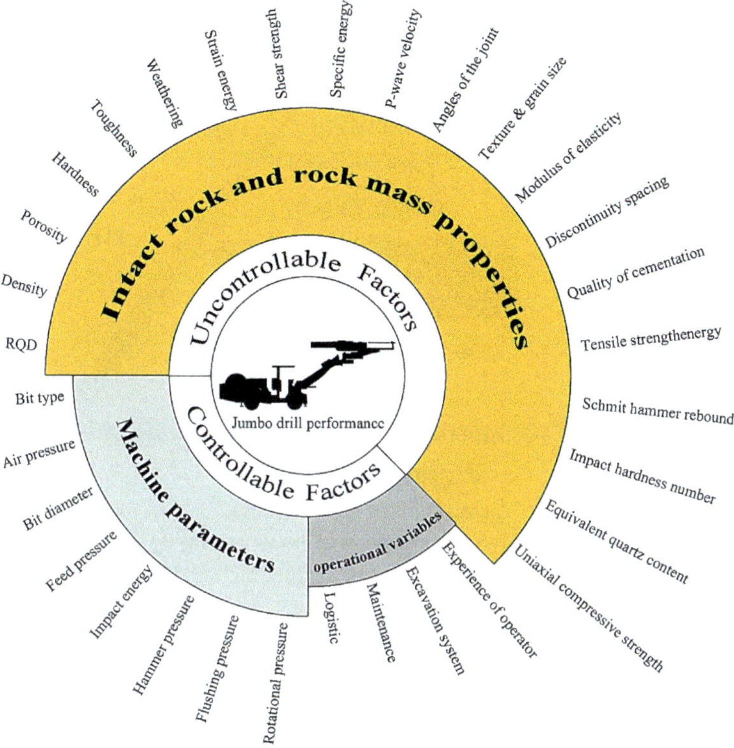

Fig. 3.13 The most important factors influencing drilling performance, Heydari et al. (2024a, b)

the Zonguldak coal field according to the applicability of roadheaders. For this, it is aimed to use the relationships given by McFeat and Fowell (1979), as shown in Fig. 3.16.

The cuttability of the rock specimen taken from the same roadways was determined using block grooving tests. The method has been widely described elsewhere (McFeat and Fowell 1979). Cutting conditions were as follows.

Depth of Cut: 5 mm
Cutting speed:150 mm/sec
Tool geometry: Chisel insert, rake angle—50
Back clearance angle: 50
Tool width:12.7 mm.

Specific energy, cutting, and normal force components acting on the cutting tool were meticulously recorded for each experimental rock using the technique mentioned above. Specific energy is defined as the work done to excavate a unit volume of rock, and it is obtained by dividing the mean cutting force by the volume of the material excavated per unit distance cut. Both the performance of a percussive drill and tunneling machine are primarily governed by rock mass properties, so it is

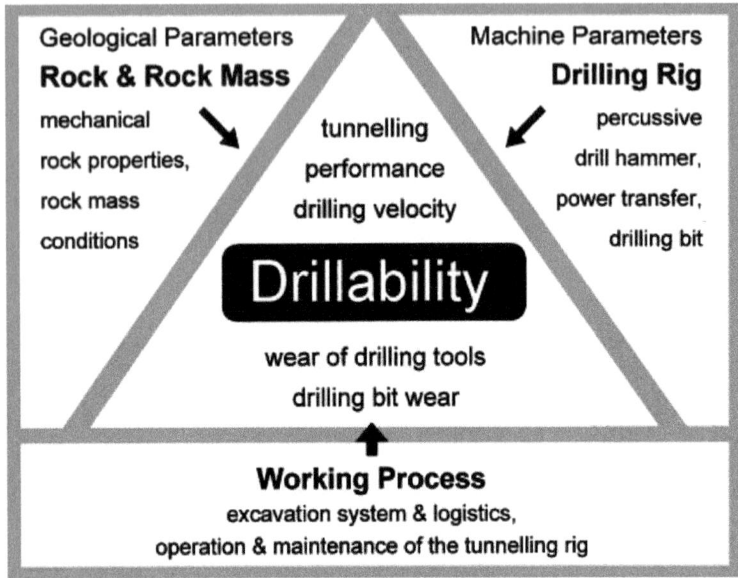

Fig. 3.14 Illustration of the main factors influencing hard rock drilling, Thuro (1977)

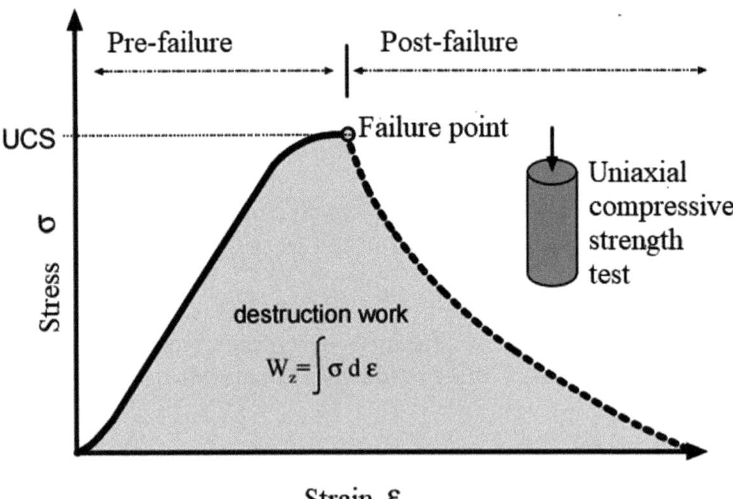

Fig. 3.15 Definition of the destruction work, Thuro (1977)

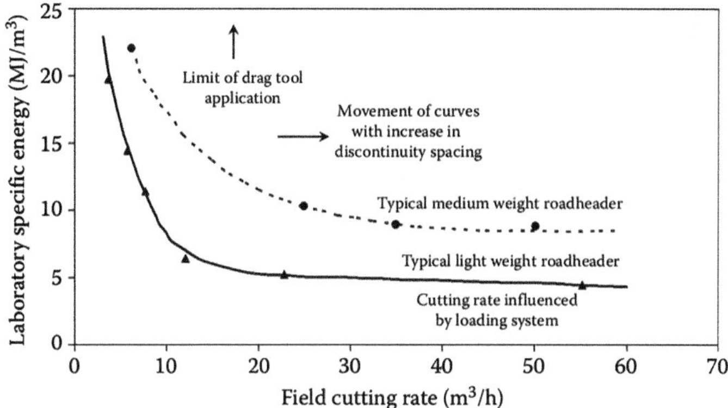

Fig. 3.16 The relationship between cutting specific energy and field cutting rates of medium (20–35 t) and heavyweight (50–70 t) roadheaders Smith McFeat and Fowell (1979)

reasonable to suggest that the drillability and cuttability of a rock mass are closely related. This study showed that pick cutting parameters can be easily predicted from percussive drilling rates. Figure 3.17 shows the relationship between the chisel pick cutting specific energy and field drilling rates of percussive drill rigs, providing reliable data for your research and work in the field.

The laboratory cutting experiments and field drilling tests showed that pick cutting forces and cutting specific energy values are closely related exponentially to percussive drilling rates. This relationship permitted the classification of the rock formations in Zonguldak Coalfield. Importantly, these relationships are only valid if the same type of drill rigs are used and operational air pressure is kept constant. The measurement of road header performance in the Middle Anatolian coalfield confirmed the above findings, allowing for an improved prediction of cutting rates and hence tunneling costs, which has direct practical implications for your in the field, Bilgin (1983).

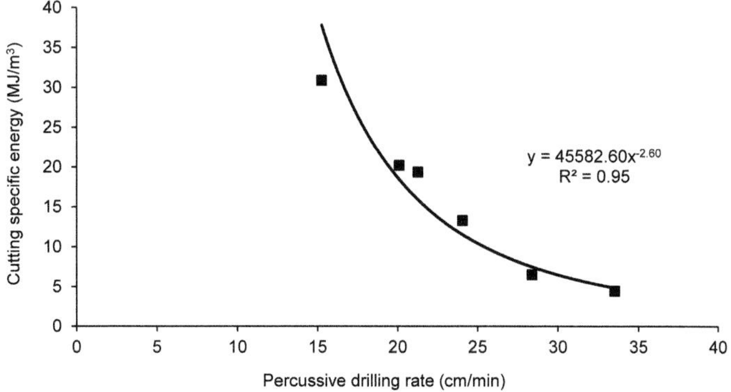

Fig. 3.17 The relationship between drilling rate and cutting specific energy, Bilgin (1983)

3.4 Full-Scale Laboratory Drilling Tests and Drilling-Specific Energy

The main objective of the research program was to select rocks having different strength properties, to carry out full-scale drilling and cutting tests, to see rock properties affecting drillability and the cuttability characteristics, and to find out the relations between these two characteristics leading to selection the proper mechanical excavators from drilling specific energy values, Bilgin et al. (2011), Balci et al. (2020). Five different rock samples were selected for the testing program. These are Travertine, Kufeki-Limestone, Limestone, Sandstone and Beige Marble. The compressive strength of the rocks varies between 9.1 and 160.7 MPa, with brittleness values between 2.9 and 20.5 (brittleness is defined as the ratio of compressive strength to tensile strength), having apparent porosity between 0.14 and 18.9%. The horizontal drill rig shown in Fig. 3.18 is used for drilling experiments. The drilling power of the machine is 132 kW with a maximum thrust force of 50 t and a maximum rpm of 90. Two PDC drill bits having 11.2 cm (4.4 inches) of diameter were used during drilling tests, Fig. 3.19. PDC cutter has 0.5 cm of height, so when choosing the maximum thrust force level, special attention was paid not to pass 0.4 cm advance per one revolution of the drill bit. The groove opened by a PDC drill bit is seen in Fig. 3.20.

Fig. 3.18 Horizontal drill rig used in the experiments, Bilgin et al. (2011), Balcı et al. (2020)

Fig. 3.19 PDC cutters used in the experiments, Bilgin et al. (2011), Balcı et al. (2020)

Fig. 3.20 Groove opened by a PDC drill bit, Bilgin et al. (2011), Balcı et al. (2020)

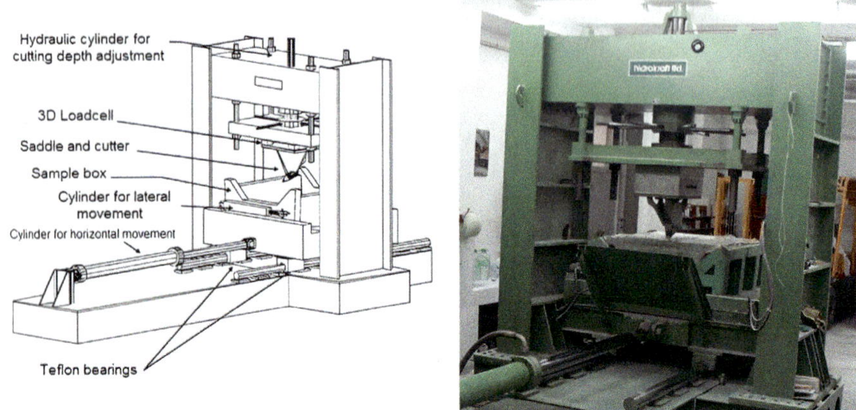

Fig. 3.21 Linear cutting rig used in the experiments, Bilgin et al. (2011), Balcı et al. (2020)

Linear Cutting Machine, seen in Fig. 3.21, is used for rock-cutting experiments. Full-scale cutting tests were realized with a conical cutter (Sandvik Q2VA-3877-2162) at a constant cutting speed of 12.5 cm/sec.

"DrillingSpecific energy" is the energy required to excavate a unit volume of rock. It is a valuable parameter that may also be taken as an index of the mechanical efficiency of a rock-working process. The following equation derived for specific energy by Teale (1965) in rotary drilling is used throughout the research study:

$$E_s = \frac{WOB}{A_B} + \frac{2\pi NT}{A_B ROP} \tag{3.1}$$

where:

E_S	Specific energy; MJ/m^3
WOB	Weight of bit, or thrust; MN
A_B	Area of the drill bit or hole; m^2
N	Rotational speed; revolution per second
T	Torque; MNm
ROP	Rate of penetration; m^3/h.

Teale also introduced the concept of minimum specific energy and/or maximum mechanical efficiency. The minimum specific energy is reached when the specific energy approaches, or is roughly equal to, the compressive strength of the rock being drilled. The mechanical efficiency is then;

$$EFF_M = \frac{E_{S\,\min}}{E_S}100 \tag{3.2}$$

$$E_s = E_{S\,\min}100 \approx \sigma \tag{3.3}$$

Es_{min} Rock strength

The maximum efficiency is reached when the specific energy approaches or roughly equals the compressive strength of the rock being drilled. The maximum efficiency is reached when

$$\frac{E_{S\ min}}{E_s} = 1 \quad (3.4)$$

The relationships between thrust and drilling rate are linear for medium-strength rocks like Travertine and K. limestone. However, there is a threshold in higher-strength rock samples. A minimum thrust force is necessary to penetrate PDC bits into the high-strength rocks; thereafter, the relationship is linear, as seen in Fig. 3.22 obtained for Sandstone. The slope of this linear portion in (m/h)/thrust is linearly related to UCS, as seen in Fig. 3.23. The test results are significant in the relationship between specific drilling rates and the rock UCS, as displayed in Fig. 3.23. A definite relationship between these two parameters exists, which means that by calculating the specific drilling rate for the field drilling program, one can estimate in-situ rock strength.

The relationships between thrust and SE are shown in Fig. 3.24. As seen from these figures, SE decreases dramatically with thrust force; after a certain level of thrust force, SE levels off, giving minimum SE. These minimum SE values will be used to calculate relevant correlations. Minimum drilling SE is correlated with UCS in Fig. 3.25. This relationship is significant for the test objectives. By analyzing

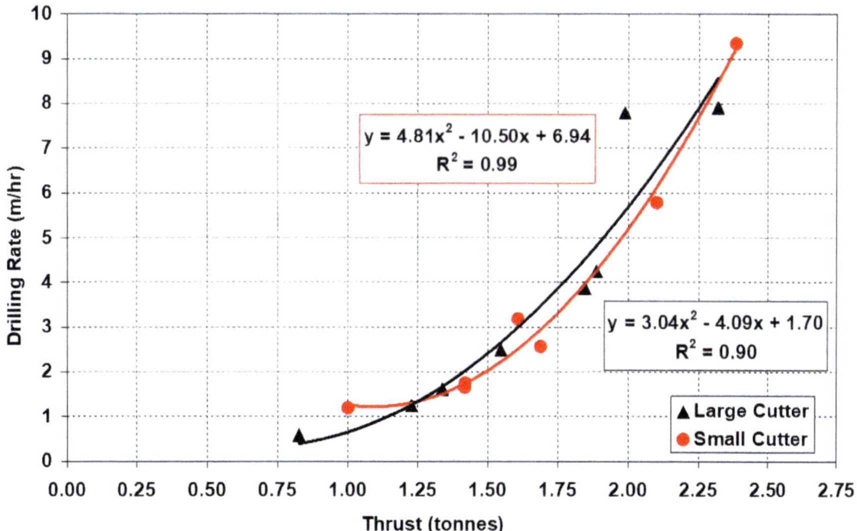

Fig. 3.22 The relationship between drilling rate and thrust in Sandstone, Bilgin et al. (2011), Balcı et al. (2020)

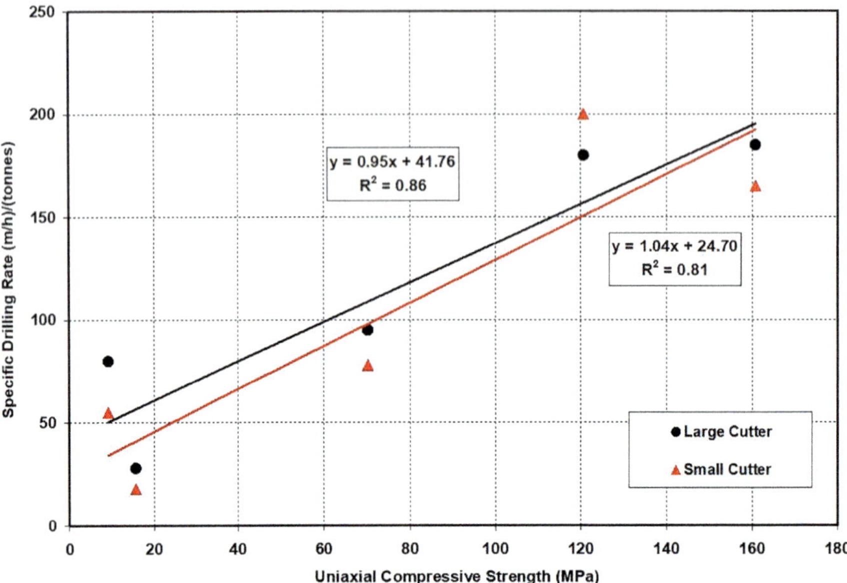

Fig. 3.23 The relationship between specific drilling rate and thrust for five rocks tested; specific drilling rate is defined as the slope of drilling rate/thrust graph, Bilgin et al. (2011), Balcı et al. (2020)

the field drilling data and calculating the in-situ drilling specific energy, one should estimate the in-situ UCS. This concept justifies the works done by Wu et al. (2024) and Zhao et al. (2024).

As a first analysis, drilling SE for two PDC and five rocks is plotted against destruction energy and cutting specific energy values for 25 and 50 mm spacing. As seen in Figs. 3.26 and 3.27, the results are auspicious for further analysis.

Drilling and cutting results and measurements of destruction-specific energy look highly promising for developing a reliable model to estimate the in-situ cutting-specific energy from drilling data. The relationship between the specific drilling rate and the rock UCS is significant based on the test results, as displayed in Fig. 3.23. A definite relationship between these two parameters exists, which means that by calculating the specific drilling rate for the field drilling program, one can estimate in-situ rock strength. By analyzing the field drilling data and calculating the in situ drilling specific energy, one should assess the in-situ UCS.

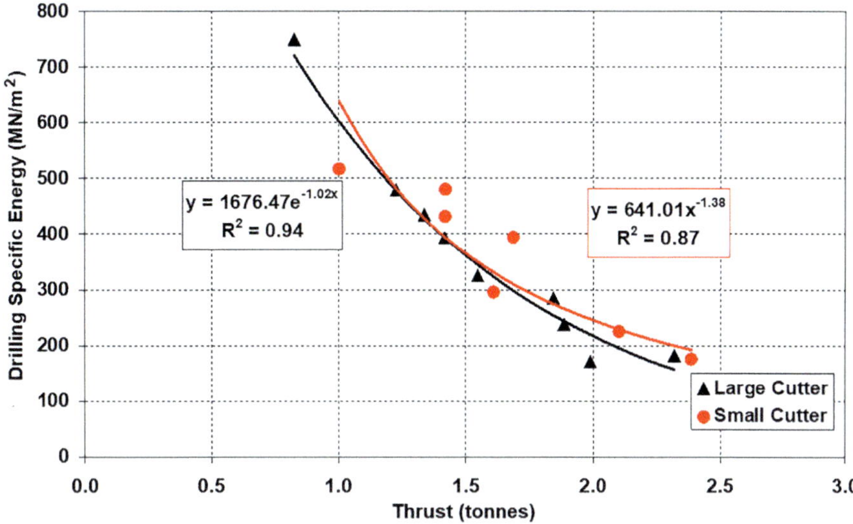

Fig. 3.24 The relationship between thrust and drilling specific energy in Sandstone, Bilgin et al. (2011), Balcı et al. (2020)

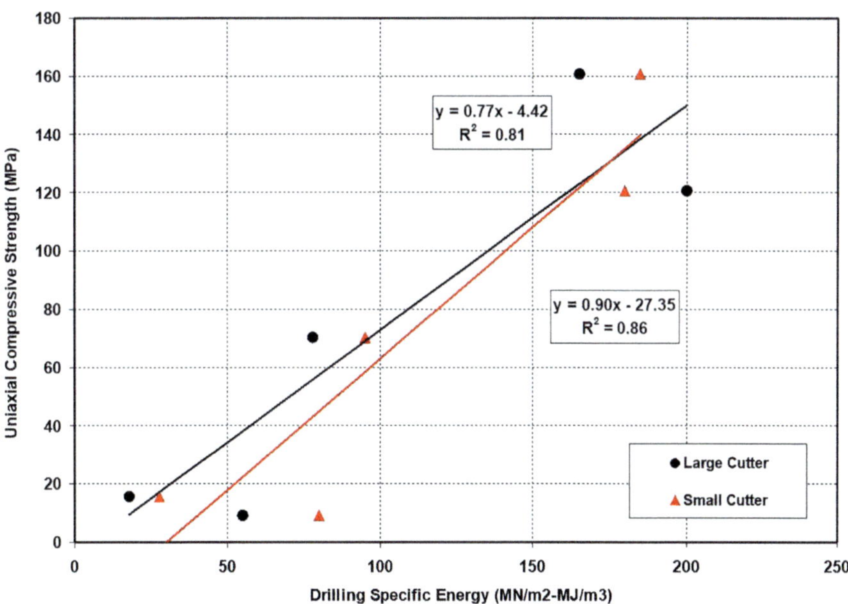

Fig. 3.25 The relationship between drilling specific energy and uniaxial compressive strength for five rocks tested, Bilgin et al. (2011), Balcı et al. (2020)

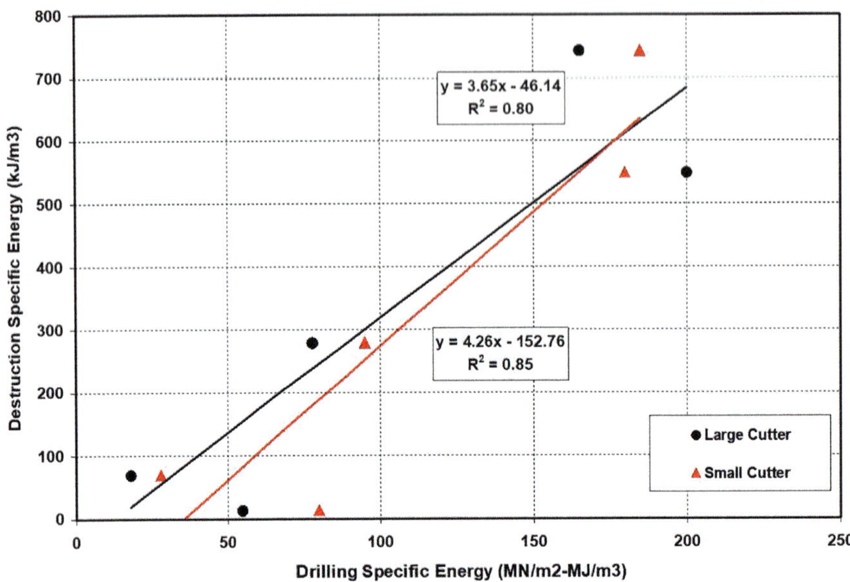

Fig. 3.26 The relationship between drilling specific energy and destruction energy/work for five rocks tested, Bilgin et al. (2011), Balcı et al. (2020)

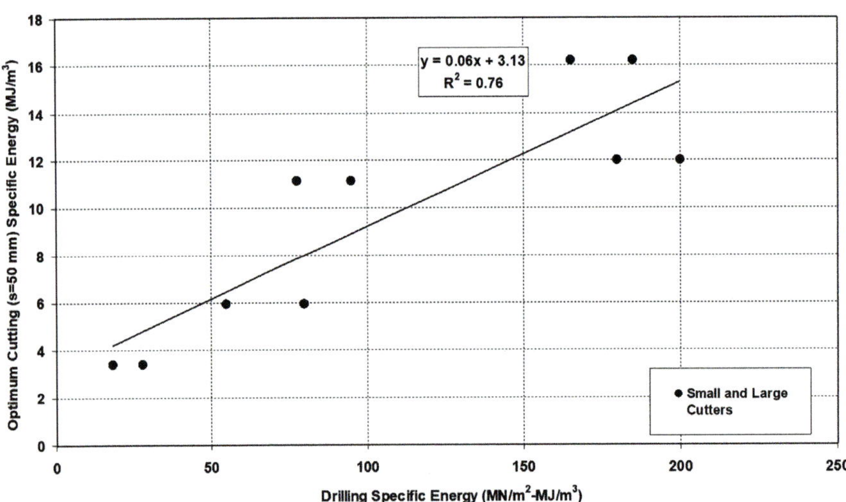

Fig. 3.27 The relationship between drilling specific energy and optimum cutting specific energy for five rocks and two small and large PDC drill bits, Bilgin et al. (2011), Balcı et al. (2020)

3.5 Some Examples of Conventional Tunneling Methods, Past Experiences

Typical examples of drill and blast tunneling methods are given in Sect. 3.7.1 for Tunnel T1 in Algeria, and another ones for Ovit and Kartal-Kozyatağı Metro Tunnels in Turkey in Chap. 9. However, a typical example of a conventional tunneling method using impact hammers as an example of a primary excavation method is given in this chapter. Almost 15 km of metro tunnels are driven in Istanbul using conventional tunneling methods, such as impact hammers, since the initial capital investment is relatively low and rock formations are highly fractured in some zones. The example given in this section is taken partly from a paper published in Tunnels and Tunneling titled "The performance prediction of impact hammers from Schmidt hammer rebound values in Istanbul metro tunnel drives," Bilgin et al. (2002). It belongs to part 1 between 4. Levent and Taksim and part 2 between Taksim and Yenikapı, with a total length of 11 km. Trakya formation of the Carboniferous age is found in the fine-grained, laminated, fractured, and inter-bedded siltstone, Sandstone, and mudstone. Some diabase or andesite dykes have also been encountered while driving the tunnels, and these affected progress rates, as they are significantly stronger and more massive (with RQD of approximately 100%) than rock excavated along the significant part of the route. Many faults and geologic discontinuities are developed in the area due to Hercinian and Alpine Orogenies. Rock Quality Designation (RQD) values vary between 0 and 100%, and uniaxial compressive strength between 30 and 150 MPa. The new Austrian Tunneling Method (NATM) was used since the tunnel diameter and ground structure vary frequently along the route. Three to four-meter-long rock bolts, wire mesh, and shotcrete were used as temporary tunnel support. Depending on tunnel diameters, the final lining is undertaken with 35–45 cm thick in-situ cast concrete. Single-track tunnel type A has a cross-section of 36 m^2 and is excavated in two steps. The method of construction is explained in Fig. 3.28.

In this Figure, the upper bench of 28 m^2 is excavated first, and the lower bench of 8 m^2 is excavated later, 30 m behind the first bench. Figures 3.29 and 3.30 summarize the overall performance of the tunnel drive in Phases 1 and 2.

As seen from these figures, the utilization of impact hammers on average is 22, and 17% of the total time is spent on mucking. Shotcrete takes almost 27% of the total time. Machine breakdowns and workers' travel time to reach the face are classified as waiting and take approximately 6–7%. The rock formation in Phase 2 was more fractured than in Phase 1, necessitating fore poling operations in Phase 2. The time spent for mucking was substantially different in phases 1 and 2 due to the difference in the muck transportation system used. Since the spacing between steel arches was 1 m, the daily advance rates were kept as the multiplier of 1 m provided that the last set of steel arches was set up in the final shift of the day. Daily advance rates change usually between 2 and 3 m. Alpine ATM 75 transverse type roadheader having a cutting power of 200 kW was used in Taksim platform tunnel 2. Eickhoff ET 250 transverse type roadheader having a cutting power of 250 kW was used in Taksim platform tunnel 1. Platform tunnels have a length of 238 m and a cross-section of

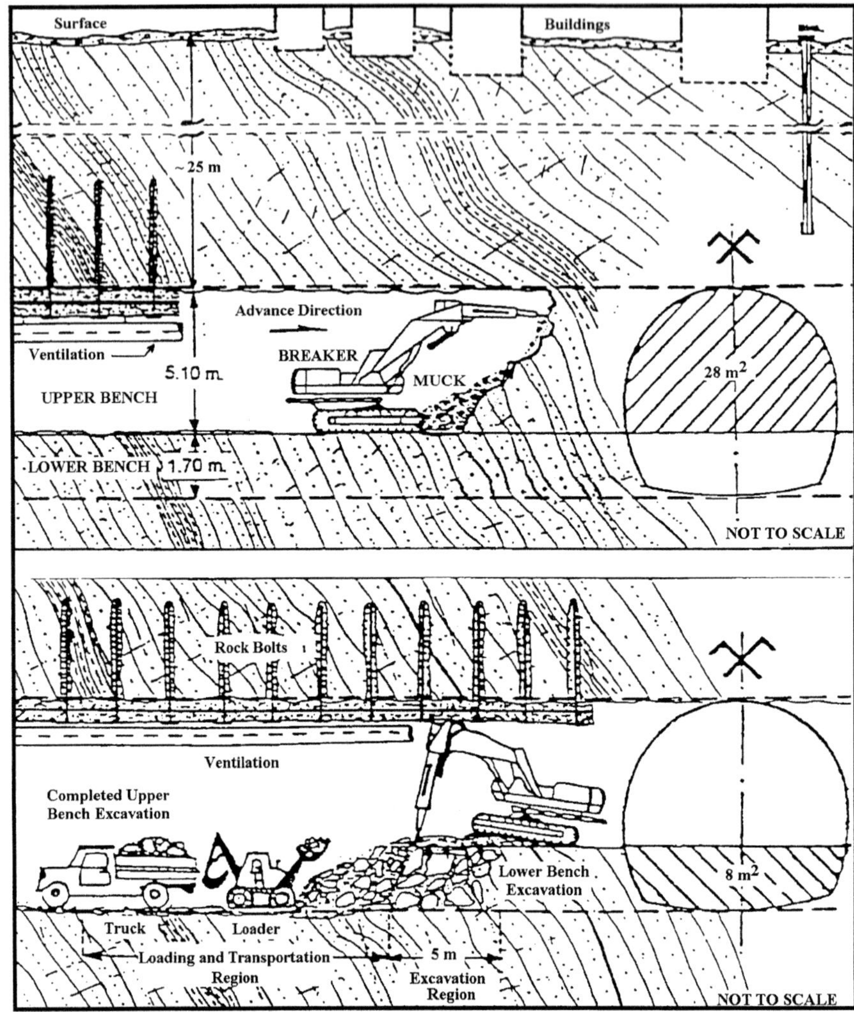

Fig. 3.28 Method of construction in Levent-Taksim metro tunnel, Bilgin et al. (2002), credit to Prof. Dr. Cüneyt Güler

64 m^2. Krupp HM 185 and Montabert BRH 250 jackhammers attached to hydraulic excavators were used in Taksim–Levent line 2 tunnels, Bilgin et al. (2002).

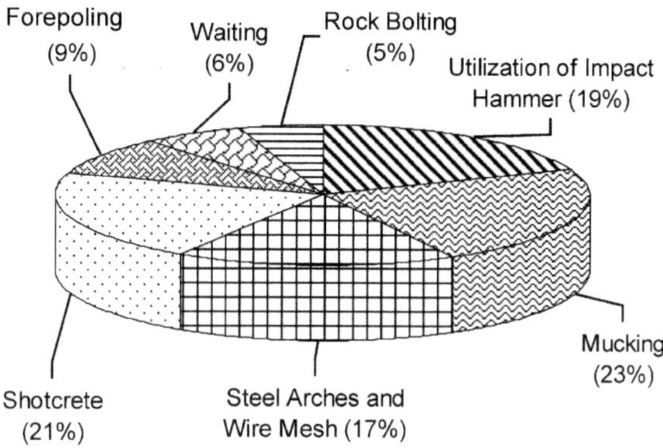

Fig. 3.29 Overall performance of the tunnel drive in phase 1, from the Author Bilgin et al. (2002)

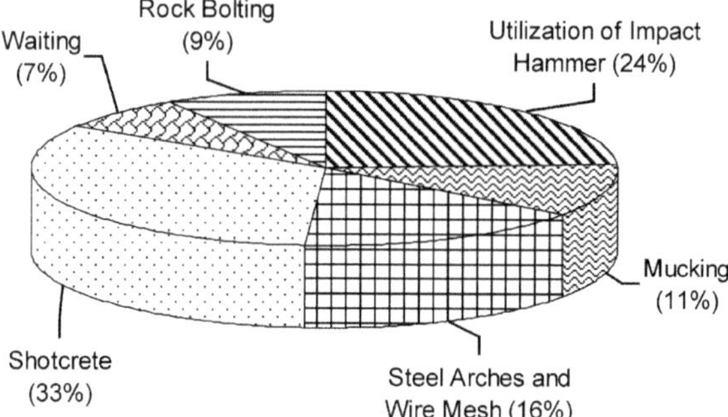

Fig. 3.30 Overall performance of the tunnel drive in phase 2, from the Author Bilgin et al. (2002)

3.6 A Summary of Blasting Technology for Tunneling

In this section, blasting technology for tunneling will be briefly discussed. We believe that the readers of this book, after consulting tree examples of drill and blast tunneling methods which will be given in Sect. 3.7.1 for Tunnel T1 in Algeria, and another ones for Ovit and Kartal-Kozyatağı Metro Tunnels in Turkey in Chap. 9, will be more acquainted with the subject.

3.6.1 Explosives

Essential elements of the tunnel blasting are explosives. Upon their detonation, the heat of around 3000–4000 °C and large quantities of high-pressure gases reaching up to 250,000 bar is rapidly released. The gases expand quickly to overcome the confining forces of the surrounding rock formation. The expansion of gases generated is very rapid, reaching velocities of 7000–8000 m/s. The entire conversion process takes only a fraction of a second and is accompanied by shock and loud noise. The primary explosives used are nitroglycerin-based dynamites or gelatins, Zou (2017). The most used explosives in the tunnels are ANFO, slurries, and emulsion explosives. ANFO is a mixture of ammonium nitrate of 94% and fuel oil of 6%. Under conditions of high moisture content, it will cake. ANFO can only be used in dry holes unless packaged in a suitable waterproof cartridge. Water gel/slurry explosive is essentially based on saturated aqueous solutions of ammonium nitrate, often with other oxidizers such as sodium nitrate and calcium nitrate, in which the fuels, sensitizers, and cross-linking agents are dispersed to avoid the segregation of the solids. Slurry explosives have a very good to excellent water resistance. They can sustain a regular detonation after 24 h under 10-m-deep water. Water gel is generally chemically stable and has a shelf life of about 1–2 years under normal storage conditions. Today, emulsion explosives have almost replaced slurry/water gel and share domination of the worldwide explosive market with ANFO. Emulsion explosives generally refer to water-in-oil, emulsion-type, water-resistant industrial explosives made by emulsification technique. One of the outstanding characteristics of emulsion explosives is their high degree of inherent safety. In addition to ANFO, the production costs of emulsion explosives are the lowest in all commercial explosives. Table 3.2 shows the composition of two types of cartridge emulsion explosives, Zou (2017).

Two parameters are often calculated from a blast design: the powder factor or specific charge (kilograms of explosives per cubic meter of blasted rock) and the drill factor (total length of drill holes per volume of blasted rock (meter/cubic meter). These indicators of the overall blasting economy permit easy comparison among blast patterns. The powder factor varies between 0.6 and 5 kg/m^3 for most typical tunnel

Table 3.2 Content of the emulsion, Zou (2017)

Emulsion explosive	EL-102-2	BME-2-2
Ammonium nitrate %	73.0	63.0
Sodium nitrate %	10.0	9.0
Ammonium perchlorate %	–	6.2
Water %	12.0	11.0
Emulsifier %	1.0	1.0–1.5
Wax %	2.5 (composite)	1.5
Vaseline %	–	2.0
Mycrocrystalled wax %	–	0.8
Density adjuster %	–	2–4
Appearance status	Cartridge with paper/plastic wrapper	Cartridge with paper/plastic wrapper

blasting. Typical drill factors vary between 0.8 and 6 m/m^3. All commercial explosives can be ignited by a blasting cap or an electric detonator, except for ANFO, slurry, and most emulsions. Sometimes, ANFO explosives are ignited with a detonator. However, primers are usually used. For this purpose, the primer should consist of an explosive with high detonation velocity. Dynamite can be successfully utilized to ignite ANFO.

3.6.2 Detonators

For safety reasons, explosives should only detonate intentionally, so charges are required to set off the explosion, which contains explosive material called detonators. A relatively small amount of energy is required for initiation. Primary explosives are often used in detonators or to trigger more considerable charges of less sensitive secondary explosives. Electrical detonators have many advantages, such as total control of the initiation time, reduced air blast and ground vibration, and better blasting results with delays, which is particularly important in tunnel blasting. However, for safety reasons, they are not allowed to tunnel in certain countries, such as Sweden, Chapman et al. (2010). Due to this disadvantage, the detonator cord was developed, which is safe for use in extraneous electricity environments, has no limitation concerning hole size, and is inexpensive. However, it causes a 'noisy' initiation and requires much cord movement. This disadvantage drove the need to find alternative initiation systems and non-electrical detonators, which were brought to the market in the early 1973s with the commercial NONEL. This non-electric initiation system, also called the shock tube initiating system, was invented by Nitro Nobel AB of Sweden. Instead of electric wires, a hollow plastic shock tube delivers the firing impulse to the detonator. The main disadvantage is its cost. Figure 3.31 shows a typical view of NONEL.

There is also another type of detonator called electronic detonator. Electronic detonators are fully programmable and intelligent high strength detonators. Each detonator contains a circuit board that can be programmed to initiate at precise millisecond timing within a firing sequence. According to Cardu et al. (2024) they are advantageous in terms of vibration reduction, increased frequencies, air blast, improved fragmentation in the muck pile, diggability, crushing cost saving (less energy used during the primary and secondary fragmentation), and control of overbreak, which allows greater profile accuracy. Basic structure of electric detonator, Nonel and electronic detonator is given in Fig. 3.32.

3.6.3 Drilling Pattern, Blasting Sequence, and Type of Cuts

The drilling pattern ensures the distribution of the explosive in the rock and the desired blasting result. Several factors must be considered when designing the drilling

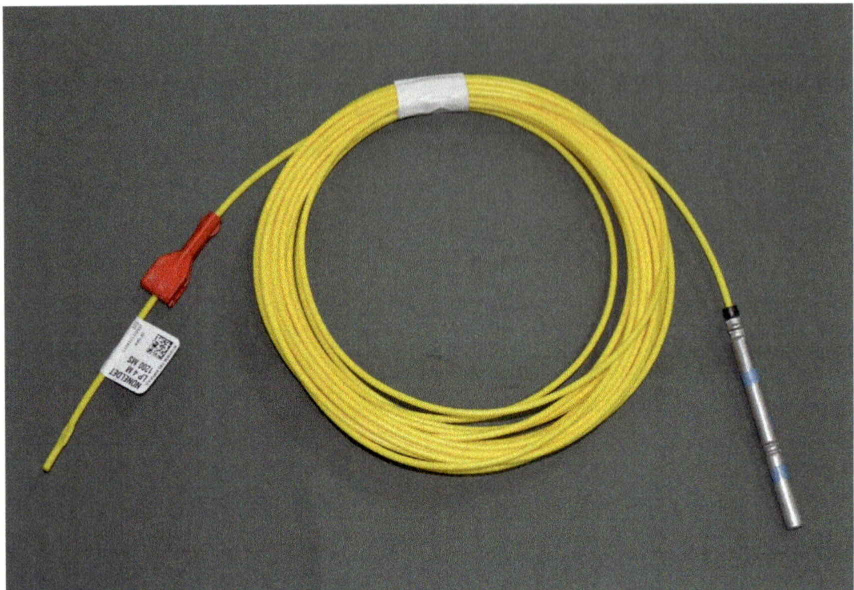

Fig. 3.31 Nonel, with kind permission of Nitromak

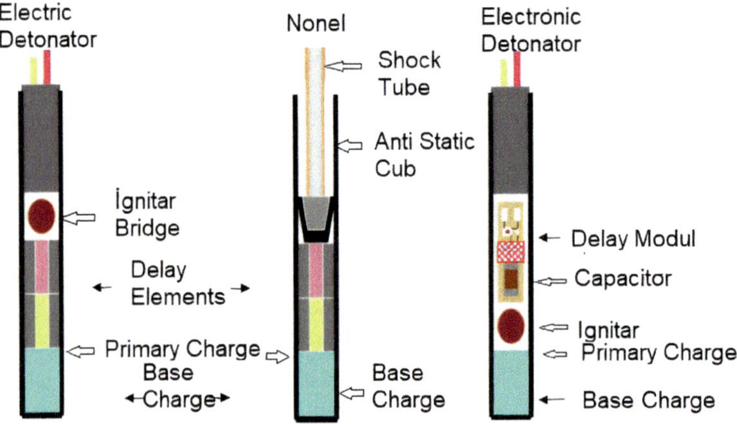

Fig. 3.32 Basic detanator structure, with kind permission of Nitromak

pattern: drilling equipment, geology, rock drillability, explosives availability, blast vibration restrictions, and accuracy requirements of the blasted wall.

A cut in the tunnel face produces this second face, which can be either a parallel hole cut, a V-cut, a fan-cut, or other ways of opening up the tunnel face. The blasting sequence in a tunnel or drift always starts from the "cut," a pattern of holes at or close to the center of the face designed to provide the ideal line of deformation. The cut's placement, arrangement, and drilling accuracy are crucial for successful blasting in tunneling. The basic layout always involves one or several uncharged holes drilled at or very near the center of the cut, providing space for the adjacent blasted holes. Uncharged cut holes are typically 76–127 mm in diameter.

The holes surrounding the cut are called stope-holes, typically between 41 and 51 mm in diameter. The contour holes lie in the perimeter of the drilling pattern to get smooth walls. Inaccurate drilling and charging of the floor holes can leave unblasted surfaces, which are difficult to remove later. The look-out angle is between the practical (drilled) and the theoretical tunnel profile. If the contour holes are drilled parallel to the theoretical line of the tunnel, the tunnel face gets smaller and smaller after each round.

The parallel cut is especially suitable for modern mechanized tunneling equipment. This cut type has also made long rounds, which is common in small tunnels. An earlier version of the parallel cut is the "burn cut," which does not use uncharged holes, relying instead on a powerful charge to burn the rock. Today, the parallel cut has replaced the burn cut. The V cut is a traditional cut based on symmetrically drilled, angled holes. It has lost some popularity with the widespread adoption of the parallel cut and longer rounds. The principle of the fan is to make a trench-like opening across the tunnel, and the charge calculations are similar to those for opening the bench. Due to the geometrical design of the cut, the hole construction is small, making the cut easy to blast. Hole drilling and charging are similar to cutting holes in the V-cut.

For detailed information concerning drill pattern design, readers should consult Chap. 9, Fig. 9.7 for the Ovit tunnel in Turkey, Chap. 11, Figs. 11.4–11.7 for Kaynarca-Kartal-Kadıköy Metro tunnels.

3.7 Vibration Due to Drill and Blast and Its Effects on Buildings and Monuments

One of the inevitable negative side-effects of excavation with blasting is blasting-induced vibrations. An overview of the blast vibration regulations and legislation in Europe is given by Kaltenegger et al. (2016). It is well-known that improving the delay design by using non-electric detonators, drilling accuracy, and choosing the right explosive type reduce blast vibrations. Particle velocity (mm/sec) is the best and the most practical description for defining potential damage to structures. However, the effects of the frequency of the vibrations are not negligible, especially if the frequency is equal to the natural frequency of the structure, which could cause resonance and damage.

Currently, the most widely accepted single measurement of ground vibration considered potentially damaging is peak particle velocity (PPV), Hosseinie and Baghikhani (2013). PPV is the speed at which earth particles move or pass a site. Through measuring vibrations with a seismograph, it has been found that this PPV can be related to the weight of the explosive charge and its distance to the recording site through the application of a simple power law formula (Eq. 3.1)

$$V = \frac{K.W\alpha}{R\beta} \tag{3.5}$$

where V is the PPV in mm/s, W is the charge weight per delay in kg, and R is the radial distance from the point of detonation in m. The constant K, α, and β used in this equation depended upon the type of blast and condition of rock mass

An essential factor for reducing blast vibrations is the charge per ignition time. Setting up a higher number of ignition delays makes it possible to optimize the amount of explosives initiated simultaneously. One alternative is using electronic detonators, which can be programmed and fired in milliseconds. More commonly, non-electronic detonators are used.

3.7.1 A Study on the Effect of Blasting Vibration on the Environment in a T1 Tunnel in Algeria

This section is intended to show typical in-situ vibration measurements due to blasting operations in a tunnel in Algeria. The northern part of the national road RN09 linking Setif and Bejaia is in a mountainous area of the country. The existing road is currently insufficient for today's traffic needs. As a result, the widening and development of the eastern road was planned recently. In the work sites, there are tunnels, viaducts, etc. One of these tunnels, T1 works located between 1 + 460 and 1 + 975, has a length of 515 m. At the north portal of T1, 25 m below the portal in the Oued Agrioun valley, there is a spring catchment from the water basin, which meets 60% of Bejaia's water demand (Fig. 3.33).

The city's governor officially asked the contractor to conduct a study to check the risk of disruption of the water source due to the vibration created by the blasting operations for the tunnel construction. Following this request, a commission of experts visited the tunneling area in December 2018. A detailed geological study and observations of the blasting operations were done to clarify the vibration's risk to the water source's disruption, Bilgin et al. (2018). This section is a summary of this field study.

This study aimed to evaluate and estimate the possible effects of using explosives in the T1 tunnel on the water sources that fill the reservoir located 25 m below the T1 tunnel providing water drinking in Bejaia. Based on these facts, a detailed geological study was conducted on-site. The rock mass in the tunnel T1 is limestone. The initial portal is located behind the water catchment supplying Bejaia and on the fault front, which separates the limestone and travertine. This location, corresponding to the foot of the Ait M'barek slide, is very close to the water catchment. The fault just behind the catchment separates the massive limestones, and the limestones/travertines are extremely carstic. This fault creates an impenetrable barrier for water underground. As a result, no underground water inflow to the fault's south side is observed along

Fig. 3.33 View of the spring catchment from the water basin of the city of Bejaia, Bilgin et al. (2018)

the route of the T1 tunnel or the valley. On the contrary, the north side of the fault is rich in terms of underground and surface water. The water catchment which feeds Bejaia comes from this fault.

The Ait M'barek landslide is an old and large-scale landslide. To stay on the secure side so as not to excavate the slip foot of the landslide and not disturb the fault filling the catchment, a new tunnel T3 is proposed at 65 m south instead of the entrance gate initially planned. A 26 m horizontal survey was carried out at the level of the new portal proposed, and no water ingress was noted. This fact demonstrates that groundwater comes from the western and northern zones. Detailed laboratory and field studies showed that the rocks around T1 and T3 have the same geotechnical characteristics. Three shots were observed in the T3 tunnel on December 9, 10, and 11. Figures 3.34, 3.35, and Table 3.3 show the face of the tunnel and a detailed shooting plan, including the number of drill holes and the characteristics of the explosives.

The Abem Vibralog (vibration monitor) device was used on December 11 to measure the vibrations to see if the shooting results were between the international limits to protect the water source. Geological studies have shown that the water source supplied to the reservoir was 65 m away from the first portal chosen in the T1 tunnel. So, the vibration measuring device was placed at this distance. Figure 3.36 shows the measuring device.

Fig. 3.34 Face of the tunnel where the experimental shooting was realized, Bilgin et al. (2018)

This device is well known worldwide for measuring the vibrations created by firing in tunnels and mines, Nateghi (2011) and Trajkoviç' et al. (2013). It is also reported that this device was used to predict the effects of the vibrations of the shots on the water system of the Gotvand dam in Iran. Figure 3.37 shows the result of the firing obtained on December 11, 2018. It is interesting to see in this figure that the fragmented rocks have dimensions that show firing efficiency.

The results of the measured vibration are as follows:

VV = 6.48 mm/sec

VL = 3.59 mm/sec

VT = 4.58 mm/sec

$$VPP = \sqrt{VV^2 + VL^2 + VT^2} \tag{3.6}$$

VPP = 8.7 mm/sec

PPV = 8.7 mm/sec

Here VV is the speed in the vertical direction

VL is the speed in the longitudinal direction

VT is the speed in the transverse direction and VPP is the maximum speed of the particles.

Load 180/12 = 15 kg/delay of 20 ms

The frequency measurements' results are:

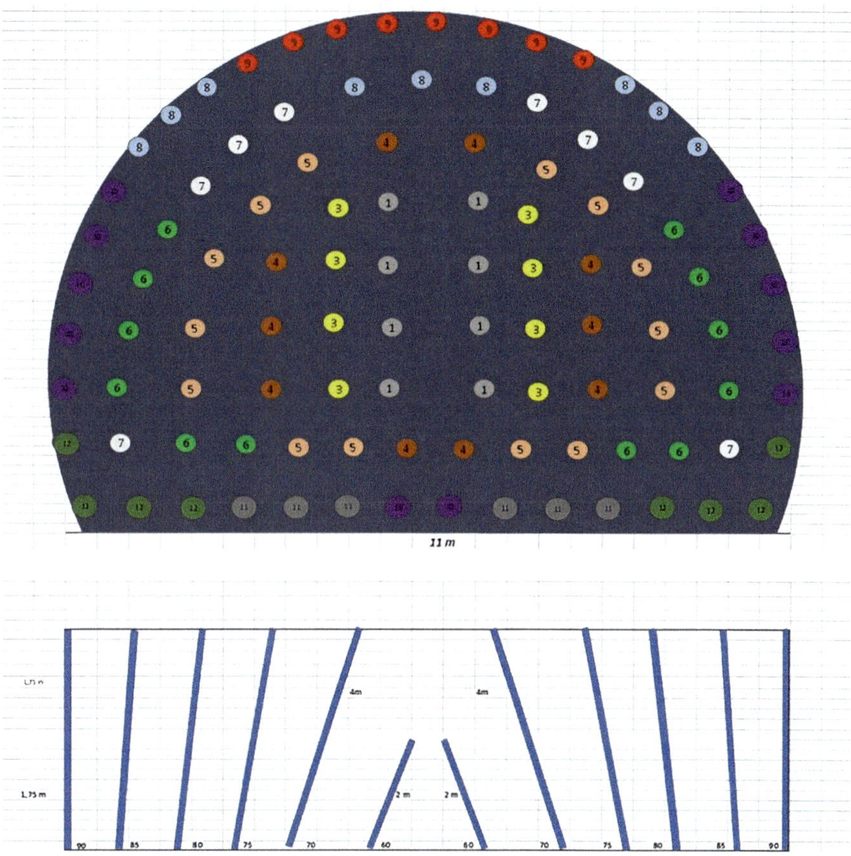

Fig. 3.35 Plan of shooting, Bilgin et al. (2018)

FV = 5.0 Hz

FL = 4.6 Hz

FT = 4.7 Hz

Here, FV is the vibration frequency in the vertical direction, FL is the vibration frequency in the longitudinal direction, and FT is the vibration frequency in the transverse direction. The results of the acceleration measurements obtained are as follows:

AV = 0.4 mm/sec^2

AL = 0.5 mm/sec^2

AT = 0.8 mm/sec^2

Here AV is acceleration in the vertical direction, AL is the acceleration in the longitudinal direction and AT acceleration in the transverse direction.

Table 3.3 Details of shooting plan in tunnel T3, Bilgin et al. (2018)

Cartridge No.	Calotte	Calotte	The number of calotte	The number of stross	Dynamite
1	5 dynamite	10 kg	8 holes	3 holes	Gelanite 30 mm
2	–	–	–	–	Gelanite 30 mm
3	10 dynamite	20 kg	8 holes	–	Gelanite 30 mm
4	8 dynamite	20 kg	10 holes	4 holes	Gelanite 30 mm
5	8 dynamite	28 kg	14 holes	–	Gelanite 30 mm
6	7 dynamite	21 kg	12 holes	–	Gelanite 30 mm
7	7 dynamite	14 kg	8 holes	–	Gelanite 30 mm
8	4 dynamite	9 kg	9 holes	3 holes	Gelanite 30 mm
9	4 dynamite	8 kg	8 holes	–	Gelanite 30 mm
10	8 dynamite	24 kg	12 holes	–	Gelanite 30 mm
11	8 dynamite	12 kg	6 holes	–	Gelanite 30 mm
12	8 dynamite	16 kg	8 holes	2 holes	Gelanite 30 mm
Total	728 dynamite	182 kg	103 holes	12 holes	

Mean calotte area 50 m^2, the name of dynamite Samex 1, the diameter of dynamite 30 mm, the weight of one cartridge 0.25 kg, the diameter of the hole 45–51 mm

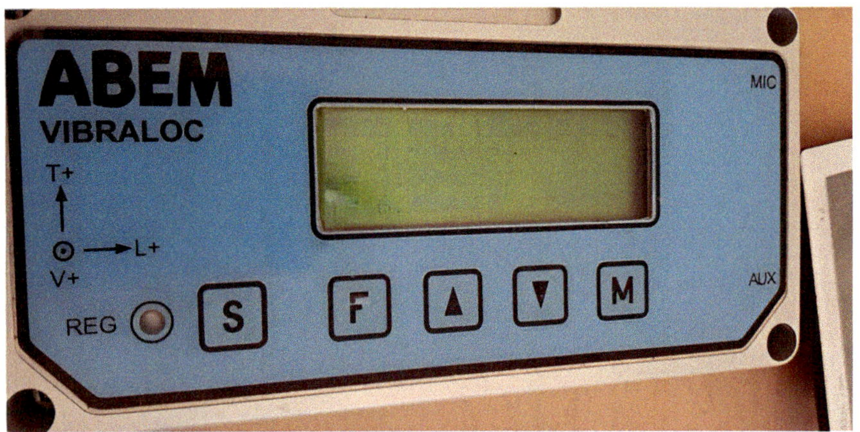

Fig. 3.36 The Abem vibration monitoring device, Bilgin et al. (2018)

3.7.2 Execution of Vibration Measurements to Verify the Suitability of the Shots Carried Out in the T1 Tunnel to Protect the Water Source Supplying Bejaia

Two international methods have been used to see the influence on the environment shots fired in the T3 tunnel.

Fig. 3.37 The results of firing obtained in tunnel T3 on December, 2018, Bilgin et al. (2018)

1. This method was developed by the U.S. Bureau of Mines and cited in Report No. 8507. Several authors use this method in their research on the effects of the shooting, Siskind et al., (1989); Nateghi, 2011; www.oricaminingservices.com, document reference: 200281; Aloui et al. (2016).

Figure 3.38 shows the diagram developed by the U.S. Bureau of Mines. Points above the red line are the points affected by vibration. Points below the red line are the points not affected by vibration. Therefore, As can be seen from this diagram, the vibration measurement result corresponds to the location suitable for protecting the water source that feeds Béjaia.

2. The criterion used by the Scandinavian countries is shown in Fig. 3.39 Weman (2015). In this figure "peak particle velocity" is the maximum particle velocity explained with the formula 3.7, "scaled distance" is a parameter calculated with the following formula,

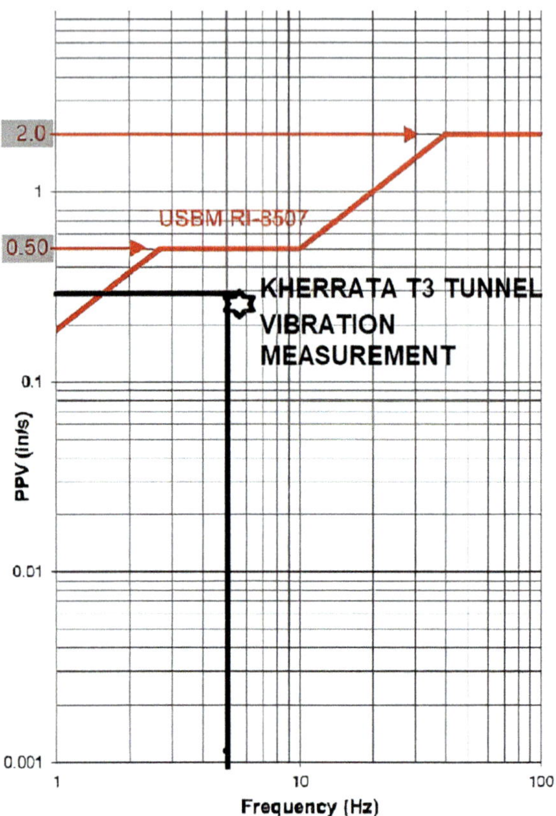

Fig. 3.38 The diagram developed by the U.S. Bureau of Mines for the evaluation of the vibration measurements, Siskind et al. (1989)

$$SD = \frac{R}{\sqrt{W}} \tag{3.7}$$

SD is scaled distance, R = distance to the firing point, here R = 65 m, W = 15 kg/ 20 ms delay. SD is calculated 16.7

Figure 3.39 also shows that the second place indicated for the entrance to tunnel T1 will be safe to protect the water source that loves Béjaia.

3.8 Impact Hammer Used in Conventional Tunneling and Performance Prediction

Hydraulic impact hammers have been used widely in the mining industry and civil engineering applications since 1960. Almost 30 km of metro tunnels were driven in Istanbul with impact hammers since the initial capital investment was relatively lower and rock formations were highly fractured in some zones; rock quality designation

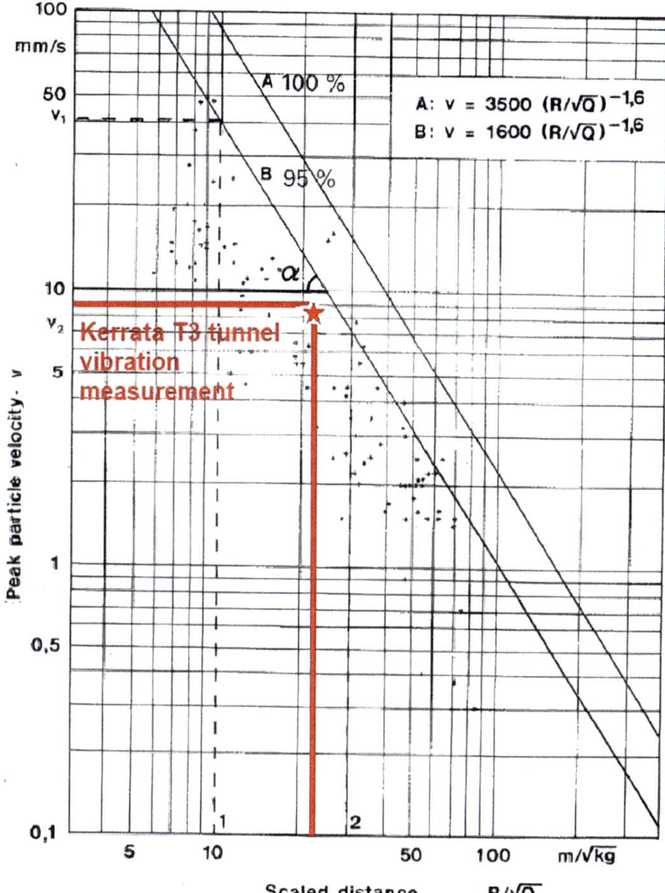

Fig. 3.39 The diagram used by the Scandinavian countries to control the safety of shots at the point vibrations were created, Weman (2015)

(RQD) values ranged from 0 to 80. The impact hammers may be mounted in any excavator and operated efficiently. Safe and efficient operations need to match the size of the carrier/excavator to the weight and power of the hammer. The excavator/carrier is more costly than the breaker, so hydraulic hammer manufacturers build hammers with high blow energies relative to their weights. An impact hammer mounted on an excavator is seen in Fig. 3.40.

The data of 600 impact hammers available in the market are analyzed, and the following equations are found, which are very useful in the proper selection of impact hammers in tunneling operations.

$$E_i = 2.4718 \times Wham - 27.774 \tag{3.8}$$

Fig. 3.40 Impact hammer mounted on an excavator

$$EW_{\max} = 0.015 \times Wham + 3.2343 \quad (3.9)$$

$$EW_{\min} = 0.0094 \times Wham + 1.3485 \quad (3.10)$$

$$P_{output} = 0.0187 \times Wham + 7.1016 \quad (3.11)$$

$$P_{input} = 0.0187 \times Wham + 11.837 \quad (3.12)$$

where:

E_i single blow energy, J
Wham operational weight of impact hammer, kg
EW_{max} maximum recommended weight of excavator, t
EW_{min} minimum recommended weight of excavator, t
P_{output} output power, kW
P_{input} input power, kW.

A contractor is always interested in predicting the machine's performance before starting a tunnel project that will define the tunnel's private economy. A research team of several staff and students collected data from several tunnels, and a statistical model for predicting the instantaneous or net breaking rate of the hydraulic impact hammers was developed. Bilgin et al. (2014) drove the following prediction equations.

$$IBR = 4.24P(RMCI)^{-0.567} \tag{3.13}$$

$$RMCI = \sigma_c \left(\frac{RQD}{100} \right)^{2/3} \tag{3.14}$$

where:

IBR	instantaneous or net breaking rate, m^3/h
P	cutting or breaking power of the hydraulic impact hammer, HP
RMCI	rock mass cuttability index, MPa
σ_c	uniaxial compressive strength, MPa
RQD	rock quality designation, %.

Detailed information on impact hammers, selection criteria, and following numerical examples with solutions was given by Bilgin et al. (2014); you may find the solution to the questions in the cited reference.

Numerical example 1

A Numerical Example to Calculate Impact Hammer Performance

A tunnel with a length of 1200 m and a cross-section of 14 m^2 will be excavated in a mudstone formation with a compressive strength of 80 MPa and an RQD value of 40%. The job organization will be arranged to have two shifts of 10 h/day. The Montabert BRH 250, having an output power of 33 HP, will be used with a suitable excavator.

(a) Calculate the net cutting rate of the impact hammer.

Instantaneous or net cutting rate may be calculated using Eqs. 3.13 and 3.14 as IBR = 17 m^3/h.

(b) Calculate how many days the tunnel may be excavated.

The job duration in (m) is calculated from the length of the tunnel, the excavated area, the number of shifts, and the number of working hours per day as 206 days.

(c) Discuss the possibility of increasing the excavation efficiency.

As recorded before, the time spent on muck transportation is around 23%. This means that if an impact hammer has a muck-gathering arm, the utilization time may be doubled, meaning that the time spent on tunnel excavation may decrease to around 100 days.

3.9 Concluding Remarks

A recent study done for Mineta Transportation Institute covering 67 long tunnels greater than 4.5 km with a total length of 1469.3 km showed the inevitable contribution of conventional tunneling within the total amount of tunnels driven in 30 different countries. Within 67 tunnels, 25.4% were driven with the conventional tunneling method, 33.7% were driven with TBMs, and 40.9% were driven with the hybrid tunneling method, i.e., conventional tunneling jointly with TBMs, Pyeon (2016). As Jennemyr mentions in his article published in 2020, drill and blast methods are often the only possible methods for short tunnels, large cross-sections, cavern construction, cross-overs, cross passages, shafts, etc. It also can be more flexible to adapt to varying profiles than a TBM tunnel that always gives a circular cross-section, especially for highway tunnels, resulting in much over-excavation about the actual cross-section needed. From 2015 to 2018, in Norway alone, 4.2 10.6 m^3 of underground rock excavation was realized by drill and blast. Most recently, drilling jumbos now have computer controls that allow an entire tunnel round to be drilled without an operator. Underground jumbo drill rigs are equipped with a Direct Control System or a computerized Rig Control System to which different levels of automation may be added. An extensive range of rock drills is available for impact power, from 16 to 40 kW. Several models of the face drill rigs are available with an optional zero-emission battery electric driveline. A recent development uses multi-function jumbos suspended from the tunnel crown, allowing multiple functions to proceed simultaneously, such as drilling and mucking. The jumbo can also be used to install lattice girders and shotcrete. The drillability tests initiated at the Norwegian University of Science and Technology (NTNU) in the early 1960s have been continuously developed and formed the foundation of various hard rock equipment capacity and performance prediction models for mechanized and conventional tunneling. Due to the national and international success of the drillability tests, several other laboratories have also set up similar equipment to perform the same tests, Yaşar et al. (2015), sakız et al. (2021).

Zare and Bruland (2013), in their paper titled "Applications of NTNU/SINTEF Drillability Indices," explain the practical application of these tests. It is interesting to note that Beniawski and his co-authors (2008) developed a rating system to estimate rock excavability (RME) which included DRI as one of the parameters of the rating system. Thuro (1977) has introduced destruction work/Energy as the best fitting parameter for drillability prediction in drill and blast tunneling, and the excavability of the rock by roadheaders. An approach to selecting roadheaders based on field drilling rates was developed due to a significant research project for the Turkish Scientific and Research Council. Detailed drillability studies were conducted in different roadways in Zonguldak coalfields, Turkey, Bilgin (1982), where drill and blast were used for roadway drives. The laboratory cutting experiments and field drilling tests showed that pick cutting forces and cutting specific energy values are closely related exponentially to percussive drilling rates. This relationship permitted the classification of the rock formations in Zonguldak Coalfield. Laboratory full-scale drilling tests showed that D-drilling and cutting results and measurements of destruction-specific energy look highly promising for developing a reliable model to estimate the in-situ cutting-specific energy from drilling data. The test results show significant relationship between the specific drilling rate and the rock UCS. A definite relationship between these two parameters exists, which means that by calculating the particular drilling rate for the field drilling program, one can estimate in-situ rock strength. By analyzing the field drilling data and calculating the in situ drilling specific energy, one should assess the in-situ UCS.

References

Aloui M, Bleuzen Y, Essefi E, Abbes C (2016) Ground vibrations and air blast effects induced by blasting in open pit mines: case of metlaoui mining basin, Southwestern Tunisia. J Geol Geophys. https://doi.org/10.4172/2381-8719.10

Balci C, Copur H, Bilgin N, Ozdemir L, Jones GR (2020) Cuttability and drillability studies towards predicting performance of mechanical miners excavating in hyperbaric conditions of deep seafloor mining. Int J Rock Mech Min Sci 130:104338

Bieniawski ZT, Celada B, Galera JM, Tardáguila I (2008) New applications of the excavability index for selection of TBM types and predicting their performance. In: ITA-AITES world tunnel congress, Agra India

Bilgin N (1982) Drillability studies carried out in Zonguldak coalfield. A research project sponsored by Turkish Scientific and research council. Istanbul Technical University, p 67

Bilgin N (1983) Prediction of roadheader performance from penetration rates of percussive drills: some applications to Turkish coalfields. In: Eurotunnel 83, Basel, pp 111–114

Bilgin N, Dincer T, Copur Ç (2002) The performance prediction of impact hammers from Schmidt hammer rebound values in Istanbul metro tunnel drivages. Tunn Undergr Space Technol 17(2002):237–247

Bilgin N, Copur H, Balci C (2011) Cuttability and drillability studies for predicting performance of mechanical miners in hyperbaric conditions. Report submitted to Nautilus Minerals Australia, Istanbul Technical University Revolving Fund Project

Bilgin N, Copur H, Balci C (2014) Mechanical excavation in Mining and Civil Industries. CRC Taylor & Francis Group, p 366

Bilgin N, Genç C, Sunal G (2018) Study done for Özgün İnşaat of drill and blast and the problems associated in Kherrata, Algeria. Istanbul Technical University, p 37

Brierely G (2015) Drill-and-blast for tunnel construction—a historical perspective. TBM Tunnel Business Magazine, 20 Apr 2015

Cardu M, Martinelli D, Todaro C (2024) Industrial explosives and their applications for rock excavation, CRC Press, p 240

Chapman D, Metje N, Stark A (2010) Introduction to tunnel construction. CRC Press, p 390

Fossati DA, Jakobsen PD, Multan MA (2017) Tunnelling experiences on using conveyor belt for mucking at the world-longest and deepest subsea road tunnel. In: Proceedings of the world tunnel congress 2017—surface challenges—underground solutions. Bergen, Norway

Gall V, Pyakurel S, Munfa N (2017) Conventional tunneling in difficult grounds. In: Proceedings of the world tunnel congress 2017—surface challenges—underground solutions. Bergen, Norway, p 10

He B, Armaghani DA, Lai SH, He X, Asteris PG, Sheng D (2024) A deep dive into tunnel blasting studies between 2000 and 2023—a systematic review. Tunn Undergr Space Technol 147:105727

Heydari S, Hoseinie SH, Bagherpour R (2024a) Prediction of jumbo drill penetration rate in underground mines using various machine learning approaches and traditional models. Sci Rep 14:8928. https://doi.org/10.1038/s41598-024-59753-6

Heydari S, Hoseinie SH, Bagherpour R (2024b) A new empirical model for prediction of jumbo drills' penetration rate in underground mines based on the rock mass characteristics, MIAB. Min Informational Anal Bull 5:17–35

Hoseinie SH, Aghababaei H, Pourrahimian Y (2008) Development of a new classification system for assessing of rock mass drillability index (RDi). Int J Rock Mech Min Sci 45:1–10

Hosseinie M, Baghikhani MS (2013) Analysing the ground vibration due to blasting at Alvand Qoly Limestone Mine. Int J Min Eng Miner Process 2(2):17–23. https://doi.org/10.5923/j.mining.20130202.01

https://en.wikipedia.org/wiki/Shotcrete. Loadedin September 2024

Jennemyr L (2020) Advancements in tunneling and underground excavation by Drill & Blast. TBM Tunnel Business Magazine, 15 April

Kahraman S (1999) Rotary and percussive drilling prediction using regression analysis. Int J Rock Mech Min Sci 36(1999):981–989

Kahraman S, Bilgin N, Feridunoglu C (2003) Dominant rock properties affecting the penetration rate of percussive drills. Int J Rock Mech Min Sci 40(2003):711–723

Kaltenegger K, Kukkonen J, Olli W (2016) Vibration control in urban drill and blast tunneling. ITAtech Activity Group Excavation, ITAtech Report No 8, p 23

Macias FJ, Dahl F, Bruland A, Käsling H, Thuro K (2017) Drillability asessment in hard rock. In: Helsinki, Finland Johansson & Raasakka (eds) 3rd nordic rock mechanics symposium October 11–12, 2017. ISBN 978-951-758-622-1

McFeat S, Fowell RJ (1979) The selection and application of roadheaders for rock tunneling. In: Proceedings of rapid excavation and tunneling conference, USA vol 1, pp 269–279

Nateghi R (2011) Minimizing negative effects of blasting on underground structures first Asian. In: 9th Iranian tunnelling symposium, November, Tehran-Iran, p 7

Pyeon JH (2016) Trend analysis of long tunnels worldwide. Mineta Transportation Institute, MTI Report WP 12-09

Sakız U, Kaya GU, Yaralı O (2021) Prediction of drilling rate index from rock strength and Cerchar abrasivity index properties using fuzzy inference system. Arab J Geosci 14:354. https://doi.org/10.1007/s12517-021-06647-w

Siskind DE, Stagg MS, Kopp JW, Dowdin HC (1989) USBM. Report of investigations 8507 structure response and damage produced by ground vibration from surface mine blasting, p 84

Teale R (1965) The concept of specific energy in rock drilling. Int J Rock Mech Min Sci Geomech Abstracts 2(1):57–73

Thuro K (1977) Drillability prediction—geological influences in hard rock drill and blast tunneling. Geol Rundsch 86:426–438

Thuro K, Plininger RJ (1999) Roadheader performance-geological and geotechnical influences. In: 9th ISRM congress Paris, 25th–28th August

Thuro K, Plinninger RJ (2003) Hard rock tunnel boring, cutting, drilling and blasting: rock parameters for excavatability ISRM 2003–Technology roadmap for rock mechanics, South African. Institute of Mining and Metallurgy, p 7

Trajkoviç' S, Lutovac S, Ravilic' M, Nikolinka DN (2013) Assessment of blast effect pit Ranci of shock waves on constructed facilities and environment, V Balkan Mine Ohrid

Weman O (2015) Vibration control in urban drill and blast tunneling. Samaa University of Applied Sciences Technology, Lappeenranta Gegree Programme of Civil Engineering

Wu S, Qiu M, Yang Z, Ji F, Yue QZ (2024) Rapid profiling rock mass quality underneath tunnel face for Sichuan-Xizang Railway. Undergr Space (2024). https://doi.org/10.1016/j.undsp.2024.02.004

www.oricaminingservices.com. Document reference: 200281, loaded September 2024

Yasar S, Capik M, Yilmaz AO (2015) Cuttability assessment using the drilling rate index (DRI). Bull Eng Geol Environ 74:1349–1361. https://doi.org/10.1007/s10064-014-0715-4

Yoshitomi Y (1999) The first in the world! Continuous belt conveyor mucking system at a tunnel excavated by blasting Tagami Tunnel of Kyushu Shinkansen. AFTES—J d'études Int PARIS—25 au 28 October

Zare S, Bruland A (2013) Applications of NTNU/SINTEF drillability indices in hard rock tunneling. Rock Mech Rock Eng 46:179–187. https://doi.org/10.1007/s00603-012-0253

Zhao R, Shi S, Yao R, Yang S (2024) Application of relationship model for the measurement while drilling data to predict rock uniaxial compressive strength for tunneling. Rock Mech Rock Eng 57:7187–7203. https://doi.org/10.1007/s00603-024-03907-5

Zou D (2017) Theory and technology of rock excavation for civil engineering. Springer, Hong Hong. ISBN 978-981-10-1988-3

Chapter 4
A Review of Mechanized Tunneling Methods

Abstract This chapter is a summary of a review made on mechanized tunneling methods starting with the Industrial Revolution based on the use of coal for power steam engines, leading to the development of roadheaders and modern tunneling machines. The role of rock-cutting experiments in developing these fantastic machines is also summarized with a brief explanation of Gripper Type TBMs, Single Shield, Double Shield, EPB, Slurry, Crossover, and variable density TBMs. Performance prediction models of roadheaders, hard rock, and soft ground TBMs are also the main topics in this comprehensive summary. The basic principles of tunnel ventilation, muck transport, rail systems, and belt conveyors are discussed. Another critical topic in this chapter is the vibration created by TBMs. The possibility of using the vibration measurements to identify the ground ahead of the tunnel and the criteria for the damage of vibration to the surrounding area and historical heritages, with the experience of one of the authors for adequate water in Istanbul, is discussed briefly.

4.1 Introduction

The Industrial Revolution, which began in Britain in the eighteenth century and later spread to continental Europe, North America, and Japan, was based on the availability of coal to power steam engines. International trade expanded dramatically when coal-fed steam engines were built for the railways. The small-scale techniques were unsuited to the increasing demand, with extraction moving away from surface extraction to deep shaft mining as the Industrial Revolution progressed. The mechanization of coal mining with cutting technology occurred as early as 1860. The first cutting machines were simple devices comprised of circular saws with picks positioned around the machine. It took nearly an entire century, however, for coal-cutting equipment to be developed to a level sophisticated enough that miners could abandon hand tools, Bilgin et al. (2019).

Mechanization in tunneling progressed parallel to coal-cutting technology. Shields with mechanical excavation equipment were first patented in 1876, but the first successful machine was designed by John Price in 1896 and steadily improved over

N. Bilgin and C. Balci, *Critical Issues in Selecting Conventional and Mechanized Tunnelling Methods*, Springer Tracts in Civil Engineering,
https://doi.org/10.1007/978-3-031-89114-4_4

the next several years. Price machines became the standard for soft-ground tunnel excavation for the next 50 years and are the ancestors of modern soft-ground tunnel boring machines, Potter (2023).

The first successful rock tunneling machine was built in 1954 by James Robbins for the Oahe Dam project in South Dakota. Previously, most attempts at rock tunneling machines used drag picks. The key innovation of Robbins' machine was the disc cutter, patented by a mining engineer, James Robbins, from the United States, Stack (Stack 1995). This invention caused a rapid increase in the number of tunneling projects worldwide due to its capability in hard rock cutting. Disc cutters also caused many new developments, especially in hard rock tunnel boring machine (TBM) technologies following laboratory full-scale cutting tests, Bilgin et al. (2014). In the 1960s, shielded hard rock TBMs were improved to excavate the fractured/jointed rock masses in the United States. The improvements in soil excavation machines began during the 1970s, especially in Japan and Germany. Many soft ground machines working with the earth pressure balance principle were developed and are still being developed. Some European countries followed the research trend in soil excavation starting in the 1990s. Scientific investigations focus on soil treatment and using slurry as a tunnel face support media. The use of mechanized tunneling in urban areas brought the problem of the vibration generated by a TBM and its effects on buildings and monuments; this topic is treated in detail in this chapter, given examples in different geological formations. A summary of mechanized tunneling with TBMs was recently done by Bilgin and Acun in (2024), so with the objective of not repeating the same topics, only a few of them will be the subject of this chapter.

Roadheaders were first developed for mechanical excavation of coal in Europe in the late 1940s. Roadheaders are partial-face machines excavating only a portion of the face at once, and a certain number of cutters are in contact with the face. The primary advantages of roadheaders over other underground excavation machines are their mobility, flexibility, and selective mining ability. Besides the general benefits of mechanical excavation, these advantages allow them to be used widely and have a critical and unique position in underground mining and tunneling operations. The comparison between the performance of TBMs and roadheaders in a Metro project is made in Chap. 5 to better understand the two mechanized systems.

4.2 Laboratory Small- and Full-Scale Cutting Tests

Rock-cutting experiments are the best choice for performance prediction of mechanical excavators since they are reliable and can define the basic specifications of these machines and design their cutter heads. A more popular small-scale rock cutting rig (SLCM) was used at the University of Newcastle Upon Tyne, England, for core cutting at 5 mm of depth of cut. Specific energy measured was then used for predicting the performance of roadheaders, McFeat-Smith and Fowell (1977, 1979) by using a standard tungsten carbide cutting tool. A similar small-scale rock cutting rig was also instrumented at Istanbul Technical University, Bilgin and Shahriar (1987), as seen in Fig. 4.1. By using this SLCM testing system, Balci and Bilgin (2007) correlated the specific energy obtained from the SLCM tests with the specific energy obtained from the full-scale linear cutting machine (FLCM) tests, which also enabled them to predict the performance of mechanical excavators. The second version of the small-scale rock cutting rig, as seen in Fig. 4.2 (also named portable liner rock cutting rig/PLCM), is being developed at ITU, Bilgin et al. (2010), Çomaklı et al. (2021). It includes a small, stiff reaction frame on which the cutter and load-cell assembly are mounted. A block sample in size up to 10 × 15 × 20 cm is cast within a metal sample box with fast-curing concrete at a certain dip angle, parallel or perpendicular to the bedding planes, to simulate the different cutting conditions on a rock sample. A servo-controlled hydraulic cylinder moves the sample box through the cutter at a preset depth of cut, cutter spacing, and constant cutting speed of 30 mm/s. The depth of cut and line spacing of the cutter can be adjusted by a mechanical device and a hydraulic cylinder, respectively. A constant cross-section (CCS) disc cutter with a diameter of 145 mm and a tip width of 4.7 mm is used throughout the cutting experiments. A triaxial dynamometer is used to record the orthogonal force components. The structural load capacity of the PLCM is 50 kN for normal force and 20 kN for rolling force. The tests performed to verify the portable linear cutting machine indicate that the performance and operational parameters of the two identical EPB TBMs, such as net cutting rate, thrust force, cutterhead torque, and power, are reliably predicted deterministically using the portable linear cutting machine test results.

The full-scale linear cutting tests measure cutter forces acting on a real-life cutter of any type, single disk, conical, radial cutter, etc., while cutting a block of rock sample cast in a sample box; the rig developed in ITU is seen in Fig. 4.3. The results of this test can be used as input for the selection, design, and prediction of the costs and performance of any type of rock-cutting machine. This test, along with deterministic computer simulation, is accepted as the most reliable and economical method for these purposes. However, the FLCM testing method used in different research institutes and universities has some disadvantages, such as requiring experienced personnel to run the tests, large blocks of rock samples, and more extended periods for testing, which explains well the invention of small linear cutting machines. The currently used FLCMs are designed to keep the depth of cut constant and to measure normal, cutting, and side forces such as the FLCM used in Colorado School of Mines, Ozdemir (1990), Istanbul Technical University, Bilgin et al. (1999, 2005, 2014), Korea Institute of Construction Technology and Seoul National University, Chang et al. (2009), Cho et al. (2013), CSIRO Earth Science and Resource Engineering, Shao et al. (2014) and Beijing University of Technology, Zhao et al. (2015) and other research institutes in China, The details of the cutting rig developed in ITU are well defined by Bilgin et al. (2014). The measured and predicted values of TBM performances using this experimental cutting rig in different case studies are proven reliable and realistic, Bilgin et al. (1999), Balci (2009).

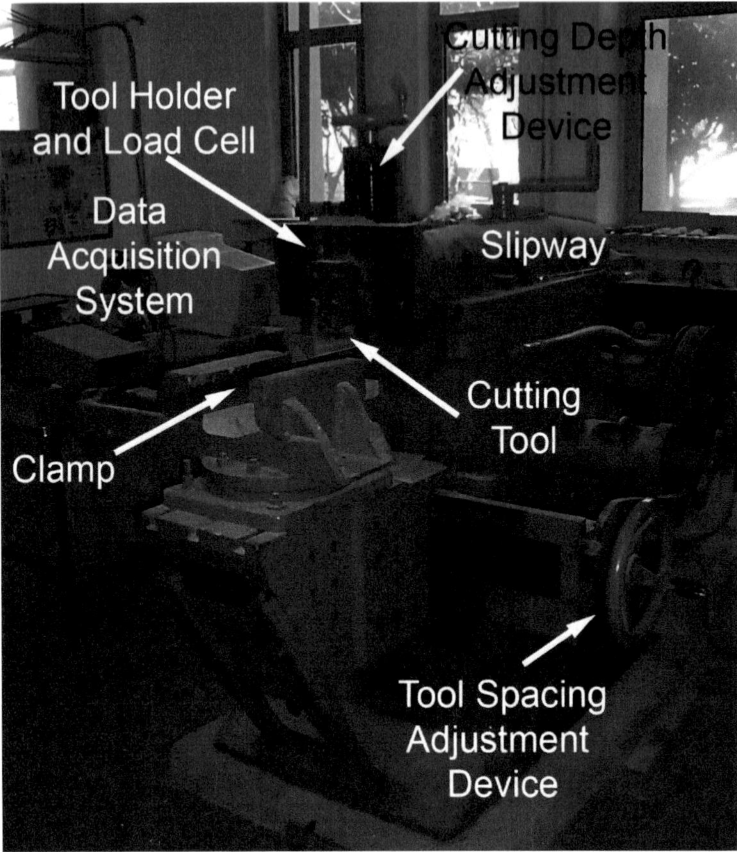

Fig. 4.1 Early design of the small-scale rock cutting rig. *Source* Bilgin et al. (2014). Mechanical Excavation in Mining and Civil Industries, CRC Press

4.3 Roadheaders

The first roadheaders were used for mining in the 1960s. By the early 2020s, more than a thousand roadheaders were used for underground civil construction. During this early phase, boom-type cutting equipment in shields or other hauling structures such as excavators also became popular. Roadheaders can be classified into two groups based on cutterhead types: axial (longitudinal) and transverse (ripping, drum), as seen in Figs. 4.4 and 4.5. Their cutterhead shapes are cylindrical and conical, with the nose section semi-spherical or flat. Both cutterheads have several advantages and disadvantages. Transversal cutterheads cut in the direction of the tunnel face. Therefore, they are more stable than roadheaders with axial cuutterheads of comparable weight and cutter head power. At transversal heads, the majority of reactive force resulting from the cutting process is directed toward the main body of the machine.

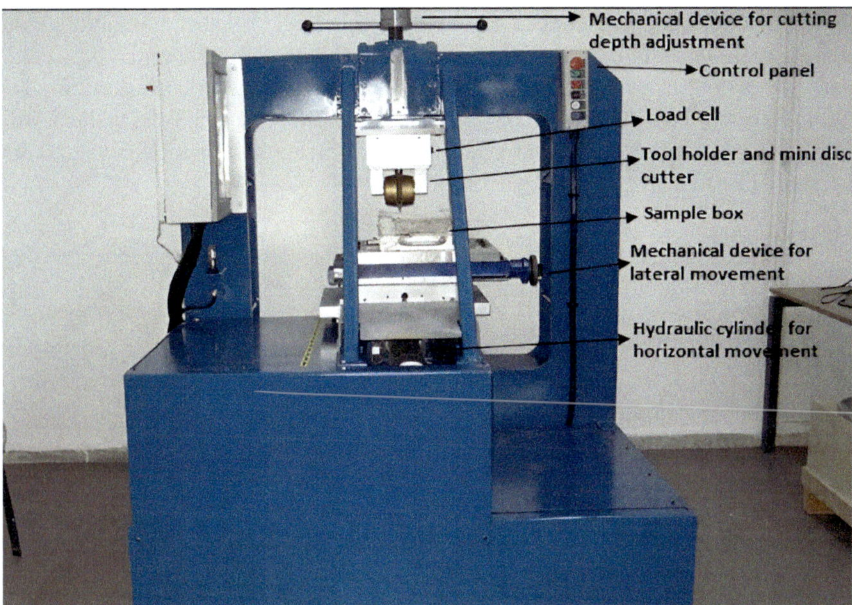

Fig. 4.2 The second version of the small-scale rock cutting rig (also named portable liner rock cutting rig developed at ITU, from the archive of the author Bilgin, N

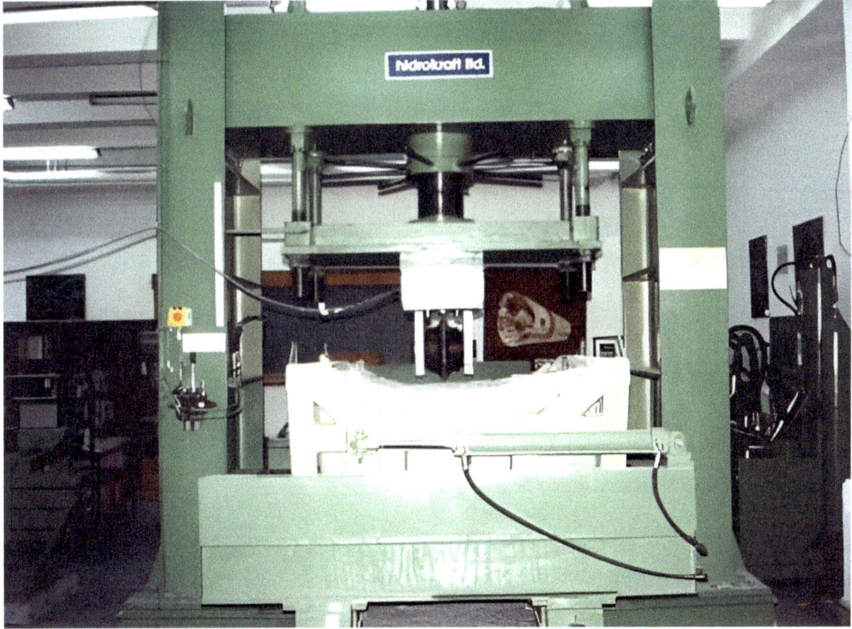

Fig. 4.3 Full-scale linear rock cutting machine in ITU, Photograph by Bilgin

On axial cutterheads, the cutter array is more accessible because both cutting and slewing motions go in the same direction. Roadheaders with transversal-type cutter heads are less affected by changing rock conditions and harder rock portions. The cutting process can better use geological discontinuities, especially in bedded sedimentary rock. Most longitudinal heads show lower figures for pick consumption, primarily due to lower cutting speed, Heiniö (1999).

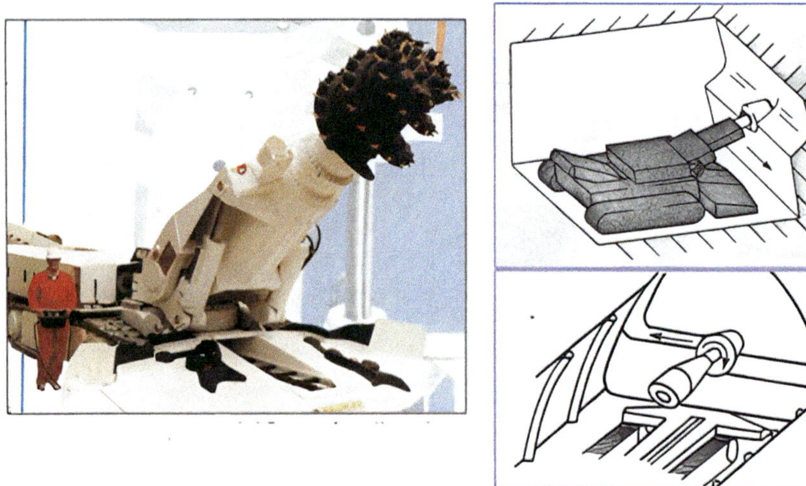

Fig. 4.4 Axial type roadheader, adopted by the author Bilgin

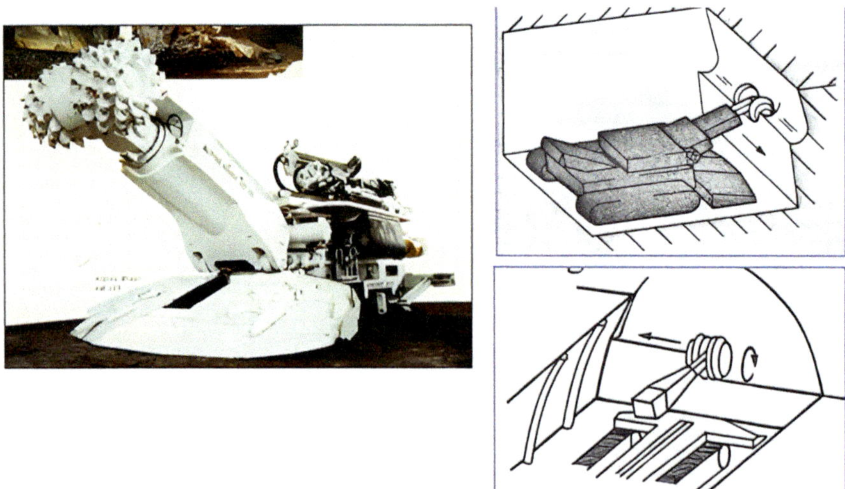

Fig. 4.5 Transverse type roadheader, adopted by Bilgin

The cutterhead is equipped with cutting tools and attached to a boom, which is movable in any direction, excavating the face. Excavated material (muck) falls into a loading apron. It is gathered and loaded into a chain conveyor in the center of the loading apron by a continuous muck loading system such as a gathering arm, star wheel, and scraper conveyor. The chain conveyor carries the loaded material through the roadheader body to the tail conveyor. From the tail conveyor, the material is transferred to any muck haulage unit, such as trucks, rail cars, belt conveyors, and shuttle cars. Roadheaders are primarily limited to excavating massive rocks up to 100–120 MPa of UCS, depending on cutterhead power and weight of the roadheader and some structural characteristics of the rock mass (Copur et al. (1998), Copur (1999). If the rock mass is highly fractured–jointed–foliated, they can excavate up to 160 MPa UCS. These UCS values should be used cautiously since they are not the only parameter for defining cuttability.

The performance prediction of roadheaders is of prime importance for estimating the project's cost and making an acceptable project schedule. For more detailed information, the reader of this book should consult the book titled "Mechanical Excavation in Mining and Civil Industries" by Bilgin et al. (2014), and Chap. 5 in this book. Besides these recommended references, a detailed literature survey on the road header performance models done by different researchers is summarized by Kahraman et al.(2023) in Table 4.1

4.4 Hard Rock and Soft Ground TBMs

Tunnel boring machines (TBMs) are an alternative to conventional tunneling methods. They have the advantages of high advance rates, limiting the disturbance to the surrounding area and producing a smooth tunnel wall without overbreak. They are suitable for constructing tunnels in urban and high-traffic regions with less surface disturbance. They are designed to operate in various ground conditions as they can dig soft ground like sand and hard rocks. They offer a continuous operation and a better working environment compared to the drilling and blasting tunneling method, Bilgin and Acun (2024)

This section explains the basic working principles of TBMs and the essential criteria for adequately selecting a TBM. After this, equipment planning for tunnel ventilation, muck/spoil transport with belt conveyors and locomotives, and material transport with multiple service vehicles will be summarized. The essential reference books are Maidl et al. (2011, 2013); Bilgin et al. (2014).

Table 4.1 The performance estimation models for roadheaders, with kind permission of Kahraman et al. (2023) and Bilgin et al. (2014)

Model no.	Model	Reference	Remarks
1	$NCR = k\frac{P}{SE}$	Rostami et al. (1994)	k value of 0.45 works for axial, 0.55 for transverse type
2	$NCR = \frac{719}{\sigma_c^{0.78}}$	Gehring (1989)	250 kW transverse type cutterhead
3	$NCR = \frac{1739}{\sigma_c^{1.13}}$	Gehring (1989)	230 kW axial type cutterhead
4	$NCR = 0.28P(0.974)^{RMCI}$ $RMCI = \sigma_c\left(\frac{RQD}{100}\right)^{2/3}$	Bilgin et al. (1990)	Axial type cutterhead
5	$NCR = 27.511e^{0.0023RPI}$ $RPI = \frac{PW}{\sigma_c}$	Copur et al. (1998)	Transverse type cutterhead, for evaporitic rocks
6	$NCR = 75.7 - 14.3ln\sigma_c$	Thuro and Plinninger (1999)	132 kW transverse type cutterhead
7	$NCR = 0.8\frac{P}{0.37\sigma_c^{0.86}}$ d = 5 mm $NCR = 0.8\frac{P}{0.41\sigma_c^{0.67}}$ d = 9 mm	Balci et al. (2004)	Developed for axial and transverse roadheaders
8	$NCR = 109.25\sigma_c^{-0.72}$ $NCR = 81.21SH^{-0.78}$	Tumac et al. (2007)	90 kW axial type cutterhead
9	$NCR = 510588\sigma_c^{-2.1779}$	Ocak and Bilgin (2010)	300 kW trans. head
10	$RMBI = e^{\left(\frac{\sigma_c}{\sigma_t}\right)}x\left(\frac{RQD}{100}\right)^3$ $NCR = 30.75RMBI^{0.23}$	Ebrahimabadi et al. (2011)	82 kW axial type cutterhead
11	$NCR = 1.79\sigma_c + 0.501\alpha + 0.636RQD - 4.839\sigma_t - 22.127$	Abdolreza and Yakhchali (2013)	82 kW axial type cutterhead
12	$NCR = -2.92I_s - 0.79A_w + 22.95$	Kahraman and Kahraman (2016)	112 kW axial type cutterhead
13	$NCR = -8.58lnNPI + 55.06$	Kahraman et al. (2019)	112 kW axial type

Nomenclature NCR is net cutting rate (m^3/h), P is cutterhead power (kW), k is energy transfer coefficient (0.45–0.55), SE is specific energy (kWh/m^3), σ_c is compressive strength (MPa), σ_t is indirect tensile strength (MPa), RQD is rock quality designation (%), RMCI is rock mass cuttability index (MPa), RPI is roadheader penetration index (kWxt/MPa), W is machine weight (ton), SH is Shore hardness, RMBI is rock mass brittleness index, α is the angle between tunnel axis and weakness planes, I_s is point load strength (MPa), A_w is water absorption ratio (%), and NPI is needle penetration index (N/mm)

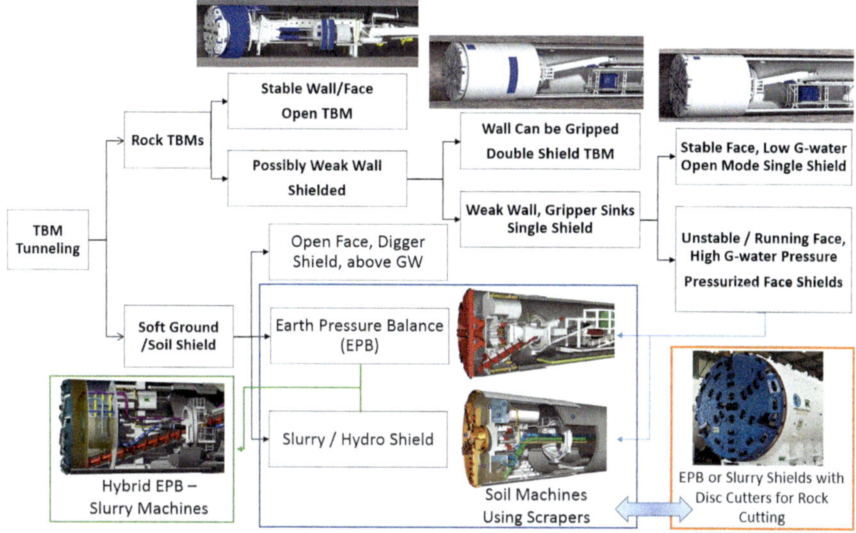

Fig. 4.6 The classification of TBMs according to ground conditions, from Rostami (2016)

4.4.1 Classification of TBMs

The classification of TBMs according to ground conditions is shown in Fig. 4.6.

4.4.2 Hard Rock TBMs

Gripper or main-bearing TBMs:

A general view of the gripper or main bearing TBM is shown in Fig. 4.7.

Gripper TBMs are designed for high advance rates in stable hard rock and are relatively less expensive than other TBMs. The front shield (1) holds the main drive (3) which drives the cutterhead (2). Hydraulic cylinders (4) allow adjusting the size and alignment of the shield. The shield is at the tip of the main beam (5) and is pushed forward by thrust cylinders (7), which lead the force into the ground through grippers (8). Hydraulic feet (9) hold the TBM while resetting the gripper. The TBM has drilling machines (6) to install rock anchors. This action is necessary to secure the geology safely.

Single Shield TBMs:

A general view of a single shield TBM is seen in Fig. 4.8.

Single shields are built to advance in solid or fractured rock. Like other shield machines, they use thrust cylinders (5) to advance and an erector (4) to build segmental lining (6). The rock chips cut by the cutterhead (1) are lifted by buckets in

Fig. 4.7 A general view of gripper or main bearing TBM, Courtesy of Herrenknecht, Brabant and Duhme (2017)

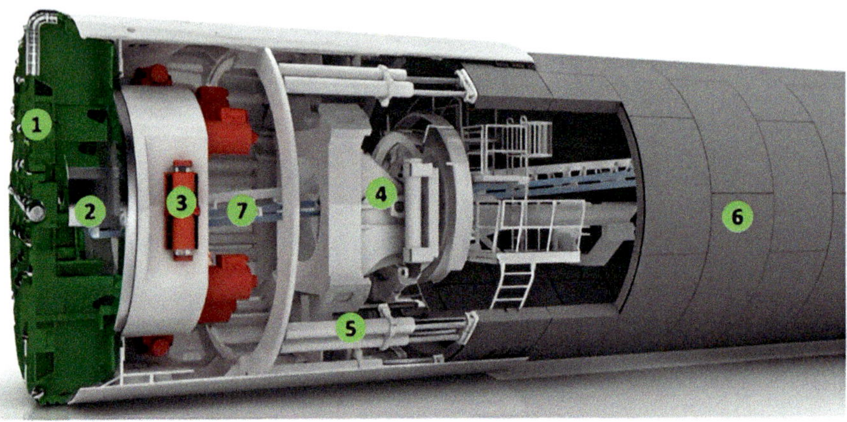

Fig. 4.8 A general view of single shield TBM, Courtesy of Herrenknecht, Brabant and Duhme (2017)

the cutterhead and dropped onto the muck ring (2). From here, they fall onto a belt conveyor (7), which removes them from the TBM. A hydraulic torque box holds The main drive within the steel structure (3). This action allows precise cutting process control, courtesy of Herrenknecht, Brabant and Duhme (2017).

Double Shield TBMs:

A general view of double shield TBM is seen in Fig. 4.9.

Double shields have been developed to increase the production rate of TBMs in fractured rock. Excavation and muck transport work in the same way as for single shields. However, their front shield (1) is equipped with a telescopic function (2) relative to the rear shield (3). It is pushed forward by the main thrust cylinders (4) from the middle shield, while the torque is transmitted by torque cylinders (7).

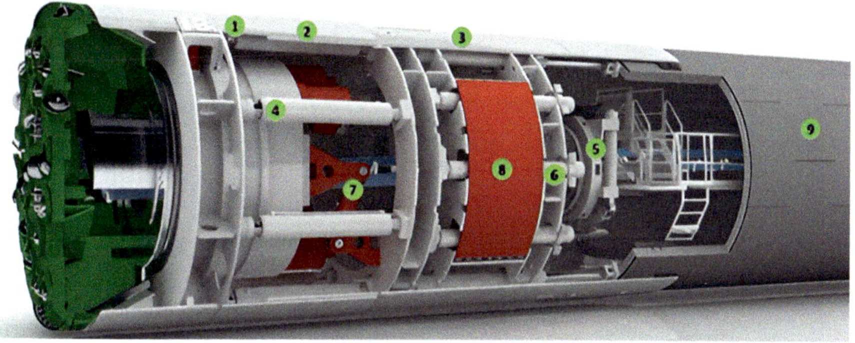

Fig. 4.9 A general view of a double shield TBM, Brabant and Duhme (2017)

The gripper leads the Thrust force into the ground (8). During the advance, the rear shield is fixed, and by using the auxiliary thrust cylinders (6), the erector (5) can erect the segment lining (9). After completing one cycle of advance and Ring-building, the gripper is repositioned, and the advance can restart. This increases productivity relative to a single shield, Herrenknecht, Brabant and Duhme (2017).

Crossover TBMs:

The machine has multi-speed gearboxes to enable the machine to advance through blocky, fractured, or mixed ground at high torque and low speed. Standing for a Crossover (X) between Rock (R) and EPB (E), the XRE has been designed and deployed by Robbins on multiple projects that feature sections of both hard rock and soft ground in the tunnel alignment. Using features of EPB and single shield complex rock machines, the XRE has successfully been used on mixed-ground projects in Australia, Turkey, Mexico, India, and other countries. Readers are advised to have more technical information on this revolutionary machine to read a paper by Harding and Alpagut (2020).

A comparison of hard rock TBMs is given in Table 4.2. This table will help the readers to understand better the understand the advantages and disadvantages of different types of hard rock TBMs.

4.4.3 *Soft Ground TBMs, EPB TBMs, Slurry TBMs, Mixshield TBMs*

The unique feature of EPB TBM is that it uses excavated soil directly as a support medium. The cutterhead is pushed onto the tunnel face and excavates the material. The soil enters the excavation chamber through openings, which mixes with the existing soil paste. Mixing arms on the cutting wheel and bulkhead mix the paste until it has the required texture. The bulkhead transfers the pressure of the thrust

Table 4.2 Comparison of hard rock TBMs, credit to Prof. Dr. Levent Özdemir, from a private communication

The comparison between hard rock TBM types		
Single shield	Double shield	Gripper type
For the poorest ground conditions Used with precast segments for thrust requirements	For poor to medium quality rock May be used with or without precast segments	For medium to high quality rock, gripper reaction is needed to advance machine
Pro's Relatively inexpensive, crew not exposed to rock, short shield, less prone to trapping, better steering ability, no need of shotcrete	**Pro's** Crew not exposed to rock, can erect lining while boring, high advance rates, can operate as a single shield in poor ground	**Pro's** High advance rates, tight turn radius, inexpensive, easy mobilization
Con's Slow advance rates, cyclic operation	**Con's** Expensive, prone to becoming stuck in high overburden, large turn radius, large thrust force required due to shield friction	**Con's** Crew more exposed to rock, ripper pressure around 4 MPa

cylinders to the soil paste. When the pressure of the soil paste in the excavation chamber equals the pressure of the surrounding soil and groundwater, the necessary balance has been achieved, Fig. 4.10.

The pressure distribution ahead of an EPB TBM cutterhead is given in Fig. 4.10, and the differences between soft-ground TBMs are in Fig. 4.11. A critical point in selecting EPB and mix shield TBMs is that the choice between them mostly depends on the granulometry of the ground, as seen in Fig. 4.12. The Slurry Shield TBM is suitable for excavating soils with high water content. This TBM is equipped with a slurry system that controls the pressure in the excavation face by injecting pressurized slurry into the cutter chamber, where the slurry is mixed with the excavated material. The mixture is pumped out of the tunnel to a separation and re circulation plant.

Variable Density TBMs:

Babler et al. (2018) emphasize that the variable density TBM can be operated as a classic slurry TBM with an air bubble system to control the face pressure and in a full EPB mode. The change between the modes can be done gradually under permanent and complete control of the tunnel face pressure without chamber interventions. This machine can also be operated using a high density in the excavation chamber, which would be too dense for a classic slurry operation but too fluid for a classic EPB operation. For all operation modes of the variable density TBM, the muck is extracted from the excavation chamber via a screw conveyor. For the first and second phases of the Kuala Lumpur Klank Vally project, Herrenknecht delivered a total of 10 variable-density TBMs. The TBMs bored through the Kenny Hill formation in EPB mode and the Kuala Lumpur Limestone and Granite formations in slurry mode

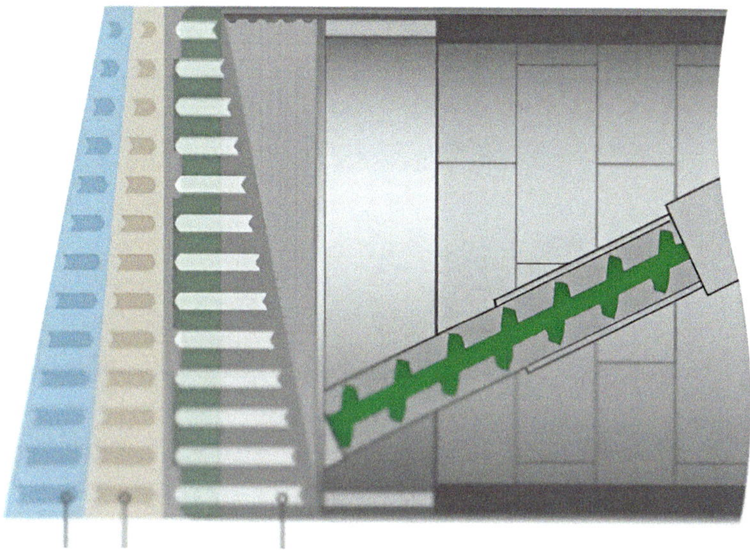

Fig. 4.10 The pressure distribution ahead of an EPB TBM cutter head, from the archive of the author Bilgin

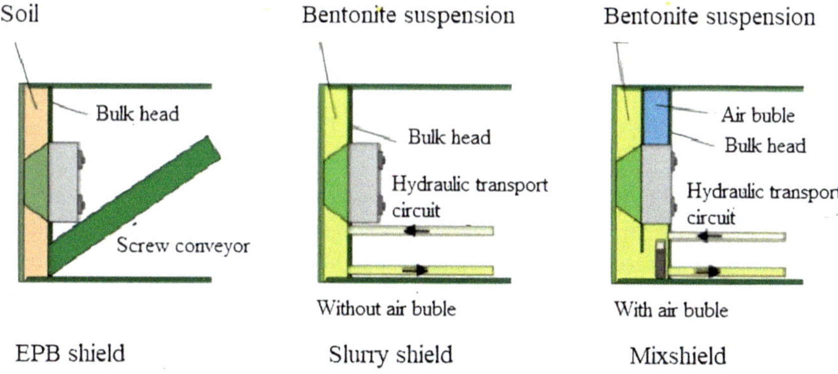

Fig. 4.11 The differences between soft ground TBMs, Burger (2016)

and high-density slurry mode. For further case studies, the readers are advised to read the references cited for this topic.

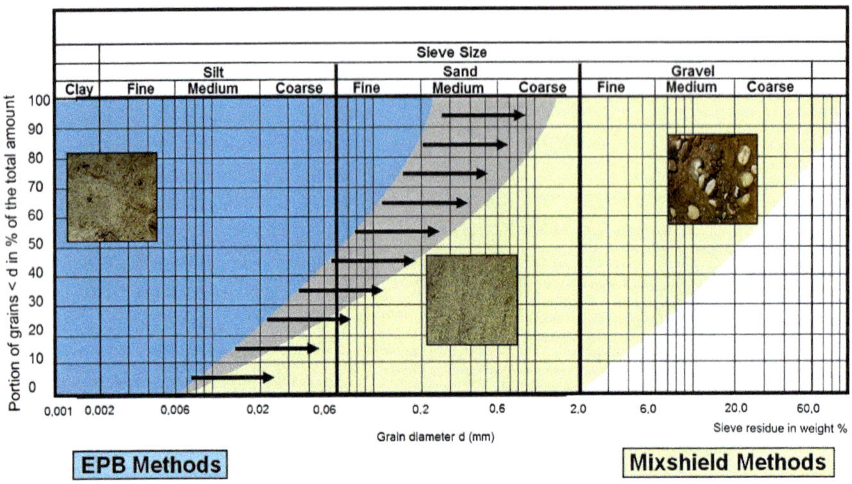

Fig. 4.12 Grain size distribution for different TBM types, Duhme (2017)

4.5 Performance Prediction of Hard Rock and Soft Ground TBM's

The following summary on the performance prediction of hard rock and EPB TBMs is taken from a recent publication by the authors Bilgin and Acun (2024).

4.5.1 Performance Prediction of Hard Rock TBMs

One of the most critical parameters in the success of mechanized tunneling depends on how TBM will behave in a given geology. In other words, what will the machine's performance be (thrust, torque, penetration, specific energy, and machine utilization factor)? Recently, there have been excellent papers on the subject, such as Farrokh (2020) and Salimi et al. (2022). However, the readers are advised to follow the references below to understand the scientific development of the subject.

Several TBM performance prediction models with few on the prediction of machine utilization time were realized in the past. Full-scale rock-cutting tests and geological and geotechnical parameters were the basis for predicting TBM performance. The works at Newcastle Upon Tyne University, Colorado School of Mines in the 1970s, and recently at ITU, Korea Institute of Construction Technology, Seoul National University, and some institutes in China are typical examples of full-scale laboratory rock-cutting experiments. Roxborough (1969), Roxborough and Phillips (1975), Bilgin (1977), McFeat Smith and Fowell (1977, 1979) carried out experiments at Newcastle Upon Tyne University, Ozdemir (1990), Rostami and Ozdemir (1993), Rostami et al. (1994, 1996), Ozdemir and Nilsen (1993), Copur (1999) did basic rock cutting tests in Colorado School of Mines. Balci (2009) and Bilgin et al. (2010, 2014) conducted full-scale tests at ITU. Chang et al. (2009) did detailed rock-cutting studies at the Korea Institute of Construction Technology and Seoul National University. The work carried out by Dollinger et al. (1998) is another example of using specially designed punch penetrating tests for estimating TBM performance. The NTNU model developed at the Norwegian University of Science and Technology is a semi-theoretical model that defines first cutter forces and later TBM performance using geological parameters, Bruland (1998). Other prediction models have recently been developed to account for a wide range of rock properties and mass conditions. The QTBM (Barton 2000) is based on an expanded Q-system, which can be used for TBM performance estimation. Hassanpour et al. (2009a, b, 2011) previously analyzed the TBM performance in relation to the field penetration index. Gong and Zhao (2007, 2009) Gong et al. (2007) introduced the boreability ındex and rock brittleness and investigated rock mass characteristics affecting TBM performance. Sapigni et al. (2002) and Ribacchi and Fazio (2005) analyzed the relationship between TBM performance and RMR. Khademi et al. (2010) and Yagiz et al. (2012, 2024a, b) created TBM performance prediction models considering the influence of the rock mass parameters. Bieniawski et al. (2007, 2008) and Bieniawski and Grandori (2007) investigated TBM borability using rock mass classification systems. Factors affecting TBM penetration rates in different rock formations were also analyzed by Nelson et al. (1983, 1985). Various models for estimating the penetration rate of hard rock TBMs were summarized recently by Farrokh et al. (2012, 2013). The problems of TBM excavation in blocky ground were investigated by Delisio and Zhao (2013) and Delisio et al. (2013).

4.5.2 *Performance Prediction of EPB TBMs*

Although much research has been done on the performance prediction of hard rock TBMs, as explained previously, the work done on the performance prediction of EPB-TBMs is rare. In this respect, the work done by Çopur et al. (2014), Namlı and Bilgin (2017) and Bilgin et al. (2017) is summarized below.

As Copur et al. (2014) emphasized, the rock excavation process is stochastic, which includes uncertainties regarding decision-makers. These authors summarized a new method for predicting the performance (instantaneous penetration rate, daily advance rate, thrust, power, torque, and specific energy) of EPB TBMs by a stochastic method implemented into a deterministic prediction model requiring input from full-scale laboratory linear rock cutting tests. Full-scale linear rock-cutting experiments using a disc cutter were performed on a block of limestone samples. Estimations were made using a commercially available Monte Carlo simulation program providing knowledge of probabilistic outcomes. Results of the suggested method were verified by measuring the field performance of an EPB TBM excavating a massive rock mass of limestone in semi-closed mode. Results indicate that the stochastic estimates allow uncertainties and variations in the experimental parameters to be seen through the estimated probability density in the function.

A model based on past experiences obtained in Istanbul was used in another work. The TBM data collected from Beykoz Utility Tunnel, Cayirbasi Water Tunnel, Kartal-Kadikoy Metro Tunnel, Pendik-Kaynarca Metro Tunnel and Uluabat Power Tunnel from Bursa, and current metro Usküdar-Umraniye-Sancaktepe-Cekmekoy projects were used to develop the model described by Namlı and Bilgin (2017). The model is based on the concept of specific energy, TBM design parameters, and machine utilization time; in this reference, a numerical example is also given on predicting the performance of EPB.

4.6 Planning of the Ventilation

The main objectives of providing ventilation systems in tunnels are to give the working crew fresh air and remove dust and poisonous gas during construction and operation. It is demanded that the concentration of various gases in the atmosphere inside the tunnel should be as follows:

(a) Methane should be measured close to the roof and not exceed 0.5% at any place inside the tunnel.
(b) Carbon monoxide should be less than 0.005%,
(c) Carbon dioxide should be less than 0.5%,
(d) Nitrogen fumes should be less than 0.0005%,
(e) Hydrogen sulfide should be less than 0.001%,
(f) Aldehyde should be less than 0.0002%,
(g) The oxygen content in the tunnel atmosphere should not be less than 19%,
(h) The ventilation system used in the Avrasya Tunnel is seen in Fig. 4.13.

In the Avrasya tunnel, the fan's diameter was 1800 mm, its power was 250 kW, and the duck's diameter was 2200 mm. The air return velocity was kept at 0.5 m/sn. In an emergency, another fan with reversible blades was used.

Jet fans used in running tunnels are seen in Fig. 4.14.

Criteria for Ventilation Computing:

To efficiently dilute dangerous gasses and dust, a return speed for air in the tunnel between 0.3 and 0.6 m/s is usually required (if a risk exists of methane gas, then 1 m/s should be used). The objective is to determine how much air in m^3/s is needed at the bottom of the tunnel for the return velocity to be sufficient and to be able to dilute gasses so that it is possible to breathe the air without it posing a health hazard. To clear diesel gas out, a minimum of 4 m^3 of air/kW/min. is required at the bottom of the tunnel (6 m^3/kW/min. for a motor without a particle filter). An estimate is made of how many kW are being used simultaneously in the tunnel, and then a computation is made to determine the total amount of air in m^3/second. Each person in the tunnel needs 3–5 m^3 of air/minute, and a computation is made to determine how many

Fig. 4.13 Ventilation system used in Avrasya tunnel/Istanbul, Bilgin and Acun (2024)

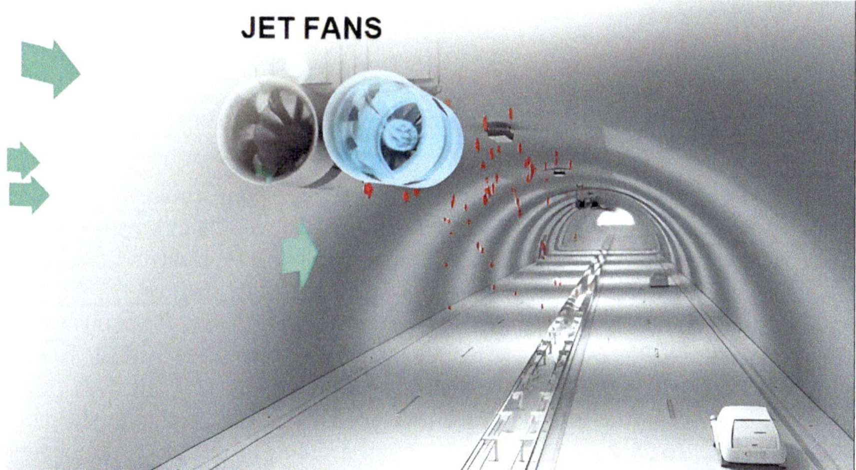

Fig. 4.14 Jet fans used in a running tunnel, Bilgin and Acun (2024)

people are usually in the tunnel simultaneously. The total amount is worked out in m^3/minute.

4.7 Transport of the Muck/Spoil with Belt Conveyors

Mucking out with a conveyor belt is a common technique for full-face tunnel boring with tunnel boring machines (TBM), and it is currently used in several projects worldwide. With TBM drives, the belt conveyor is extended continuously while the TBM progresses without stopping. The advantages of using belt conveyors are reduced costs in long tunnels, environment, safety, and considerable energy-saving compared to gas oil consumption in rail transport. Material excavated from the TBM is continuously moved in the tunnel and/or from the excavation shaft; there is no interference on the road between vehicles going out (mucked material) and in (segment transportation, injection material, and personnel). There is no auto-repair garage for assistance and maintenance of the locomotives, no tankers or re-fueling plants, optimization and minimization of the spare part area, and less personnel to handle the muck as a whole (drivers, mechanics, maintainers, safety systems inspectors, etc.

The disadvantages of rail transport include waiting for locomotives, setting up switches, derailing locomotives, and the need to up size the ventilation if diesel locomotives are used. With the belt conveyor system, the only time tunneling stops is to add a belt. Some dimensions of the belts used in Turkish Tunnels are seen in Table 4.3.

4.8 Transport of the Muck/Spoil with Locomotives

It is always the contractor's decision, but if the tunnel is longer than 1.5 km long, belt conveyors are used. If it is shorter, then locomotives with muck cars are usually preferred. With small-diameter TBMs, although there is an opportunity to develop belt conveyors smaller than 60 cm wide, they are not used in practice. On the other hand, some contractors prefer locomotive transport because of their lower investment cost compared to belt conveying. A locomotive used in one of the Turkish Tunnels is given in Fig. 4.15.

Finding the size and specifications for a locomotive may require the consideration of many factors. The first and most crucial step is deciding on the necessary locomotive weight to develop the tractive effort to move the train load. The maximum grade and the load are the input for this calculation. Locomotives are limited to shallow gradients, and the frictional grip between the driving wheels and the rails also limits the train size that can be hauled. Diesel locomotives are used in tunneling operations since in battery locomotives, batteries need to be charged very often, complicating the system. A robust, functional, and maintenance-friendly hydrodynamic locomotive with a strong and maintenance-friendly design exists on the market. These types are available in the weight range from 18 to 45 t with corresponding engine power from 125 to 360 kW in gauges up to 1000 mm, suitable for tunnel diameters from 5 to 15 m and gradients up to 3%. California switch is indispensable to ease the passage of the train. A portable or fixed platform rides on a track in a tunnel and is used for passing

Table 4.3 Some dimensions of belt conveyors used in Turkish tunnels, Bilgin and Acun (2024)

Projet	D	L1	L2	W	V	Q	P
	m	m	m	cm	m/sec	t/h	kW
Atakoy—Ikitelli Metro, EPB-TBM	6.6	5500	5367	800	3.5	450	750
İstanbul Airport Line 1, EPB-TBM	6.6	4200	4200	800	3.5	450	400
İstanbul Airport Line 2, EPB-TBM	6.6	3560	3650	800	3.5	450	400
Goztepe Metro Line 1, EPB-TBM	6.6	6400	6400	800	3.5	450	675
Göztepe Metro Line 2, EPB-TBM	6.6	6400	6400	800	3.5	450	675
Götepe Metro Line 3, EPB-TBM	6.6	6240	6240	800	3.5	450	475
Göztepe Metro Line 3, EPM-TBM	6.6	6240	6240	800	3.5	450	475
Esme Railway, XRD	13.8	3400	3400	1000	3.5	1600	710
Kemerburgaz—Hasdal (× 2), EPB	6.6	5090	3500	800	3.5	450	400
Kemerburgaz İhsaniye (X2), EPB	6.6	5855	5855	800	3.5	300	600
Istanbul Airport Ihsaniye Line 1 (X2), EPB-TBM	6.6 EPB	9860	9860	800	3.5	300	875
Istanbul Airport Ihsaniye Line 2 (X2), EPB	6.6	7220	7220	800	3.5	300	875
Istanbul Airport Hasdal, EPB	6.6	5330	5330	800	3.5	300	600
Kargı Power tunnel*, DS TBM	9.8	11,700	11,900	950	3.5	1500	370
Bahçe-Nurdağı Railway, SS	8.0	9700	88,500	1000	3.5	650	600

L1 is the tunnel length, *L2* is the length of the belt, *W* is the width of the belt, *V* is the speed of the belt, *Q* is the capacity of the belt, *P* is the power of the drive unit, *XRD* Robbins crossover machine, *EPB* Earth Pressure Balance, *DS* Double shield, *SS* single shield. *The belt used in Kargı Project has a curve radius of 1524 m, 4 carrying side boosters 370 kW, 2 return side boosters 75 kW, and a radial stacker at the portal

cars and trains. It has space for two or more trucks, crossovers, and switches and has sliding and tapered end rails that ride on mainline rails. Optimizing the transport systems by increasing the number of trains and adding the California switch increases TBM utilization time. One of the biggest problems is getting people and materials in and out of the tunnel; the problem may be reduced by using California switches, as seen in Fig. 4.16.

Fig. 4.15 Muck transport system used in Yamanlı energy tunnel/Turkey, with courtesy of MSD Company

Some examples of using locomotives in Turkey:

(a) Kozyatağı Kadıköy Metro Tunnels:

Two twin tunnels were excavated in Istanbul between Kozyatağı and Kadıköy with two TBMs of 6.57 km in diameter. Due to a high degree of inclination of % 3, a Tandem Locomotive system was selected. Five cars of 12 m^3 (with 20 t, with muck in) and 25 t of weight were used. Fixed California switches were preferred due to the small distances between stations.

(b) Yamanlı Energy Tunnel:

The Yamanli II HEPP Project's Energy Tunnel, which is 6618 m in length, was constructed with a double shield TBM with a 4.305 mm excavation diameter. Schoema locomotives of 20 t and 100 kW of power were used, along with muck cars of 8 m^3 with a self-discharger. One fixed and one moving California switch were also used.

Fig. 4.16 Car changing with California Switch, Bilgin and Acun (2024)

4.9 Multiple-service Vehicles (MSVs) for Material Transport

They are designed to carry heavy loads up to 200 t in narrow environments. They may work in inclines and declines up to 20%. Tunnels are used on the job site to transport all operating materials needed for work in the tunnel, like precast segments, grout tanks, material platforms, personnel modules, etc. Unlike rolling stock, MSVs do not require rails, as seen in Fig. 4.17, saving a lot of time and money in setting up and cleaning rails on the job site. They have extremely high-performing braking distances. They can move to both sites. The characteristics of some multi-service vehicles to supply TBM tunneling are given in Table 4.4.

Fig. 4.17 A typical MSV carrying out precast segments, with courtesy of the Company TMS

Table 4.4 Some characteristics of some multi-service vehicles to supply TBM tunnelling, Bilgin and Acun (2024)

Parameter	MSV single	MSV double	MSV triple
Load capacity, t	10–65	40–140	120–200
Width, m	1.0–1.2–1.5–1.7–1.8	1.5–1.7–1.8	1.9
Number of wheels	4.6 or 8	8–20	24
Drive weals	4.6 or 8	8–20	12–24
Drive, kW	100–150–200–315	200–315–400	315–400
Max. speed, km/h	25	20	16
Max. incline, %	25	20	15
Steering system	Electronic	Electronic	Electronic

4.10 Vibration Due to TBM and Its Effects on Buildings and Monuments and Relations with Lithology

Tunnel Boring machines have been used extensively for 30 years, especially in urban areas. Using these mechanical excavators can generate vibrations, causing damage to neighbouring buildings and disturbing the residents or the equipment. This problem is particularly challenging in hard rock and mix faces, where TBMs are increasingly large in diameter and shallow in depth. Vibrations of cultural objects due to different sources (blasting, mechanical excavation, transport, etc.) have recently been raising more and more interest from the heritage community. However, the cutting-induced vibration can help clarify the TBM-ground interaction and provide a valuable ground identification approach. In this section, the research works realized in this respect will be summarized, and later, the results of a study carried out on the effect of the vibrations of 4 different EPB-TBM on the historical aqueducts in Istanbul "Uzun Kemer and Eğri Kemer" will be discussed. One of the authors of this book was a team leader in this work realized for the contractor, Kolin-Şenbay, who did the job for the Istanbul New Airport Metro Project, Bilgin et al. (2017).

4.10.1 Past Studies on the Vibration Generated by TBMs

Carnavale et al. (2000) did work on the West Water Supply Tunnel, a 28.3 km long, 4.9 m diameter, pressurized tunnel outside Boston, Massachusetts. This project used two identical Robbins gripper-type Tunnel Boring Machines (TBMs). The main rock formations were granodiorite, granite, and schist. The monitoring aimed to detect TBM-induced vibrations and study the potential impacts on residences above and adjacent to the tunnel alignment. The monitoring is conducted at the ground surface as the TBM passes underneath the sensors at depths between 61 and 137 m. The secondary concern was the annoyance factor that could surface if TBM vibrations were perceptible to the humans and animals above the tunnel alignment. The peak particle velocities of TBM vibrations measured were found perceptible by humans. However, using the USBM criteria, Siskind et al. (1980) they found that most of the TBM steady-state peak particle velocities and the frequencies of the associated vibrations measured for this project should not cause structural damage. The vibration levels measured at these locations are lower than those in the adjacent natural ground.

Gao et al. (2021) measured ground vibrations due to subway moving loads in China's two distinct geological conditions (soft soil ground and granite ground). Vibration measurements are performed on the tunnel ground after the vibration generated by another running subway. It was found that the vibration acceleration of the granite ground was approximately 5 times higher than that of soft soil ground under the subway train moving load. Furthermore, the maximum vibration level in granite ground is nearly 90 dB, seriously affecting adjacent structures, sensitive machinery, and passengers.

A 9.60 m long and 6.68 m diameter EPB shield was selected to excavate in the varying ground conditions in Metro Line R2 of Jinan, 21.5 km long in the east–west direction through the most congested part of the city in China. The overburden of the tunnel was around 17.5 m. Field measurement of TBM vibration in changing ground conditions was carried out with accelerometers mounted on the TBM bulkhead, Liu et al. (2022). The vibration characteristics and patterns under different ground conditions were compared by signal processing, and their relationships to the operating parameters were investigated. In homogeneous soft ground, the vibration magnitude was low and stable, while the high frequency and strong vibration occurred under mixed-face ground conditions; the vibration waveform in the mixed face had an apparent periodicity and was consistent with the cutter head rotation speed.

Wang et al. (2023) measured the vibration generated during the construction of subway tunnels of the Qingdao Metro with a double-shield tunnel boring machine of 6.3 m diameter. Most sections of the tunnel passed through areas with old historical houses, factory buildings, and shops, which caused multiple complaints from residents. The overburden changed between 12.3 and 25.8 m, and the tunnel passed from slightly weathered granite. Nine boreholes with a depth of 23.18 m were drilled in the monitoring section for vibration measurements, and vibration sensors were installed at three positions in each borehole. It was found that the vibration could be reduced by reducing the field penetration index, the TBM thrust, cutter head torque, and cutter head speed. In China, vibration is listed as one of seven significant public hazards, and its occurrence can have a major impact on the standard of living in urban environments, causing dizziness, fatigue, and psychological depression. In severe cases, it can lead to permanent pathological damage to the human body. Consequently, China has formulated standards such as the "Urban Regional Environmental Vibration Standard" and the "Urban Regional Environmental Vibration Measurement Method" to address the vibration problem in urban areas. It was also found that TBM vibration characteristics were sensitive to changes in working face conditions. The larger the complex rock ratio, the larger the amplitude and the more concentrated the frequency.

Rallu et al. (2023) measured vibration on four French tunnels and micro-tunnel projects in predominantly rocky terrain. The data collected were compared with the limited data available in the literature. The amplitude of the velocities measured in the TBMs is of the order of 0.1 mm/s in clay soils, 1 mm/s in sand and gravel, and 10 mm/s in rocky soils. On the ground surface, the particle velocities between 10 and 30 m in front of the cutting face are about 10 times lower than those inside the TBM. The four measurement campaigns in various geotechnical contexts (fine soil, gravelly soil, and rock) revealed that 80% of the energy of the vibration signals measured in the TBMs is 40 Hz.

Ates and Copur (2023) measured vibrations generated by Earth Pressure Balance EPB-TBMs during mechanized excavation. A unique vibration data recording system is developed and installed onto four EPB TBMs, excavating in three different project sites. Vibration data on a 6 km tunnel excavation is recorded in three orthogonal directions. The effects of geological conditions, including soils and hard rocks, EPB TBM operating parameters (thrust, torque, face pressure, cutterhead rotational speed, penetration rate), cutter damages/breakages, positions, and types of accelerometers, and TBM diameters on vibration are investigated. Following the detailed analysis of vibration characteristics (time domain parameters) and geotechnical data, it is seen that geological and lithological changes significantly influence the vibration patterns generated during the excavation. The TBM operational parameters and cutter conditions also affect the vibrations. It is realized that the broken disc cutters create specific high-amplitude signals. The results of this study indicate that the accelerations measured on the EPB TBMs can exceed $\pm$ 10 g, particularly in hard rock formations, and it is possible to detect the geological/lithological changes using vibration measurements. Vibrations increase with increasing rock strength, being between 0.27 g on a weathered dyke (with 15 MPa UCS) and 1.0 g on a fresh dyke (with 210 MPa UCS).

Aslan et al. (2024) placed geophones and accelerometers along tunnel route 18 of the Grand Paris at the Orly site Express in France: on the surface, in the ground near the TBM, and inside the TBM itself. This metro line employed an EPB-TBM with a diameter of 9.15 m. Two boreholes, T1 and T2, were used at the construction site, spaced 10 m apart and having diameters of 80 mm. The lengths of these boreholes were 26 m and 38.8 m for T1 and T2, respectively. On the surface, seven measuring devices were installed. Within the TBM's manlock, two measuring devices were positioned at 2.5 and 4.5 m distances from the cutterhead. The first measurements were conducted in January 2023 to collect data in normal mechanical noise conditions. The second campaign was executed in April 2023 upon the arrival of the TBM at the measuring side. The velocity magnitude inside the TBM is found approximately 10 times larger than that within the ground. This situation indicates that not all the energy the TBM generates is transferred to the surrounding area. Surface velocity vibrations (10 − 2 mm/s) were significantly lower than those inside the TBM (100 times); Velocity amplitudes within the TBM were ten times higher than those within the ground. Energy dissipation was confirmed by analyzing the highest singular values of spectral density matrices from all sensors, with three levels of energy recorded: approximately 10–2 (mm/sn^2) inside the TBM, 10–5 (mm/sn^2) within the soil, and 10–2 (mm/sn^2) on the ground.

4.10.2 Investigation on the Effect of TBM Vibrations on "Eğri Kemer and Uzun Kemer" Aqueducts During Istanbul Airport Metro Project Construction

The effect of vibration that will occur during tunnel excavation on structures is an important issue, and different researchers have studied this in detail: Carnevale et al. (2000), Johnson and Hannen (2015), and SKM (2011).

The technical specifications for constructing of the Istanbul Airport Metro Project are prepared in respect to the environmental issues and conservation of historical heritages. It states that the intensity of the ground vibration must not exceed the maximum horizontal particle velocity of 5 mm/sec. Furthermore, the vibration intensity in other infrastructure lines and structures will be less than the vibration that causes damage, loss of function, and pressure, ensuring the safety of the project.

In Fig. 4.18, it is clearly seen how the vibration of an excavation machine may affect a building. The peak particle velocity and the frequency of the vibration in the X, Y, and Z directions are important parameters that may affect the buildings. Figure 4.19 recommends the vibration limits in this respect given by BS 7385-2 (British Standard) which is superimposed on the USBM RI 8507 (US Bureau of Mines Research Report limits).

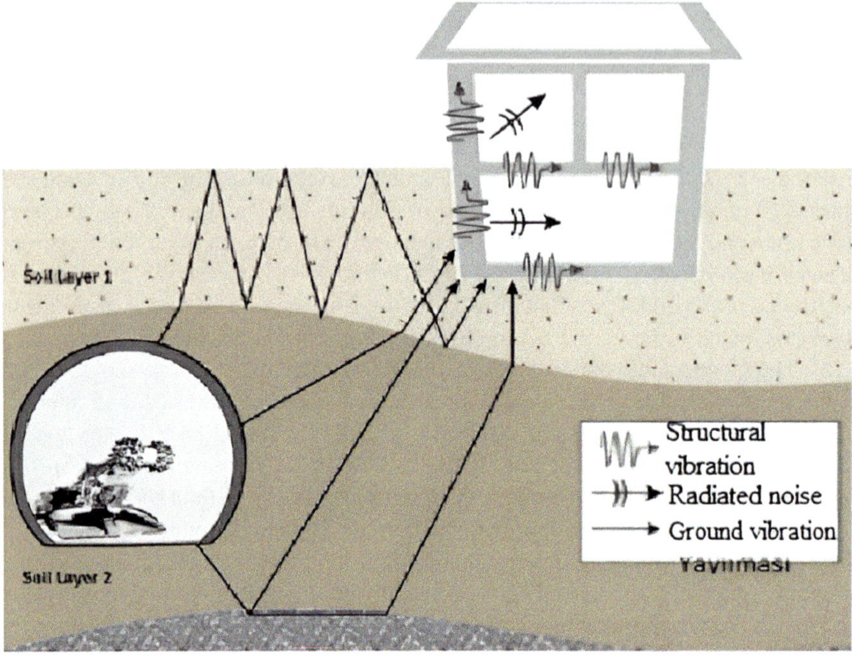

Fig. 4.18 Vibration noise transmission paths during mechanical excavation, SKM (2011)

The research on the subject was conducted meticulously, using interpretation of the vibration values generated during the excavation of Herrenknecht, Lovat and Terratec EPB-TBMs in Istanbul Airport Metro Construction. The characteristics of these TBMs are given in Table 4.5, Bilgin et al. (2017).

Vibration data were collected with commercially available single-channel seismographs. Monitoring locations were chosen on the surface in the areas close to the TBMs. These machines were excavated in the Trakya Formation, which consists of interbedded sandstone-mudstone and siltstone and the sand and silty clay in the Mahmutbey-Mecidiyeköy Metro line. Table 4.6 presents the average of the measurements, with each value being the average of at least 150 measurements. The chart clearly shows that all values are well within the safety limits, with none exceeding the peak particle velocity of 5 mm/sec, the accepted limit value that could damage historical buildings.

The variations of peak particle velocity and frequency of vibration with different TBMs in different geological formations are plotted from Figs. 4.20, 4.21, 4.22, 4.23, 4.24, 4.25, 4.26, 4.27, 4.28 and 4.29. As is seen in Figs. 4.20, 4.22, and 4.24, the peak particle velocity changes between 0.3 and 1.4 mm/s, and the frequency of the vibration changes between 0.2 and 1.4/1.8 Hertz in highly fractured Trakya Formation with RQD changing between intervals of % 0–60. However, these values reach 0.4–1.2 mm/sand 0.5–1.5 Hz in more homogeneous sandy and clayey formations. It is interesting to note that the breakdown of the moving machines is related to some

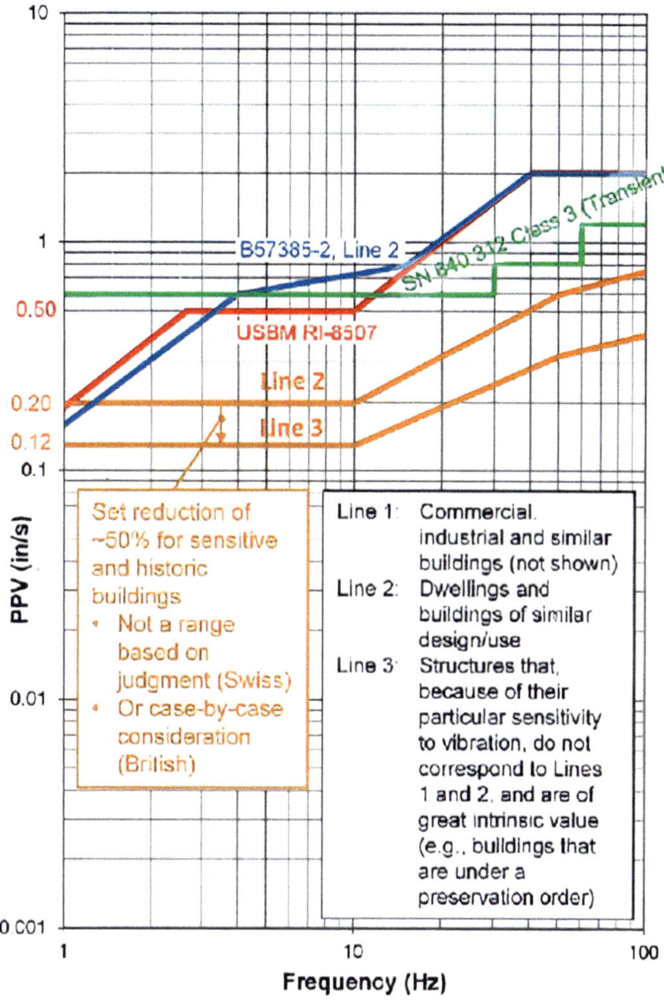

Fig. 4.19 Vibration limits recommended by BS 7385-2 are superimposed on the USBM RI 8507 limit, Johnson and Hannen (2015)

Table 4.5 The characteristics of TBMs used in vibration measurements

Parameter	Herrenknecht	Lovat	Terrateck
Diameter, m	6.55	6.57	6.55
Thrust, kN	32,000	55,000	40,000
Torque, kNm	4400	4500	5440
Disc Nr.xDiam	43 × 15.5”	41 × 17”	43 × 17”
Cutterhead rpm	0–3.8	0–4	0.3
Opening ratio	20	29	37

Table 4.6 Vibrations created by three different TBM in different formations during the Mecidiyeköy-Mahmutbey Metro excavation (each value is the average of at least 150 readings), Bilgin et al. (2017)

TBM	Formasyon	Peak particle velocity in X direction	Peak particle velocity in Y direction	Peak particle velocity in Z direction	Resultant peak particle velocity	Resultant frequency Hz
Herrenknecht	Trakya Sandstone Mudstone	0.41	0.37	0.48	0.73	0.88
Lovat	Trakya Sandstone Mudstone	0.45	0.49	0.42	0.69	0.92
Terrateck	Trakya Sandstone Mudstone	0.30	0.4	0.35	0.64	0.85
Terrateck	Clay	–	–	–	0.65	0.9
Herrenknecht	Sand	–	–	–	0.73	1.02

extent to the magnitude of the vibration, which gives a clue to a higher degree of breakdowns in TBMs working in fractured and hard rocks, Figs. 4.19, 4.20, 4.21, 4.22, 4.23, 4.24, 4.25, 4.26, 4.27 and 4.28 give the peak particle velocities of all three TBMs in the Ttakya Formation and the frequency change at this velocity. Likewise, these values are far below those that can damage the aqueducts (Fig. 4.30).

The research studies carried out by Carnavale et al. (2000), Bilgin et al. (2017), Liu et al. (2022), Wang et al. (2023), Ateş and Çopur (2023) and Aslan et al. (2024) are summarized in Table 4.7.

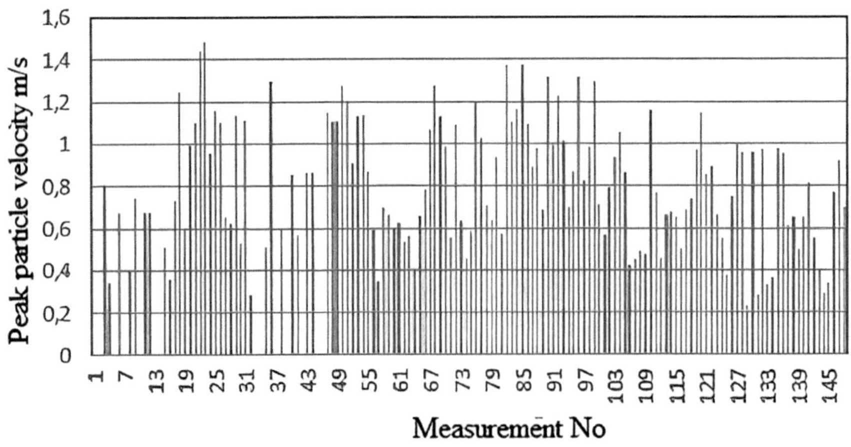

Fig. 4.20 The variation of peak particle velocity on the surface with Herrenknecht TBM in Trakya Formation, Bilgin et al. (2017)

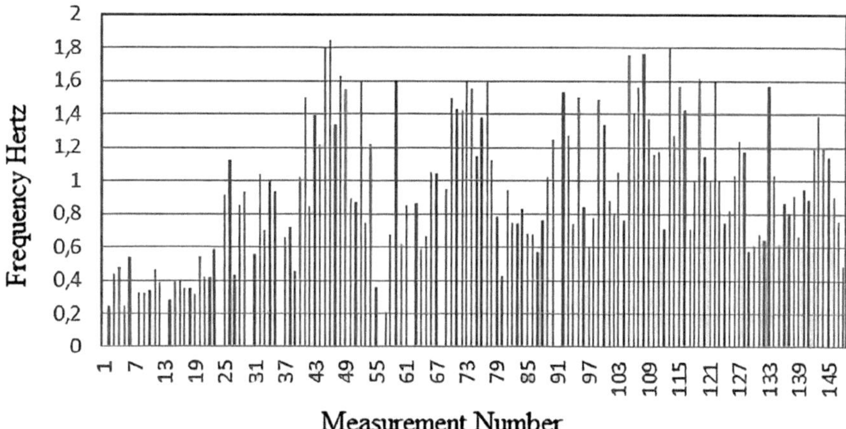

Fig. 4.21 The variation of frequency on the surface with Herrenknecht TBM in the Trakya Formation, Bilgin et al. (2017)

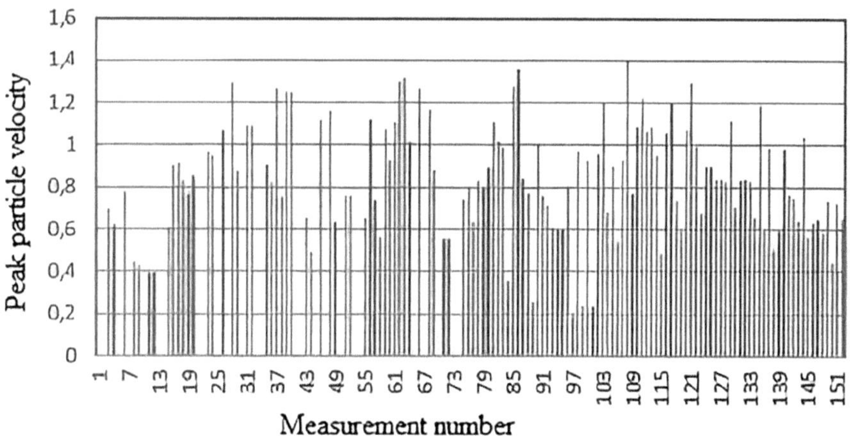

Fig. 4.22 The variation of peak particle velocity on the surface with Lovat TBM in Trakya Formation, Bilgin et al. (2017)

Table 4.7 tells us that vibration generated by TBM excavation is sensitive to changes in the geological characteristics of the ground and the type and operational parameters of the TBMs. The type of TBM affects the vibration generated. Velocity in the cutterhead is 10 times higher than that on the surface.

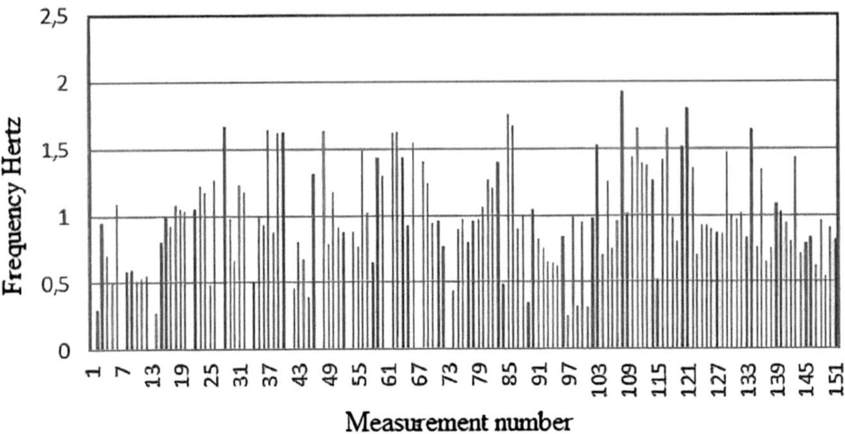

Fig. 4.23 The variation of the frequency on the surface with Lovat TBM in Trakya Formation, Bilgin et al. (2017)

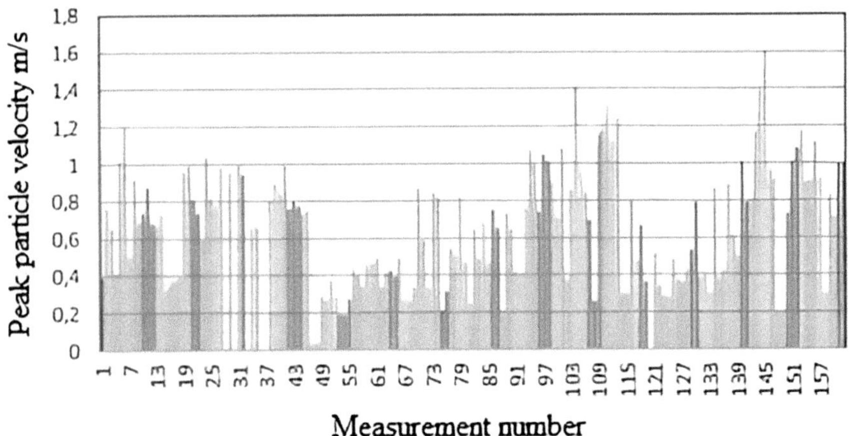

Fig. 4.24 The variation of peak particle velocity on the surface with Terrateck TBM in the Trakya Formation, Bilgin et al. (2017)

4.11 Concluding Remarks

Rock-cutting experiments are the best choice for developing mechanized cutting technology and predicting performance since they are reliable, can define the basic specifications of mechanical excavators, and can design cutter heads.

Roadheaders can be classified into two groups based on cutterhead types: axial (longitudinal) and transverse. Both cutter heads have several advantages and disadvantages. Transversal cutterheads cut in the direction of the tunnel face. Therefore,

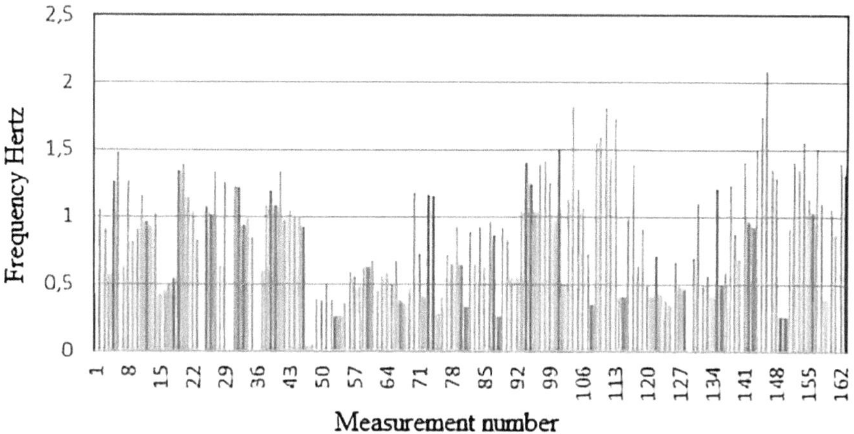

Fig. 4.25 The variation of the frequency on the surface with Terrateck TBM in Trakya Formation, Bilgin et al. (2017)

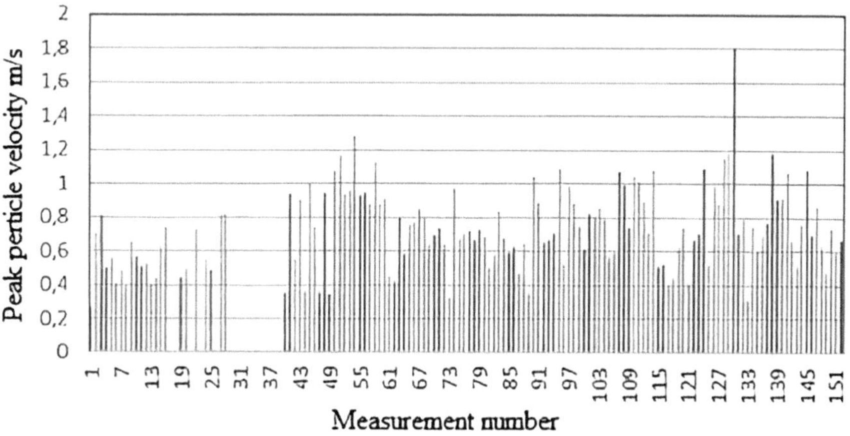

Fig. 4.26 The variation of peak particle velocity on the surface with Terrateck TBM in clayey Formation, Bilgin et al. (2017)

they are more stable than roadheaders with axial letterheads of comparable weight and cutter head power. At transversal heads, the majority of reactive force resulting from the cutting process is directed toward the main body of the machine. Roadheaders with transversal-type cutter heads are less affected by changing rock conditions and harder rock portions.

TBMs have the advantage of high advance rates in hard rock and soft ground, limiting the disturbance to the surrounding area. Gripper TBMs are designed for high advance rates in stable hard rock and are relatively less expensive than other TBMs. Single shields are built to advance in solid or fractured rock. Like other shield

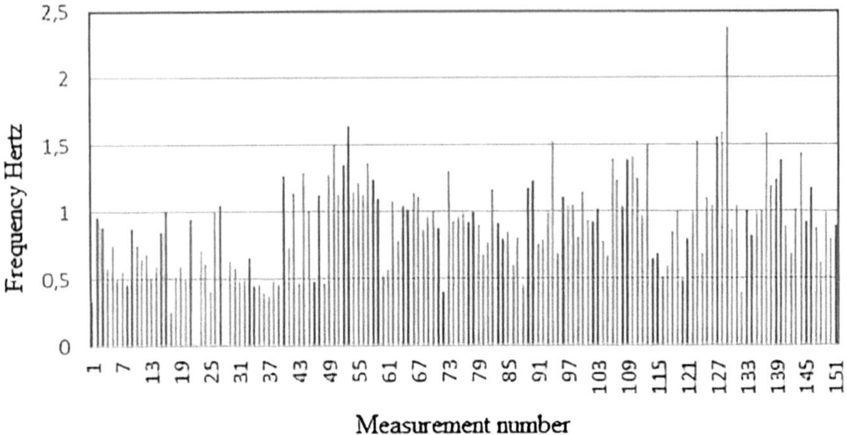

Fig. 4.27 The variation of the frequency on the surface with Terrateck TBM in clayey Formation, Bilgin et al. (2017)

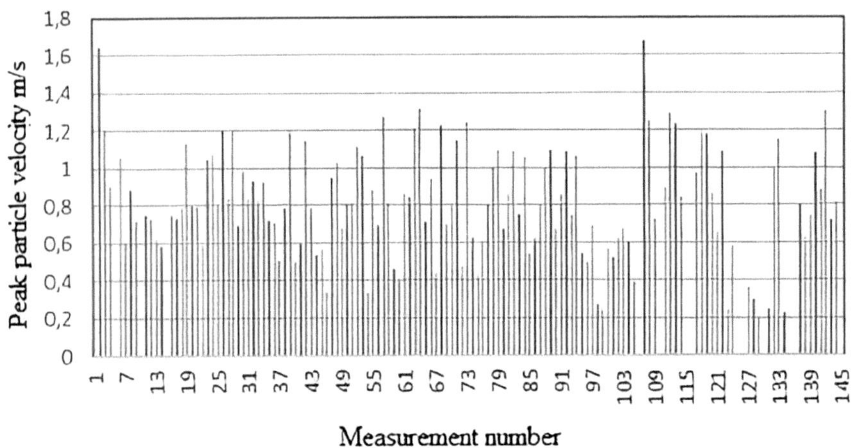

Fig. 4.28 The variation of peak particle velocity on the surface with Herrenknecht TBM in sandy Formation, Bilgin et al. (2017)

machines, they use thrust cylinders to advance and an erector to build segmental lining. Double shields have been developed for continuous boring and to increase the production rate in fractured rock. The machine may advance while the precast segments have been erected. Crossover TBMs have multi-speed gearboxes to enable the machine to advance through blocky, fractured, or mixed ground at high torque and low speed. EPB TBMs, Slurry TBMs, and variable density TBMs are soft-ground TBMs. The unique feature of EPB-TBMs is that they use the excavated soil directly as a support medium. When the pressure of the soil paste in the excavation chamber equals the pressure of the surrounding soil and groundwater, the necessary balance is

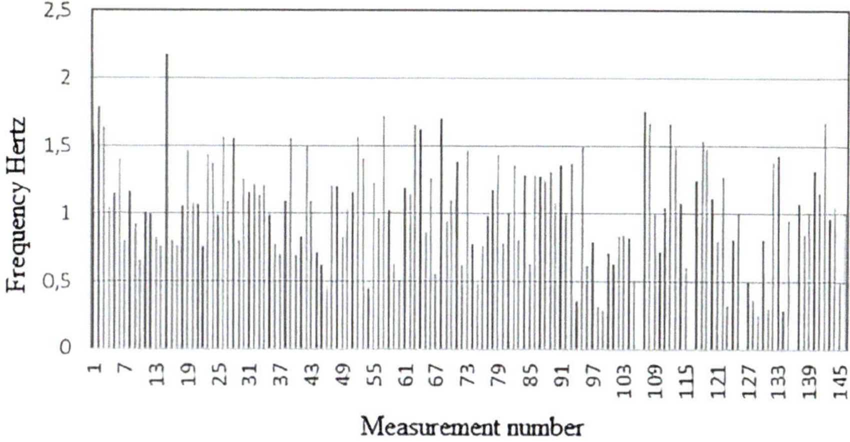

Fig. 4.29 The variation of frequency on the surface with Herrenknecht TBM in sandy Formation, Bilgin et al. (2017)

Fig. 4.30 A water aqueduct from Ottoman Time in Istanbul from the archive of the author Bilgin

Table 4.7 Comparison of the results of different authors obtained in TBM vibration measurements

Parameter	Carnavale et al. (2000)	Bilgin et al. (2017)	Liu et al. (2022)	Wang et al. (2023)	Rallu et al. (2023)	Ateş and Çopur (2023)	Aslan et al. (2024)
TBM, Diameter	Gripper, 4.9 m	3 EPB 6.6 m	EPB 6.7 m	Double sh. TBM, 6.3 m	M. tunnel 1.3 m, EPB 9.8 m	In 3 project sites, EPB 6.6 m	EPB 9.15 m
Rock	Granodiorite, granite, schist	Sandstone-mudstone, clay, sand	Silty clay mixed face		Sandy gravel, granite	Sandstome, modstone, clay	Marl, gypsum
Overburden	61–137 m	20–40 m	15–17 m	12–26 m	–	10–30 m	30 m
Measurements made on:	The surface	Surface	Cutterhead	Nine boreholes in front of TBM	Cutterhead and surface	Cutterhead	Cutterhead, inside soil and on the surface
Objective of the study	1. Impact on the buildings 2. Impact on human being	Impact on historical monument	Ground relation	Impact on historical houses	General effect of vibration	Ground relation	Research
Peak particle velocity	0.06–0.21 mm/s	0.4–1.4 mm/s	–	–	On surface 0.1 m/s in aluvium 1 mm/s in granite	–	–
Amplitude of the velocity	–	–	–	–	0.1 mm/s in clay,1 mm/s in sand, 10 mm/s in rocky soi	–	In TBM 10 mm/sec in soil 0.1 mm/s, on surface 0.5 mm/s
Acceleration	–	–	0.1–10 g	0.01–0.03 g	0.1 g in granite, 0.03 g in alluvium	1 g on fresh dyke, 0.27 g on weathered dyke	10^{-2} (mm/s)^2 inside TBM, 10^{-6} (mm/s)^2 on surface

(continued)

Table 4.7 (continued)

Parameter	Carnavale et al. (2000)	Bilgin et al. (2017)	Liu et al. (2022)	Wang et al. (2023)	Rallu et al. (2023)	Ateş and Çopur (2023)	Aslan et al. (2024)
Noise dB	–	–		53–65 dB	–	–	
Frequency of vibration	3–99 Hz	0.2–1.8 Hz	100–625 Hz	9–90 Hz	10 Hz	–	Within the ground 85 Hz, 175 Hz on surface
Results	1. Perceptible by human 2. Vibration is related to fault structure	Monuments are not affected by vibration	Vibration is sensitive to ground	Houses are affected	Velocity in cutterhead is 10 times higher than on the surface	Vibration is related to the type of ground and TBM operational parameters	Surface velocity 0.02 mm/s 100 times lower than inside TBM

achieved, and the machine may advance safely without causing surface deformations. The Slurry Shield TBM is suitable for excavating soils with high water content. This TBM is equipped with a slurry system that controls the pressure in the excavation face by injecting pressurized slurry into the cutter chamber, where the slurry is mixed with the excavated material. The mixture is pumped out of the tunnel to a separation and re-circulation plant. The variable density TBM can be operated as a classic slurry TBM with an air bubble system to control the face pressure and be in full EPB mode. This machine can also be operated using a high density in the excavation chamber that would be too dense for a classic slurry operation. Still, it would be too fluid for a classic EPB operation.

The primary elements of planning the mechanized tunneling operations are ventilation, muck transport, material, and the crew. To efficiently dilute dangerous gasses and dust, a return speed for air in the tunnel between 0.3 and 0.6 m/s is usually required (if a risk exists of methane gas, then 1 m/s should be used). It is always the contractor's decision for muck transport, but generally, if the tunnel is longer than 1.5 km long, belt conveyors are used. If it is shorter, then locomotives with muck cars are usually preferred. On the other hand, some contractors prefer locomotive transport because of their lower investment cost compared to belt conveying. Multiple-service vehicles (MSVs) are necessary for material transport to carry heavy loads, like precast segments, grout tanks, and crew modulus up to 200 t in narrow environments. Unlike rolling stock, MSVs do not require rails.

The vibration generated by TBM excavation is sensitive to changes in the geological characteristics of the ground, operational parameters, and type of the TBMs. The vibration is perceptible by humans, and it may affect the building and surrounding areas depending on tunneling conditions. Velocity in the cutterhead is 10 times higher than that on the surface.

References

Abdolreza YC, Yakhchali SH (2013) A new model to predict roadheader performance using rock mass properties. J Coal Sci Eng 19(1):51–56

Aslan Y, Rallu A, Branque D, Berthoz N, Chatzigogos C, Makrypidi T (2024) Monitoring and analyzing vibrations from tunnel boring machine in urban areas. In: ECSMGE24 XVIII European conference on soil mechanics and geotechnical engineering, Proceedings of the XVIII ECSMGE 2024, geotechnical engineering challenges to meet current and emerging needs of society. ISBN 978-1-032-54816-6, https://doi.org/10.1201/9781003431749-263

Ates A, Copur H (2023) Investigation of parameters affecting vibration patterns generated during excavation by EPB TBMs. Tunn Undergr Space Technol 138:105185

Babler K, Battistoni F, Burger WA (2018) Variable density TBM—combining two soft ground TBM technologies. TBM Tunnel Business Magazine, March 22

Balci C (2009) Correlation of rock cutting tests with field performance of a TBM in a highly fractured rock formation: a case study in Kozyatagi-Kadikoy metro tunnel, Turkey. Tunnell Undergr Space Technol 24:423–435. https://doi.org/10.1016/j.tust.2008.12.001

Balci C, Bilgin N (2007) Correlative study of linear small and full-scale rock cutting tests to select mechanized excavation machines. Int J Rock Mech Min Sci 44:468–476. https://doi.org/10.1016/j.ijrmms.2006.09.001

Balci C, Demircin MA, Copur H, Tuncdemir H (2004) Estimation of optimum specific energy based on rock properties for assessment of roadheader performance. J S Afr Inst Min Metall 104(11):633–642

Barton N (2000) TBM tunnelling in jointed and faulted Rock. Balkema, Rotterdam

Bieniawski ZT, Celada B, Galera JM (2007) Predicting TBM excavability—part I. Tunnels Tunnell Int 32–35 (September)

Bieniawski Z, Grandori R (2007) Predicting TBM excavability—part II. Tunnels Tunnell Int 15–18 (December)

Bieniawski ZT, Celeda B, Galeras JM, Tardaguila I (2008) New applications of the excavability index for selection of TBM types and predicting their performance. In: Proceedings of the world tunnel congress. Akra, India, pp 1618–1629

Bilgin N (1977) Investigations into the mechanical cutting characteristics of some medium and high strength rocks Ph.D thesis. The University of Newcastle Upon Tyne, p 332

Bilgin N, Acun S (2024) Practical management of tunneling with tunnel boring machines. CRC Press Taylor and Francis Group, p 217

Bilgin N, Shahriar K (1987) Development of an instrumented testing system and its application to TTK Amasra Coal Region. Istanbul Technical University; Mining Engineering Department. Tubitak Project No: MAG674; (in Turkish)

Bilgin N, Seyrek T, Erdinc E, Shahriar K (1990) Roadheaders clean valuable tips for Istanbul Metro. Tunnels Tunnell 22:29–32

Bilgin N, Balci C, Acaroglu O, Tunçdemir H, Eskikaya S, Akgül M, Algan M (1999) The performance prediction of a TBM in Tuzla–Dragos sewerage tunnel. In: The world tunnel congress'99, Oslo, 31 May–3 June, Balkema, Rotterdam, pp 817–822. ISBN: 90 5809 0639

Bilgin N, Feridunoglu C, Tumac D, Çınar M, Palakci Y, Özyol YO (2005) The performance of a full face tunnel boring machine (TBM) in Tarabya (Istanbul), WTC. In: Erdem & Solak (eds) Underground space use: analysis of the past and lessons for the future. Taylor & Francis Group, London. ISBN: 04 1537 452 9

Bilgin N, Balci C, Tumac D, Feridunoglu C, Copur H (2010) Development of a portable rock cutting rig for rock cuttability determination. In: Zhao J, Labiouse V, Duth JP, Mathier JF (eds) Proceedings of the European rock mechanics symposium (EUROCK 2010), June 15–18, Lausanne-Switzerland, pp 405–408

Bilgin N, Copur H, Balci C (2014) Mechanical excavation in mining and civil industries. CRC Press, Taylor and Francis Group, London

Bilgin N, Akay S, Engin A, Ercan C (2017) The effect of metro construction on "Eğri—Uzun Kemer Aqueducts" The evaluation of geological structure, possible surface deformations and vibrations created by TBMs

Bilgin N, Copur H, Balci C, Tumac D (2019) Strength, cuttability, and workability of coal. CRC Press/Taylor & Francis Group, p 249

Brabant JD, Duhme R (2017) Hard rock TBM tunnelling—technical developments and recent experience. In: The world congress on advances in structural engineering and mechanics (ASEM17), 28 August–1 September, 2017, Ilsan (Seoul), Korea

Bruland (1998) Hard rock tunnel boring doctoral dissertation. Norwegian University of Science and Technology, Trondheim

Burger W (2016) Slurry, hybrid and large TBMs. Herrenknecht AG Schwanau, Tunneling Short Course, Boulder CO, September 12–15, 2016

Carnevale M, Young G, Hager J, Carnevale MC (2000) Monitoring of TBM-induced ground vibrations. In: Ozdemir (ed) North American Tunneling '00, Balkema, Rotterdam. ISBN: 90-5809-162-7

Chang SH, Choi SW, Bae GJ, Jeon S (2009) Performance prediction of TBM disc cutting on granitic rock by the linear cutting test. Tunnell Undergr Space Technol 21:271. https://doi.org/10.1016/j.tust.2005.12.131

Cho JW, Seakwon J, Jeong H, Chang SH (2013) Evaluation of cutting efficiency during TBM disc cutter excavation within a Korean granitic rock using linear-cutting-machine testing and photogrammetric measurement. Tunn Undergr Spacce Technol 35:37–54. https://doi.org/10.1016/j.tust.2012.08.006

Comakli R, Balci C, Copur H, Tuma, D (2021) Experimental studies using a new portable linear rock cutting machine and verification for disc cutters. Tunnel Undergr Space Technol 108:103702. https://doi.org/10.1016/j.tust.2020.103702

Copur H (1999) Theoretical and experimental studies of rock cutting with drag bits toward the development of a performance prediction model for roadheaders PhD thesis. Colorado School of Mines, 361p

Copur H, Ozdemir L, Rostami J (1998) Roadheader applications in mining and tunneling industries. In: Annual meeting of American Society for mining, metallurgy and exploration (SME), Orlando, Florida, March 10–12, Preprint Number: 98–185

Copur H, Aydın H, Bilgin N, Balci C, Tumac D, Dayanc C (2014) A stochastic method for predicting performance of EPB TBMs, working in semi-closed/open mode, ITA-AITES WTC, Brasil

Delisio A, Zhao J (2013) Review of the TBM performance in blocky rocks with potential face stability issues. In: World tunnelling congress 2013, Geneva, pp 1179–1186.ta

Delisio A, Zhao J, Einstein HH (2013) Analysis and prediction of TBM performance in blocky rock conditions at the Lötschberg Base tunnel. Tunn Undergr Space Technol 33:131–142

Dollinger GL, Handewith JH, Breeds .CD (1998) Use of punch tests for estimating TBM performance. Tunnell Undergr Space Technol 13(4):403–408

Duhme R (2017) Urban tunneling in settlement risk areas. In: The 2017 world congress on advance in structural engineering and mechanics (ASEM17) 28 August 1 September, 2017, Lisan (Seoul), Korea

Ebrahimabadi A, Goshtasbi K, Shahriar K, Cheraghi Seifabad M (2011) A model to predict the performance of roadheaders based on rock mass brittleness index. J South Afr Inst Min Metall 111:355–364

Farrokh E (2020) A study of various models used in the estimation of advance rates for hardrock TBMs. Tunnel Undergr Space Technol 97:103219

Farrokh E, Rostami J, Laughton C (2012) Study of various models for estimation of-penetration rate of hard rock TBMs. Tunn Undergr Space Technol 30:110–123

Farrokh E, Rostami J, Askilsrud OG (2013) Down time analysis of hard rock TBM case histories. In: World tunnelling congress 2013, Geneva

Gao M, Song YS, Wang Y, Chen QS (2021) Effect of geological conditions on vibration characteristics of ground due to subway tunnel: field investigation. Geotech Geol Eng 39:4689–4696. https://doi.org/10.1007/s10706-021-01767-2

Gehring KH (1989) A cutting comparison. Tunnels Tunnel 21:27–30

Gong QM, Zhao J (2007) Influence of rock brittleness on TBM penetration rate in Singapore granite. Tunn Undergr Space Technol 22:317–324

Gong QM, Zhao J (2009) Development of a rock mass characteristics model for TBM penetration rate prediction. Int J Rock Mech Min Sci 46:8–18

Gong QM, Zhao J, Jiang YS (2007) In situ TBM penetration tests and rock mass boreability analysis in hard rock tunnels. Tunn Undergr Space Technol 22:303–316

Harding D, Alpgut Y (2020) Tunneling through 48 fault zones and high water pressures on Turkey's Gerede water transmission tunnel. In: ITA-AITES world tunnel congress, 12–15 May, Kuala Lumpur

Hassanpour J, Rostami J, Khamehchiyan M, Bruland A (2009a) Developing new equations for TBM performance prediction in carbonate-argillaceous rocks: a case history of Nowsood water conveyance tunnel. Geomech Geoeng 4:287–297

Hassanpour J, Rostami J, Khamehchiyan M, Bruland A, Tavakoli HR (2009b) TBM performance analysis in pyroclastic rocks: a case history of Karaj water conveyance tunnel. Rock Mech Rock Eng 1–19

Hassanpour J, Rostami J, Zhao J (2011) A new hard rock TBM performance prediction model for project planning. Tunn Undergr Space Technol 26:595–603

Heiniö M (1999) Rock excavation handbook. Sandvik Tamrock Corporation, p 363

Johnson AP, Hannen WR (2015) Vibration limits for historic buildings and art collections. APT Bull J Preserv Technol 46:2–3

Kahraman E, Kahraman S (2016) The performance prediction of roadheaders from easy testing methods. Bull Eng Geol Env 75:1585–1596

Kahraman S, Aloglu AS, Aydın B, Saygın E (2019) The needle penetration index to estimate the performance of an axial type roadheader used in a coal mine. Geomech Geophys Geo-Energ Geo-Resour 5:37–45

Kahraman S, Dibavar B, Rostami M, Fener M (2023) Performance prediction of roadheaders using the rock mass cuttability classification. Arab J Geosci 16:686

Khademi HJ, Shahriar K, Rezai B, Rostami J (2010) Performance prediction of hard rock TBM using Rock Mass Rating (RMR) system. Tunn Undergr Space Technol 25:333–345

Liu MB, Liao SM, Men YQ, Hing HT, Liu H, Sun LY (2022) Field monitoring of TBM vibration during excavating changing stratum: patterns and ground identification. Rock Mech Rock Eng 55:1481–1498. https://doi.org/10.1007/s00603-021-02714-6

Maidl B, Schmid L, Ritz, W, Herrenknecht Sturge M (2011) Hardrock tunnel boring machines. Wiley, p 356. ISBN-13: 978-3433016763

Maidl B, Herrenknecht M, Maidl, U, Wehrmeyer G, (2013) Mechanized shield tunnelling 2012. Ernst & Sohn, Berlin, p 490. ISBN: 978-3-433-02995-4

McFeat-Smith I, Fowell RJ (1977) Correlation of rock properties and the cutting performance of tunnelling machines. In: Proceedings of the conference on rock engineering, Newcastle upon Tyne, pp 581–602

McFeat-Smith I, Fowell RJ (1979) The selection and application of roadheaders for rocktunnelling. In: Proceedings of the rapid excavation and tunnelling conference, Atlanta, vol 1, pp 261–279

Namli M, Bilgin N (2017) A model to predict daily advance rates of EPB-TBMs in a complex geology in Istanbul. Tunn Undergr Space Technol 62(2017):43–52

Nelson P, O'Rourke TD, Kulhawy FH (1983) Factors affecting TBM penetration rates in sedimentary rocks. In: Proceedings, 24th US symposium on rock mechanics. Texas A&M, College Station, TX, pp 227–237

Nelson PP, Ingraffea AR, O'Rouke TD (1985) TBM performance prediction using rock fracture parameters. Int J Rock Mech Min Sci Geomech Abstr 22:189–192

Ocak I, Bilgin N (2010) Comparative studies on the performance of a roadheader, impacthammer and drilling and blasting method in the excavation of metro station tunnels in Istanbul. Tunn Undergr Sp Tech 25:181–187

Ozdemir L (1990) Recent developments in hard rock mechanical mining technologies. In: Proceedings of the 4th Canadian symposium on mining automation, September 16–18, Saskatoon, pp 143–165

Ozdemir L, Nilsen B (1993) Hard rock tunnel boring prediction and field performance. In: Rapid excavation and tunneling conference (RETC) proceedings, Boston, USA (Chapter 52)

Potter P (2023) The evolution of tunnel boring machines, construction physics, Oct 06, 2023. https://www.construction-physics.com/p/the-evolution-of-tunnel-boring-machines

Rallu A, Berthoz N, Charlemagne S, Branque D (2023) Vibrations induced by tunnel boring machine in urban areas: In situ measurements and methodology of analysis, its lists available at science direct. J Rock Mech Geotech Eng 15(2023):130–145

Ribacchi R, Fazio AL (2005) Influence of rock mass parameters on the performance of a TBM in a gneissic formation (varzo tunnel). Rock Mech Rock Eng 38:105–127

Rostami J (2016) Performance prediction of hard rock Tunnel Boring Machines (TBMs) in difficult ground. Tunnel Undergr Space Technol 57:173–182

Rostami J, Ozdemir L (1993) A new model for performance prediction of hard rock TBMs. In: Bowerman LD et al (eds) Proceedings of rapid excavation and tunneling conferences, Boston, MA, USA, pp 793–809

Rostami J, Ozdemir L, Neil DM (1994) Performance prediction: a key issue in mechanical hard rock mining. Min Eng 11:1263–1267

Rostami J, Ozdemir L, Bjorn N (1996) Comparison between CSM and NTH hard rock TBM performance prediction models. ISDT, Las Vegas

Roxborough FF (1969) Rock cutting research for the design and operation of tunneling machines. Tunnels Tunnel 125–128

Roxborough FF, Phillips HR (1975) Rock excavation by disc cutter. Int J Rock Mech Min Sci Geomech Abst 12:361–366

Salimi A, Rostami J, Moormann C, Hassanpour J (2022) Introducing tree-based-regression models for prediction of hard rock TBM performance with consideration of rock type. Rock Mech Rock Eng 55:4869–4891. https://doi.org/10.1007/s00603-022-02868-x

Sapigni M, Berti M, Bethaz E, Busillo A, Cardone G (2002) TBM performance estimation using rock mass classification. Int J Rock Mech Min Sci 39(6):771–788

Shao W, Li X, Sun Y, Huang H (2014) Linear rock cutting with SMART*CUT picks. Appl Mech Mater 477–478:1378–1384

Siskind DE, Stagg MS, Kopp JW, Dowding CH (1980) Structure response and damage produced by ground vibration from surface mine blasting. USBM report of investigations 8507. USBM, Washington

SKM - Aurecon CRR Joint Venture (2011) Cross river rail: environmental impact statement cross river rail

Stack B (1995) Encyclopedia of tunnelling, mining and drilling equipment. Muden Publishing. ISBN 10095877112X

Thuro K, Plinninger RJ (1999) Roadheader excavation performance—geological and geotechnical influences. In: The 9th ISRM congress, theme 3: rock dynamics and tectonophysics/rock cutting and drilling, Paris, pp 1241–1244

Tumac D, Bilgin N, Feridunoglu C, Ergin H (2007) Estimation of rock cuttability from shore hardness and compressive strength properties. Rock Mech Rock Eng 40(5):477–490

Yagiz S, Rostami J, Ozdemir L (2012) Colorado school of mines approaches for predicting TBM performance. In: ISRM EUROCK. Rock Engineering and Technology for Sustainable Underground Construction

Yagiz S, Hassanpour J, Rostami J (2024a) Introducing the concept of fracture index for performance prediction of hard rock TBMs New challenges in rock mechanics and rock engineering. CRC Press, pp 1513–1519

Yagiz S, Yazitova A, Smirnov G, Thyagarajan VM (2024b) Appraisal of boreability characteristics of rocks to estimate the TBM advancement. In: ARMA US rock mechanics/geomechanics symposium, 2

Wang Z, Jiang Y, Shao X, Liu C (2023) On site measurement and environmental impact of vibration caused by construction of double shield TBM tunnel in urban subway. Sci Rep 13:17689. https://doi.org/10.1038/s41598-023-45089-0

Zhao XB, Yao XH, Gong QM, Ma HS, Li XZ (2015) Comparison study on rock crack pattern under a single normal and inclined disc cutter by linear cutting experiments. Tunnell Undergr Space Technol 50:479–489. https://doi.org/10.1016/j.tust.2015.09.002

Chapter 5
A Unique Example from Kozyatağı–Kadıköy Metro Tunnel: Comparison of the Performance of Drill and Blast, Roadheader, Impact Hammer, and TBM

Abstract Mechanical excavation with tunnel boring machines is the most widely used tunneling method in Metro tunnels in Istanbul. However, drilling and blasting, impact hammers, and roadheaders are also used in some cases, especially in the platform tunnels. This chapter gives a unique opportunity to compare the performances of TBM, impact hammer, roadheader and conventional drilling and blasting methods in a selected Metro Project. Although blasting in urban areas is restricted due to vibrations and difficulties transporting blasting agents, it becomes necessary in extreme conditions. Roadheaders have been proven more efficient than impact hammers in terms of machine utilization time and production rate. However, high pick consumption of roadheaders makes using impact hammers in similar conditions necessary. An attempt is made to modify the equation used in the past to predict the net breaking rate of impact hammers, and it is proved that a reduction factor in the breaking rate, taking into account the thickness of the dykes improved the validity of the proposed model. One crucial point emerging from this study is that the dip and strike of the joints of 45° in the direction of cutting of the roadheaders are found to favour the cutting rate of roadheaders given higher values than predicted performance values. The dip and strike of the joints having an angle of 45° in favour of the cutting direction improves the net cutting rate of roadheaders.

5.1 Introduction

In a metro project planned to be realized in complex geology, like in Istanbul, different excavation methods will inevitably be used in the same project. The main tunnels are usually opened with EPB-TBMs. As seen in Table 5.1, in the last five metro projects of 194,121 m in length, 156,127 m of tunnels were opened with EPB-TBM's, station tunnels or crosscuts of 37,994 m in length were opened with drill and blast method, impact hammers or roadheaders as in Kozyatağı-Kadıköy Metro project. Table 5.1 clearly shows that conventional tunnelling methods are inevitable parts of mechanized tunneling since there is always room for conventional tunnelling in a metro project planned to be realized with TBMs due to the short length of crosscuts

N. Bilgin and C. Balci, *Critical Issues in Selecting Conventional and Mechanized Tunnelling Methods*, Springer Tracts in Civil Engineering,
https://doi.org/10.1007/978-3-031-89114-4_5

Table 5.1 The length of tunnels opened with conventional tunnelling methods and with TBMs in five recently terminated metro projects in Istanbul

Metro project	Tunnels with conventional methods, m	TBM tunnels, m	Total m
Gayrettepe New Airport	8368	68,282	76,650
Halkalı Airport	6961	55,617	62,578
Bakırköy Kirazlı	4820	12,100	16,920
Ataköy İkitelli	10,200	16,200	26,400
Başakşehir Kayaşehir	7645	3928	11,573
Total, m	37,994	156,127	194,121

and station (platform) tunnels. Before the introduction of TBMs in Istanbul in Moda and Baltalimanı tunnels, the main excavation methods were with impact hammers and roadheaders leading to the accumulation of data and development of performance prediction models of these machines, Bilgin et al. (2014, 2016). Time by time, the uniaxial compressive strength of andesite dykes surpasses 200 MPa necessitating the use of explosives. Therefore, tunnel project planners or practising engineers who are responsible for the proper and economic excavation of the project should be aware of the difficulties which may be encountered when driving tunnels in complex geology and the use of performance prediction models for the proper time scheduling of the projects, This chapter is written to introduce the readers understanding the performance of TBMs, roadheaders, impact hammers, and drill and blast tunnelling methods within the same geology of Göztepe Station of Kozyatağı–Kadıköy Metro Project and compare their performances. It includes the results of a re-evaluated and enlarged published paper by Ocak and Bilgin, (2010).

5.2 Kozyatağı–Kadıköy Metro Project, the Geology and Geomechanical Properties of the Formations

The population of Istanbul is approximately 16 million, and one-third of this population lives in the Anatolian area. The project Kaynarca-Kadıköy metro is planned to meet the enormous transportation demand between the East and West of Istanbul. Integrated with the Marmaray Project, the project connects the European and Anatolian sides at Ibrahimağa station. The project is planned to be realized in two phases, one from Kozyatağı to Kadıköy and the second phase from Kaynarca to Koztağı. This chapter covers the study concerning the first phase.

The project Kadıköy–Kartal metro system starts from Kadıköy Square, joins the station of the Marmaray Project at Ibrahimağa, and extends up to Kaynarca. The metro project includes 16 stations in total. The project comprises twin tunnels, and the distance between the two tunnels is nearly 32 m. The excavation diameter of the main tunnels is 6.62 m, and the inner diameter is 5.70 m. Tunnels between Kozyatağı–Kadıköy are generally excavated in Trakya and Kartal formations (Fig. 5.1). Trakya formation of carboniferous age consists of sandstone–siltstone–claystone–mudstone–shale in sequences. Limestone and conglomerate layers are also rarely observed. Diabase and andesite dykes having some 10 m thickness are frequently encountered. In the east of the tunnel alignment, the Kartal formation of the Devonian age is found. The Kartal formation consists of fine-grained, laminated, fractured and interbedded siltstone, limestone, sandstone, and shale. Water ingress is common in both Kartal and Trakya formations. Many faults dykes and geological discontinuities exist in the area due to Alpine Orogenies, creating alteration along discontinuity surfaces, the rock material in the contact zone between dykes and the main rock is highly weathered. The number of joint sets per meter occurring along the length of the tunnel can be classified as class 3 (one joint set plus random) as defined by Brown (1981), and the number of discontinuities per meter as described by Brady and Brown (1985) changes between 15 and 20. The overburden varies between 20 and 45 m, and the distance between twin tunnels is around 32 m. The geotechnical properties of the two rock formations show a wide range of variance, which is summarized in Table 5.2. Figure 5.1 shows the A, B, and C study zones. The geological cross-section of the studied area is given in Fig. 5.1.

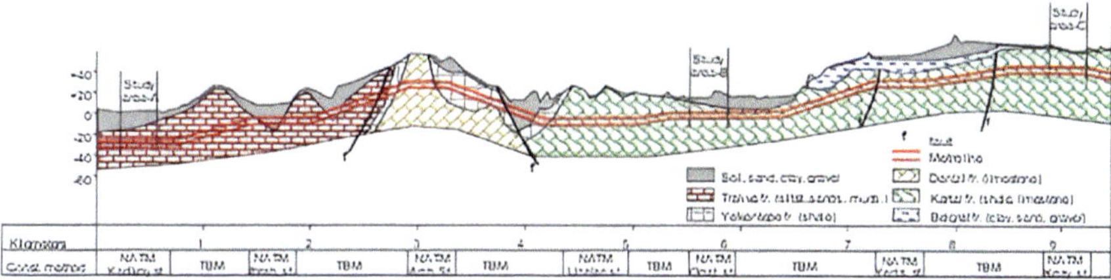

Fig. 5.1 The geological cross of the metro line between Kozyatağı–Kadıköy, from the archive of the author, Bilgin

Table 5.2 Physical and mechanical characteristics of the studied areas

Formation	γ, kN/m^3	C, kPa	σ_c, MPa ($\pm$ sd)	RQD $\pm$ sd
Kartal	26.4	142	34.3 ($\pm$ 21.5)	48.5 ($\pm$ 11)
Trakya (W1–W2)	26.5	139	50.2 ($\pm$ 38.1)	20.3 ($\pm$ 20)
Tuzla	26.4	210	58.0 ($\pm$ 12.3)	26.7 ($\pm$ 19.5)
Baltalimanı	26.5	238	62.8 ($\pm$ 13.1	30.6 ($\pm$ 18.8

γ is density, C is cohesion, σ_c is the compressive strength

5.3 TBM, Roadheder, and Impact Hammer Used in the Project

The characteristics of TBM's, roadheader and impact hammer used in the project are given below.

5.3.1 *TBMs Used in the Metro Project*

Two identical Herrenknecht TBM's were chosen for Line 1 and Line 2, these machines can work in open and closed modes The general view of the TBM's is seen in Fig. 5.2, and technical data is given in Table 5.3

Fig. 5.2 A view of the cutterhead of TBM used, from the archive of the author, Bilgin

Table 5.3 The main characteristics of the TBM

Machine diameter	6.75 m
Rotational speed	1.65–5.5 rpm
Number of disc cutters	28
Cutterhead power	4 × 315 kW
Cutterhead torque 1	5200 kNm at 1.6 rpm
Cutterhead torque 2	1515 kNm at 5.5 rpm
Maximum thrust	42,575 kN at 350 bar
Opening ratio	%35

In the beginning, TBM worked without a screw conveyor. It was noticed that big blocks were coming from the face, causing a lot of problems, such as clogging the cutterhead. The mean advance rate was 3 m/day at the beginning of the operation, and this value increased up to 7 m/day after the modification of the cutterhead installing grizzly bars, reducing the area of openings. However, it is interesting to note that the daily advance rate increased to 10.5 m (day) after installing a screw conveyor within the TBM chamber and passing from open mode to Earth Pressure Balance (EPB) mode. In the meantime, in different tunnels in Istanbul, the same remedial works were carried out for the Beykoz and Marmaray Tunnels, increasing the daily advance rates to a great extent.

5.3.2 Roadheader Used in the Project

Transverse type roadheader of model WAV 178 Westfalia was used in the station tunnels of upper and lower bench headings. The view of the machine is given in Fig. 5.3 and the characteristics of the machine are in Table 5.4.

5.3.3 Impact Hammer Used in the Project

Hydraulic Hammers are widely used in the tunneling, and several km of tunnels were excavated in Istanbul Levent metro tunnels by hydraulic hammers mounted on the excavators. The hammers used in Kozyatağı–Kadıköy metro station tunnels were Atlas Copco MB 170-type hydraulic hammers. All hammers are the same and have the following characteristics: weight 1.7 t, oil flow rate 130–160 lt/min, working pressure 160–180 bar, impact frequency 320–600 impact/min, impact energy 4170 J. The selection of carriers is essential for the success of the operations since the stability of the carrier plays a vital role in the efficiency of the system, Tuncdemir (2007), Volvo EC210B of 7 t, Sumitomo SH200LC OF 9.5 t, and Caterpillar 3066 T of 7.5 t

Fig. 5.3 Transverse type roadheader used in the project, from the archive of the author, Bilgin

Table 5.4 Characteristics of Westfalia WAW 178 model

Weight, t	75
Cutting power HP/kW	402/300
Max. excavation area m^2	61
Cutting head diameter, cm	88.8
Cutter type	Conical
Cutter number	$82 \times 2 = 164$
Cutting head type	Transverse

in weights were used with attached hammers. A general view of a hydraulic hammer is given in Fig. 5.4.

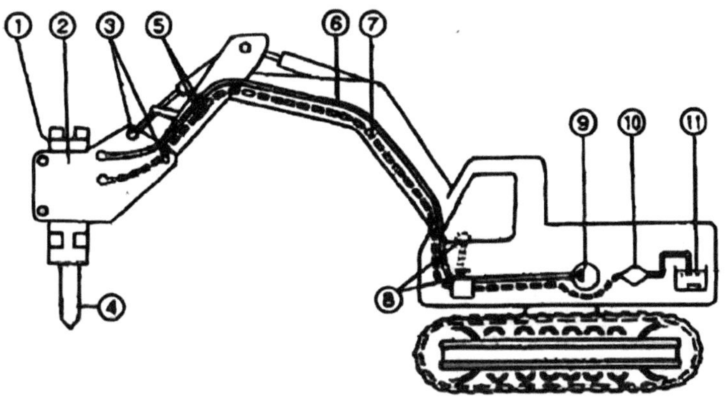

1. Hydraulic hammer	5. Shut off valves	9. Hydraulic pump
2. Bracket	6. Pressure line	10. Oil filter
3. Boom pins	7. Return line	11. Oil tank
4. Tool	8. Hammer control valve	

Fig. 5.4 Schematic view of a hydraulic hammer mounted on an excavator, after Tuncdemir (2007)

5.4 Main Factors Affecting the Performance of Tunneling Machines

TBMs are mainly affected by the geotechnical characteristics of the ground. Hard rock, soft ground, or groundwater level will affect the type of TBMs used as gripper type TBMs, single shield TBMs, double shield TBMs, slurry, or EPB TBMs. Weathering degrees of the rock formations, faults, transition zone, and plasticity of the ground affecting clogging of the cutters are among the most critical parameters affecting the performance of TBMs. Swelling clays are the main reason for squeezing ground, and it is the nightmare of the tunnelers. The main advantage of TBMs is that they are continuous excavation systems. However, excavation with roadheaders, impact hammers, and drill and blast methods is not continuous. Due to this fact, the tunneling rates using these machines are generally lower than those of TBMs. If the ground is not favourable, roof bolts, foreboding, wire mesh, shotcrete, and steel arches will be among the factors that decrease the advance rates. The cross-section of the platform tunnel T1 opened with a roadheader, impact hammer and drill, and blast method is seen in Fig. 5.5.

Platform tunnels are carried out in three stages: top heading, bench one, and bench two. The width of the tunnel excavated by the traditional tunnelling method is 10.44 m, its height is 8.48 m, and its excavation area is 74 m^2. Before starting a round, forepolling (or pipes) is applied. Forepolling is either steel bars of 32 mm in diameter and 4 m in length or pipes of 38 mm in diameter and 4 m in length with a

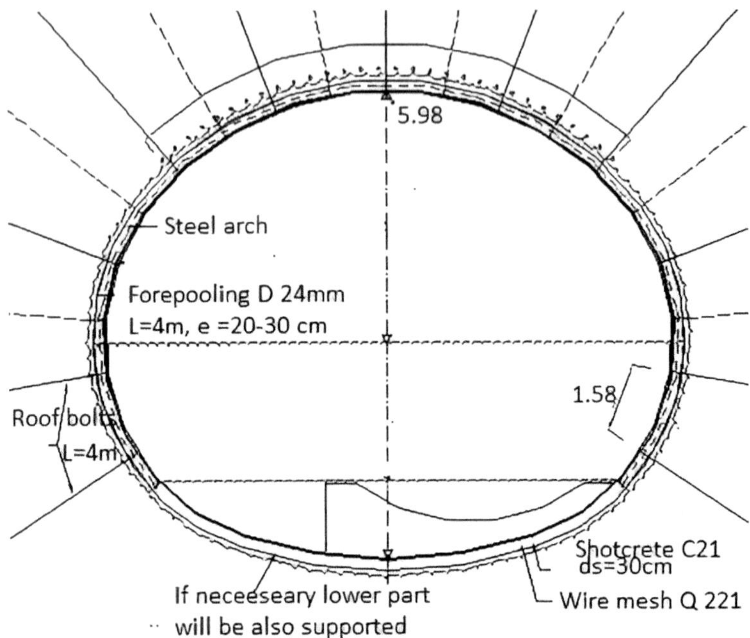

Fig. 5.5 The cross-section of the platform tunnel T1 opened with a roadheader, impact hammer and drill, and blast method; the tunnel face has an area of 74 m^2, according to the archive of the author Bilgin

thickness of 3.25 mm, they are installed along the outer lines of the excavation line as a protective layer before the excavation, Fig. 5.6. The shotcrete is applied after installing wire mesh and lattice girder in Kadikoy-Kartal metro tunnel excavations. 2–3 days after the shotcrete application, as a last step of the primary support, rock bolts with cement injections are installed. The bolts are specially produced from 26 mm steel 4 or 6 m long. The diameter of the hole into which it is placed is 41 mm. The proportion of water/cement in the injection mixture is 0.35. Elements of tunnel support are seen in Fig. 5.6.

5.5 The Performance of TBM

A detailed analysis was carried out to investigate the effect of geotechnical properties on the performance of a TBM. Table 5.5 summarizes the results of multiple regression analysis between dependent and independent variables corresponding to 0–1 bar face pressure since it is well-known that all EPB-TBM parameters are affected by tunnel face pressure, Bilgin and Yüksel (2023). The conclusion emerging from this statistical analysis was that geotechnical factors affecting the instantaneous cutting rate and penetration were the geological strength index as defined by Marinos et al.

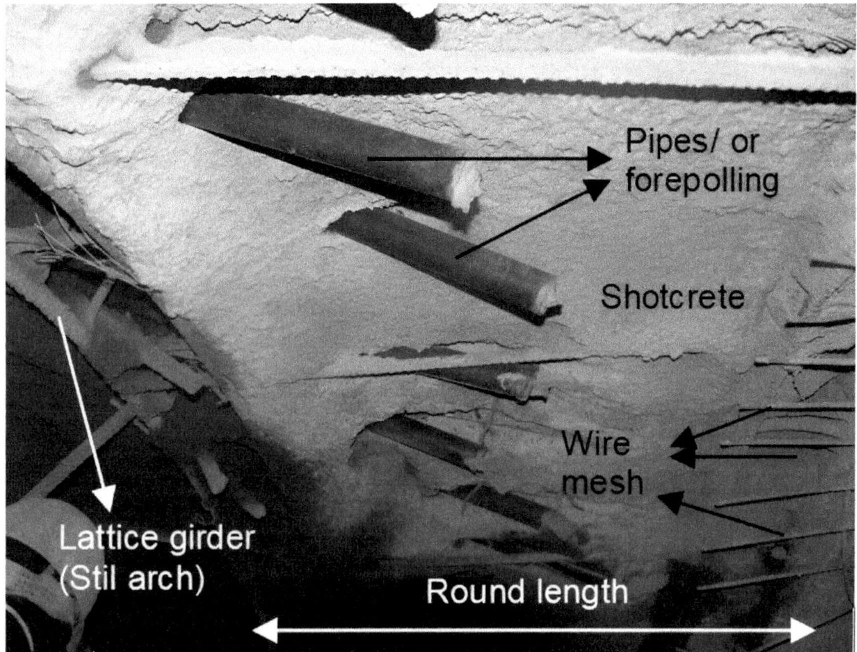

Fig. 5.6 Elements of tunnel support, from the archive of the author, Bilgin

(2005), Cerchar abrasivity index, and destruction energy as defined by Thuro and Spaun (1996) and Thuro and Plinninger (2003). The thrust index and torque index were predicted best from the compressive strength times Cerchar abrasivity index. Nizamoglu (1978) was the first to notice this in 1978 during his field studies on Robbins and Wirth TBMs performances for his Ph.D. thesis, and he called this product an index of nocivité or harmfulness.

In the above equations, TF/p is thrust index in kN/(mm/rev), T/p is torque index in kNm/(mm/rev), SE is specific energy in kWh/m^3, ICR is the instantaneous cutting rate in m^3/h, p is penetration in mm/rev, σc is compressive strength in MPa, CAI is Cerchar abrasivity index, E elasticity modulus in MPa, GSI is geological strength

Table 5.5 Performance predictor equations of TBM

$ICR = 0.896GSI - 37.416CAI + 189.804\left(\frac{\sigma_c^2}{2E}\right)$, R2 = 0.84
$p = 0.141GSI - 5.842CAI + 29.373\left(\frac{\sigma_c^2}{2E}\right)$, R2 = 0.83
$\frac{TF}{p} = 0.1615(\sigma_c.CAI)^2 - 17.668(\sigma_c.CAI) + 737.37$, R2 = 0.80
$\frac{T}{p} = 0.0366(\sigma_c.CAI)^2 - 3.4861(\sigma_c.CAI) + 203.15$, R2 = 0.79
$SE = (35.6 \times 10^{-2})E - 0.0381(\sigma_c.CAI)$, R2 = 0.69

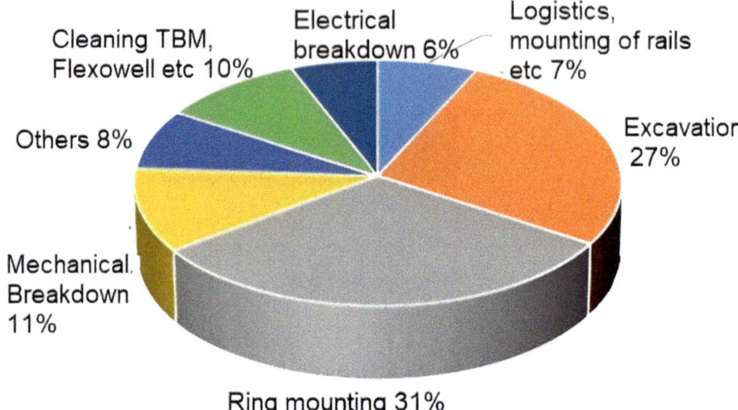

Fig. 5.7 Overall performance of tunnelling with TBM, from the archive of the author Bilgin

index, RQD is rock quality designation, $\sigma c^2/2E$ is destruction work as defined by Thuro and Spaun (1996), Thuro and Plinninger (2003).

Instantaneous cutting rate, with machine utilization time, are the parameters used to estimate daily advance rates. Machine utilization time is strictly related to job organization, the skill of the TBM operators, and the design of the tunnel operation parameters. A machine utilization time of 45% is a good number, justifying the efficiency of the operational parameters used in the mechanized tunnelling. The major criticism of mechanized tunneling with TBMs in Kozyatağı Kadıköy Metro tunnels is that 27% of a round is spent on excavation, as seen in Fig. 5.7, which is too low compared with the other projects. The main reason for this is that the muck is transported with a rail system, and the muck is transported from the shaft to the surface with a Floxewell vertical band. Sticky muck needed to be cleaned within the flyers of the vertical band, with a ratio of 10% in the overall performance of tunnelling with EPB-TBMs.

5.6 The Performance of Tunneling with Drill and Blast, Roadheaders and Impact Hammer

Platform tunnels opened using the drill and blast tunnelling method have a cross-section area of 74 m^2, as illustrated in Fig. 5.8. Drill holes were drilled using double boom Atlas Copco Rocket Boomer 282. Fracture characteristics of the rock mass were dictated by the blasting agent as Powergel Magnum 365 and Trimex (with a density of 1.20 g/cm^3, ideal blasting velocity of 6437 m/s, and ideal blasting agent of 121,400 atm. The wedge cut and parallel-burn cut methods were used with blasting holes of parallel cuts having a diameter of 89 mm and other blast holes of 41 mm in diameter. Nonel Exel MS/LP blasting cabs of 47 different blasting intervals were also used, Alan (2018). A pie chart showing the work done in drill and blast tunnelling is given in Fig. 5.8.

The Roadheader and impact hammer used in the project are described in Sects. 5.3.2 and 5.3.3. Pie charts obtained for these machines are given in Figs. 5.9 and 5.10

Machine utilization time is as important as the net cutting/breaking rate in determining the excavation efficiency and economy since daily advance rates are directly related to these factors for a given tunnel section. The performances of mechanical excavators and drilling and blasting methods are summarized in Figs. 5.8, 5.9, and 5.10. These figures provide a unique opportunity to compare roadheaders, impact hammers, and drilling and blasting methods for a given job. As seen from these

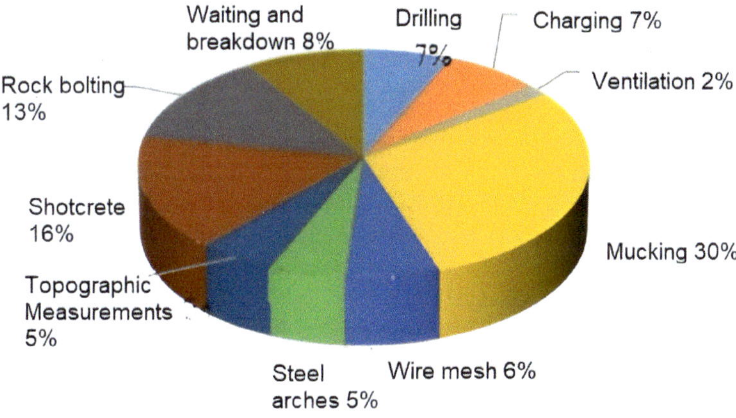

Fig. 5.8 Pie chart showing the works done in drill and blast tunnelling, from the archive of the author Bilgin

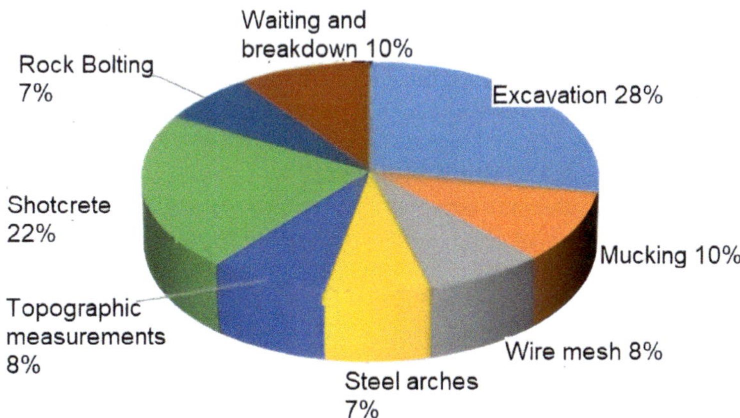

Fig. 5.9 Pie chart obtained when using a roadheader, from the archive of the author Bilgin

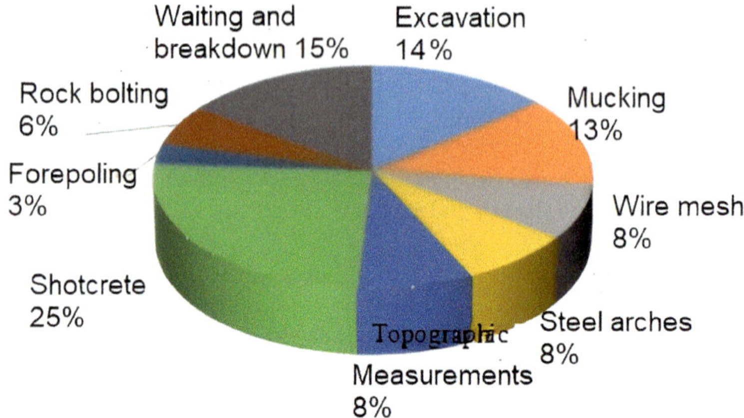

Fig. 5.10 Pie chart obtained when using an impact hammer, from the archive of the author Bilgin

figures, machine utilization time for roadheaders is 28%, almost twice compared to impact hammers, and time spent for excavation in drilling and blasting is in the same order as the time spent in the excavation by impact hammers with a value of 15.4% (drilling, blast hole charging, blasting, and ventilation). Mucking is another factor affecting job duration. In fractured rock formations like Kartal and Trakya formations, sometimes roadheaders excavate the rock over their loading capacities with gathering arms, and extra time is necessary to load the muck. The time spent for mucking was 9.8% for roadheaders, 13.3% for impact hammers, and as high as 29.4% for drilling and blasting methods. Controlling muck size in fracture rock formations with conventional excavation methods is challenging. Some big muck samples always create transportation problems, and due to the disturbed zone, the number of bolts used for a unit length of the tunnel is always higher than mechanical excavation methods. Time spent for rock bolting in the drilling and blasting method is twice as compared to the other excavation methods, with a value of 13%.

The following Eqs. 5.1, 5.2, and 5.3, were already developed to predict the net production rates of roadheaders and impact hammers. The validity of these equations was also tested for Kozyatağı–Kadıköy Metro tunnels. However, the variety of the igneous dykes in thickness served to develop Eq. 5.4. K index, which is a function of the thickness of the dykes is given in Table 5.5. As a statistical test result, the field values and predicted values are compared, which are summarized in Table 5.6. The dip and strike of the joints of 45° in the direction of cutting favour the cutting rate of roadheaders given higher values than predicted performance values.

$$IBR = 4.24 \times P \times (RMCI)^{-0.567} \tag{5.1}$$

$$ICR = 0.28 \times P \times (0.974)^{RMCI} \tag{5.2}$$

Table 5.6 A guide to estimating reduction factor K used to predict net braking rates of impact hammers related to the thickness of the dyke

The thickness of dykes (cm) (for UCS > 120 MPa)	Reduction factor K for net breaking rate
10–20	0.8–0.90
20–30	0.7–0.8
30–40	0.6–0.7
40–50	0.5–0.6
> 50	D&B is recommended

Table 5.7 The relationship between net field production rate and the predicted production rate for impact hammers and roadheaders

Impact hammer Eq. 5.4	y = 1.98x − 12.02	$r^2 = 0.63$
Roadheader	y = 3.46x − 40.97	$r^2 = 0.80$

In this table y is the field value in m^3/h; x is the predicted value in m^3/h

$$IBR = K \times 4.24 \times P \times (RMCI)^{-0.567} \tag{5.3}$$

$$RMCI = \sigma_c.\left(\frac{RQD}{100}\right)^{2/3} \tag{5.4}$$

In the above equations, NBR is the net breaking rate (for impact hammers) in m^3/h; Pi is the power output of the impact hammer in HP, it is found by multiplying the oil flow rate by working pressure; RMCI is the rock mass cutability index; UCS the uniaxial compressive strength in MPa; RQD the rock quality designation, %; NCR the net cutting rate (for roadheaders), m^2/h; P the cutting power for roadheaders in HP., Bilgin et al. (2014) (Table 5.7).

5.7 Criticism of Using TBMs, Drill and Blast, Roadheader and Impact Hammers

Criticism will be made upon daily advance rates based on a tunnel diameter of 6.75 m as given in Table 5.8.

Table 5.8 Production rates obtained in Kozyatağı Kadıköy Metro tunnels with different tunnelling methods

Method of excavation	Daily advance rate m/day
TBM	10.5
Drill and blast	5.5
Roadheader	6.4
Impact hammer	1.3

Although drill and blast, roadheader and impact hammers were used in the platform tunnels of 74 m^2 area, production rates were converted for a tunnel of 6.7 m diameter, which is the excavation diameter of the TBM. An advance rate of 10.5 m/day obtained for Kozyatağı Kadıköy is too low. Looking at Fig. 5.7, 7% of overall time is spent on mounting rails and 10% cleaning the Flexowell (vertical belt). A job organisation with belts for muck transport would increase the time spent for excavation from 27% to 40–44%, increasing the daily advance rate consecutively.

Using high-power impact hammers can also increase the daily advance rates of impact hammers almost twice since, as seen from Eq. (5.3), the net breaking rate of hammers is related to the power of impact hammers. Another alternative is to use impact hammers mounted on carriers having gathering arms, eliminating the time spent mucking.

5.8 Concluding Remarks

TBMs are the most widely used in Metro tunnels in Istanbul. However, drilling and blasting, impact hammers, and roadheaders are also used in some cases, especially in metro station tunnels (platform tunnels). In Istanbul, in the last five metro projects of 194,121 m in length, 156,127 m of tunnels were opened with EPB-TBM's, Station tunnels, or crosscuts of 37,994 m were opened with drill and blast method, impact hammers or roadheaders as in Kozyatağı Kadıköy Metro project. In this project, daily advance rates for TBM, drill and blast roadheaders and impact hammers were 10.5 m, 5.5 m, 6.4 m, and 1.3 m, respectively. This chapter concisely criticizes the performance of these tunnelling methods and modifies some of the performance prediction models that have already been developed using accumulated data.

References

Alan E (2018) Determining "k" and "β" coefficients of the formation that excavated by blasting on Kadıköy-Kartal metro route, the effect of detonator delay time on amplitude. In: Proceedings of the 4th international symposium on underground excavations for transportation, organized by Turkish tunnelling society and chamber of mining engineers of Turkey, Istanbul Branch

Bilgin N, Yüksel A (2023) The effect of EPB face pressure on TBM performance parameters in different geological formations of Istanbul. Tunnell Undergr Space Technol 138(2023):105184. https://doi.org/10.1016/j.tust.2023.105184

Bilgin N, Çopur H, Balcı C (2014) Mechanical excavation in mining and civil industries. CRC Press, Taylor and Francis Group, London

Bilgin N, Çopur H, Balcı C (2016) TBM excavation in difficult ground conditions. Case studies from Turkey. Ernst & Sohn GmbH & Co

Brady BHG, Brown ET (1985) Rock mechanics for underground mining. Chapman and Hall Publisher, p 570 (ISBN 0 412 47550 2)

Brown ET (1981) Rock characterization testing and monitoring. In: ISRM suggested methods. Pergamon Press, pp 211 (ISBN 0 08 027309-2)

Marinos V, Marinos P, Hoek E (2005) The geological strength index: applications and limitations. Bull Eng Geol Environ 64:55–65

Nizamoglu YN (1978) Contribution to the performance prediction of TBMs and analysis of the wear of disc cutters, Ph.D thesis (in French), Institute Nationale Polytechnique de Lorraine, Ecole Nationale Superieur de la Metallurgie et de L'industrie des Mines de Nancy, France, p 139

Ocak I, Bilgin N (2010) Comparative studies on the performance of a roadheader, impact hammer and drilling and blasting method in the excavation of metro station tunnels in Istanbul. Tunn Undergr Space Technol 25:181–187

Tuncdemir H (2007) Impact hammer applications in Istanbul metro tunnels. Tunnell Undergr Space Technol 23:262–264

Thuro K, Plinninger RJ (2003) Hard rock tunnel boring, cutting, drilling and blasting: rock parameters for excavatability. Proc ISRM 2003:1227–1233

Thuro K, Spaun G (1996) Introducing 'destruction work' as a new rock property of toughness referring to drillability in conventional drill- and blast tunnelling. In: Barla G (ed) Proceedings Eurock'96 conference, vol 2, Rotterdam, Brookfield. Balkema, pp 707–713

Chapter 6
The Cost of the Initial Investment Affecting the Selection of Tunneling Methods

Abstract This chapter gives the initial cost investment for conventional and mechanized tunneling methods. The data provided is for a tunnel of 6.6 m in diameter and 10 km in length. The ratio of the equipment cost in TBM tunneling to the price of the equipment in conventional tunneling is almost 10. The most expensive equipment in mechanized tunneling is EPB-Tunneling, with a value of 7,500,000 \$. However, the most costly equipment in conventional tunneling is 750,000 \$.

6.1 Introduction

The initial investment constitutes purchasing the tunneling equipment and mobilization costs. The equipment used in both tunneling methods, mechanized or conventional, is entirely different, making a significant difference in the initial investment. Conventional Tunneling startup costs are often lower. It also requires a shorter time to start the construction. In contrast, mechanized tunneling requires a higher in-front investment for the machinery. TBMs are manufactured after they are ordered; it takes several months to build these machines. In TBM's excavation or boring, the diameter of TBM is essential to determining its price. As a general rule, as the diameter increases, so does the price tag. An often-quoted rule for a Tunnel Boring Machine is that it costs US \$1 million per meter in diameter. Thus, an 8-m diameter TBM would cost around US \$8 million. But in reality, the price increase will not be linear, and exceptions exist. A recent example of the price of a large-diameter TBM is the one bought by the Santa Clara Valley Transportation Authority (VTA) in the USA for the BART Silicon Valley Phase II Project. For a 16.46-m diameter Herrenknecht TBM, VTA didn't pay US \$16.46 m, but a considerable amount of US \$76 m. Although the deal may include some other services (training, etc.), it is clear that the typical "US \$1 million per meter in diameter" rule does not apply in large-diameter TBMs, McAuley (2023). In addition to the fact that it takes time to procure a brand new TBM, purchasing a used TBM is not a fast option. Even if a used TBM is found to meet the project requirements, refurbishment work will be required, which can easily take several months. Contrarily, the most expensive equipment in conventional

N. Bilgin and C. Balci, *Critical Issues in Selecting Conventional and Mechanized Tunnelling Methods*, Springer Tracts in Civil Engineering,
https://doi.org/10.1007/978-3-031-89114-4_6

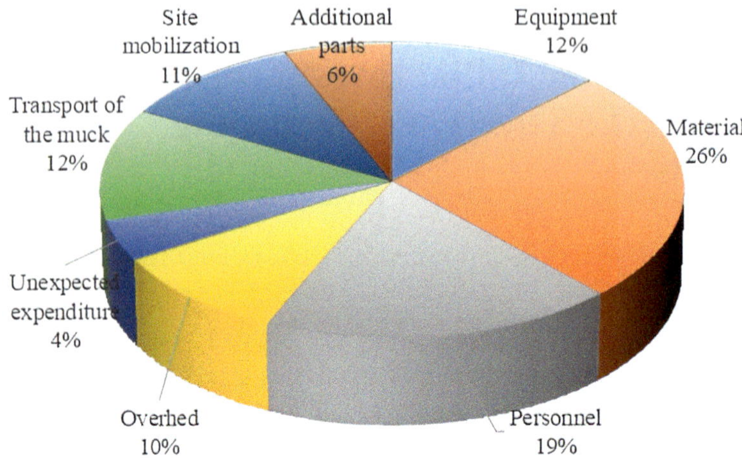

Fig. 6.1 Distribution of the cost in conventional tunneling. *Source* Wagner (2004)

tunneling would be a jumbo drill rig. Epiroc E2C of 125 kW of 2 booms and 35 t would cost about US $0.75 million.

Mobilization costs in tunneling also play an important role in initial investment. It refers to the expenses incurred to prepare a tunneling site, which includes equipment transportation, setting up site offices, securing permits, and performing other preliminary activities necessary to commence construction. The time for the transportation of TBM to the working site is much higher than that of a Jumbo drill rig, and getting ready for both machines favours the drill rig. İhsaniye job site for Istanbul New Airport Metro may be cited as a typical example of job side dimensions. The area covers 42,000 m^2, with 5000 m^2 for the dormitory, 200 m^2 for the chiller, 1300 m^2 for the muck area, and 450 m^2 for grout central, 400 m^2 for ware house, 500 m^2 for segment stock, Bilgin and Acun (2024). Job site dimension in conventional tunneling is less than that of TBM tunneling.

Comparative cost estimation in conventional tunneling and TBM tunneling is given in Figs. 6.1, 6.2 and 6.3. These figures show that equipment cost in conventional tunneling is 12% of the total cost. And the share of the equipment cost in TBM tunneling is between %24 and 35%,

6.2 The Initial Investment Cost in the Conventional/Drill and Blast Tunneling Method

Table 6.1 gives the cost of the initial investment in conventional tunneling. The price is for a 6.6 m diameter tunnel for a length of 10,000 m.

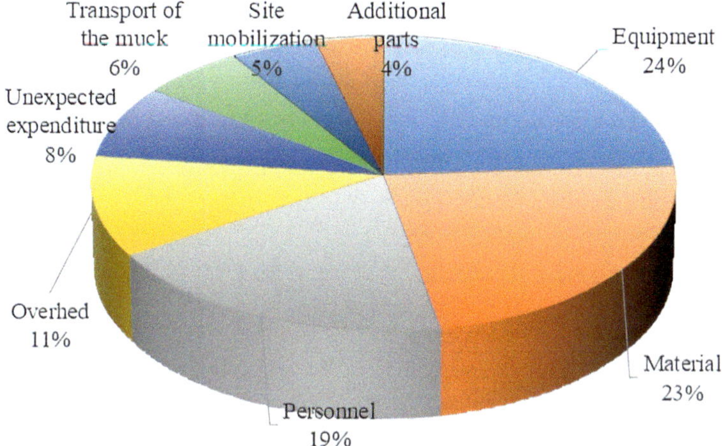

Fig. 6.2 Distribution of the cost in TBM tunneling. *Source* Wagner (2004)

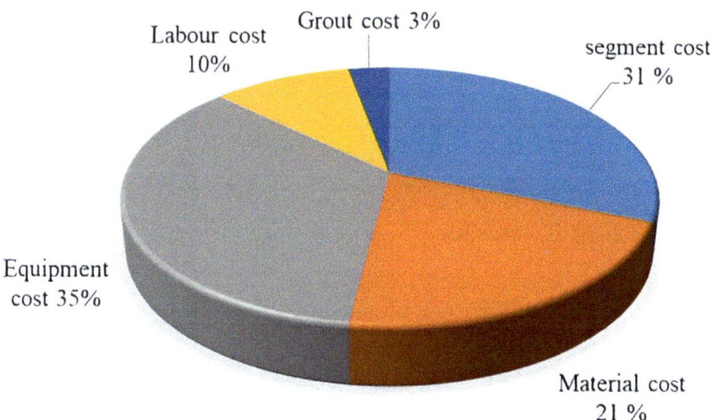

Fig. 6.3 Distribution of the cost in TBM tunneling. Bilgin and Acun (2024)

6.3 The Cost of the Initial Investment in Mechanical Tunneling Method

As mentioned above, the equipment provided in Table 6.2 is for for a tunnel of 6.6 m diameter and a tunnel length of 10,000 km.

Table 6.1 The cost of the initial investment in the D&B tunneling method, our thanks are due to Dr. Ali Yüksel from Yapi Merkezi for providing the information

Equipment	Brand/Technical specification	Price (approx), $
Excavator, with Jack hammer	Hitachi 220, power of 163 HP, backhoe 1 m^3	230,000
Mini excavator	JCB, Cat 428	75,000
Loader	Hidromek WL 640, 324 HP, 26.3 t, 4.2 m^3	225,000
Dump trucks × 4	Ford Trucks 3542D DC Euro-6	105,000
Telehandler	Manitou 1440;75 HP, engine power rating,	100,000
Wet shotcrete machine	Tünelmak Adroit 450H/W;100 HP, 11 t	135,000
Dry shotcrete machine	Meyco GM, 7 kW 9 round hole motor	25,000
Driller (Jumbo)	Epiroc, E2C 125 kW; 2 boom rings, ~ 35 t	750,000
Grout pump	DSI MAI 400 NT	9000
Truckmixer	Ford Trucks 4142 M	120,000
Ventilation fan and fan tube	Teknima, 75 kW, 1400 mm Diameter	24,000
Submersible pumps × 2	Atlas Copco, 2–4 inches	6000
Generator × 2	Atlas Copco, 500 kVA	45,000
Track with crane	FASSI 450/VOLVO FH 480 6*4	105,000
Tunnel ınner lining form	–	50,000
Stationery concrete pump	Tatmak	70,000
Total $		2,074,000

6.4 Concluding Remarks

This chapter gives the initial cost investment for conventional and mechanized tunnelling methods. The data provided is for a tunnel of 6.6 m in diameter and 10 km in length. The ratio of the equipment cost in TBM tunneling to the price of the equipment in conventional tunneling is almost 10. The most expensive equipment in mechanized tunneling is EPB-TBM with a value of 7,500,000 $. However, the most costly equipment in conventional tunneling is 750,000 $.

Table 6.2 The cost of the Initial Investment in TBM tunneling method, our thanks are due Dr. Ali Yüksel from Yapi Merkezi for providing the information

Specific machinery for tunnel	Purpose of use	Technical specification	Price (approx), $/piece
EPB-TBM	Excavation	D 6.6 m EPB type	7,500,000
Logistic-Haulage (MSV)4–5	Transportation of segments and other materials	50 ton carriage cap	657,000
Horizontal conveyor system, 10 km	Transportation of excavated material	With belt storage (315 + 110) kW	7,000,000
Ventilation fans and fan tubes	Ventilation of tunnel	D1000 mm-2 × 75 kW motor	55,000
Segment moulds 4.-5	Production of precast segments	According to segment geometry	120,000
Segment production plant and equipment	Production of precast segments	According to segment production	4,000,000
Grout plant	Grout production of segments backfill	20 m^3/h	500,000
Cooling system (Chiller)	Cooling for TBM hydraulics		120,000
Portal cranes, if there is a shaft	Vertical transportation of segments and other materials	40 ton and 15 ton hook capacity	600,000
Forklift	Handling of segments	15–20 ton lifting capacity	350,000
Total			20,902,000

References

Bilgin N, Acun, S (2024) Practical management of tunneling with tunnel Boring machines. CRC Press, Taylor and Francis Group, ISBN: 978-1-032-41622-9, p 217

McAuley R (2023) https://tunnelcontact.com/pages/view/37495/how-much-does-a-tunnel-boring-machine-tbm-cost, uploaded in November 2024, contribution written Robin McAuley in 2023

Wagner H (2004) The governance of cost in tunnel design and construction. In: First International South American tunneling symposium, p 6

Chapter 7
The Length of the Tunnel Affecting the Selection of Tunneling Methods

Abstract This chapter first discusses the work done in the past on the effect of tunnel length, tunnel diameter, geology, type of tunnel, and environment on the construction cost. The scatter between dependent and independent variables shows the complexity of predicting the tunnel construction cost and the choice between conventional and mechanized tunneling methods. A comparative cost analysis is made for drill and blast and mechanized tunnelling methods for a given geology and tunnel diameter to make the solution to the problem more comprehensive. The initial cost investment for both methods presented in Chap. 6 is used in the comparative study.

7.1 Introduction

The advantages of TBM tunneling include high advance rates, safety, minimal surface disturbances, minimal overbreak, and consistency in tunnel diameter. However, drill and blast tunneling surpass TBM tunnelling with a lower initial investment; unlike TBM tunneling, which requires a higher initial investment to procure the TBM before construction, the drill and blast method requires a lower investment at the start of the project. As explained in the previous chapter, the initial investment in conventional tunneling may be five times less than that of TBM tunnelling. The drill and blast method can adapt to varying geological conditions, making it suitable for tunnels in complex rock formations. The drill and blast method can be more cost-effective than TBMs for shorter tunnel lengths and smaller projects. The question arises regarding the limit for similar geological conditions and the same tunnel diameter. Katuwall defines this limit as 1.5–4.5 km depending on the geology, tunnel geometry, environment, and project timelines, Katuwall (2023).

Regarding this question, the objective of this chapter is to give an example of how to calculate this limit value depending on the actual values of tunneling conditions and using actual initial cost investment. However, the construction costs of tunnels are very variable, and it isn't easy to give representative construction cost values per tunnel length because these ratios may vary in essential proportions (average of 1 to 5) according in particular to:

N. Bilgin and C. Balci, *Critical Issues in Selecting Conventional and Mechanized Tunnelling Methods*, Springer Tracts in Civil Engineering,
https://doi.org/10.1007/978-3-031-89114-4_7

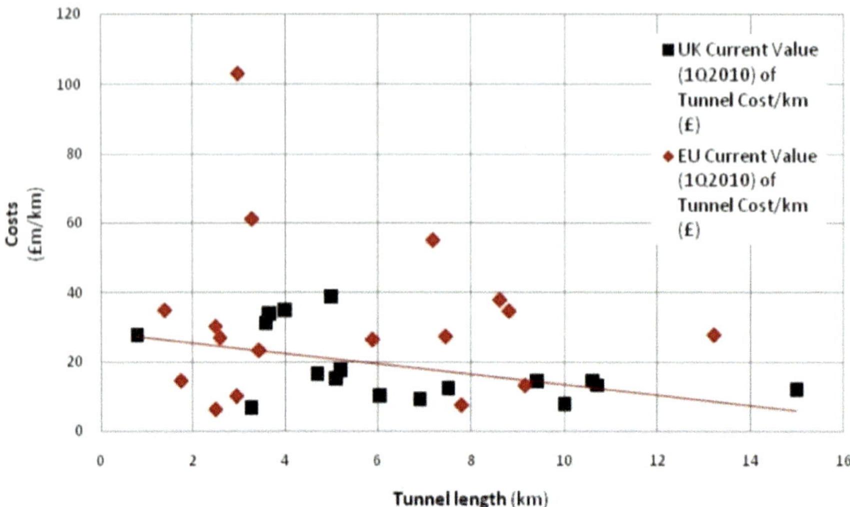

Fig. 7.1 The effects of the tunnel length on unit construction costs, HM Treasury (2010) based on the source taken from the UK cost questionnaires and British Tunnelling Society

(a) Unit construction cost ($/km) decreases with tunnel length, HM Treasury (2010), Benardos et al. (2013), Paraskevopoulou and Boutsis (2020), Ahmed (2021), Kondrachova et al. (2021). This relation is illustrated in Fig. 7.1.
(b) Tunnel unit construction cost increases with tunnel diameter, HM Treasury (2010), Benardos et al. (2013), Paraskevopoulou and Boutsis (2020), Ahmed (2021), Kondrachova et al. (2021). The typical relation between tunnel outside diameter and unit construction cost is given in Fig. 7.2.
(c) Construction cost decreases with increasing parameters like the geological strength index (GSI), Petroutsatou et al. (2006), Benardos et al. (2013), Membah and Asa (2015), Sayadi et al. (2015), and Paraskevopoulou and Boutsis (2020). Figure 7.3 shows that construction costs decrease with GSI values.
(d) Tunnel types (wastewater, water, railway, highway, etc.) affect the cost of tunnel construction, Rostami et al. (2013).
(e) Environment, the geographical location of the tunnel, the tunnel environment may lead to expensive protection arrangements for the mitigation of its impact, urban or non-urban, Membah and Asa (2015).

Looking at the costs in Figs. 7.1, 7.2 and 7.3, one should remember that they are only construction costs. However, Flyvbjerg et al. (2008) noted total project cost is much higher looking at European projects, as the total capital costs per route-kilometer (including stations and rolling stock) lie between US $50–100 million (2002 prices) and in US projects, the range is US $50–150 million.

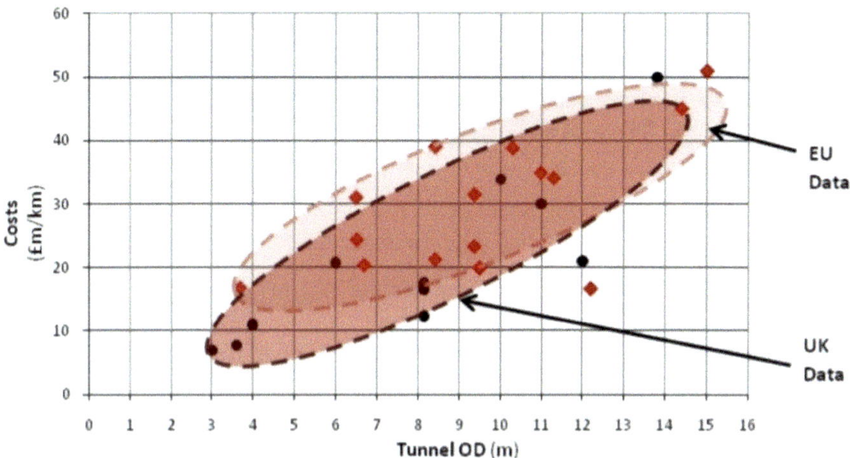

Fig. 7.2 The effects of the tunnel outside diameter on unit construction costs, HM Treasury (2010) based on the source taken from the UK cost questionnaires and British Tunnelling Society

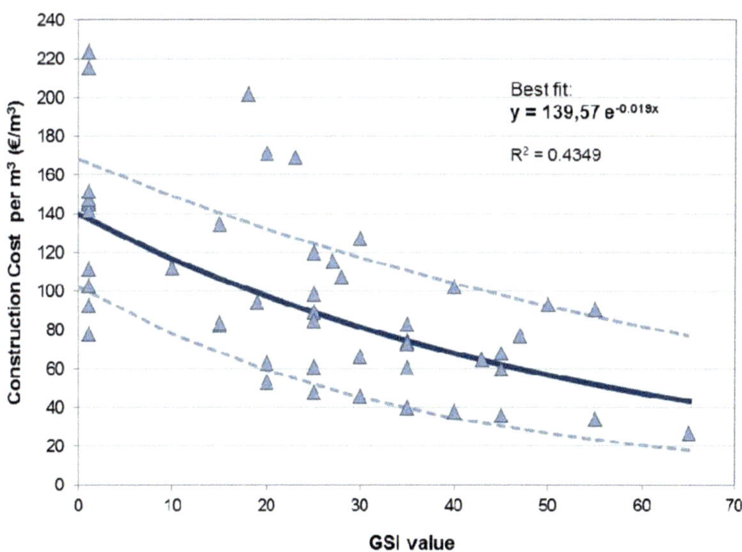

Fig. 7.3 Tunnel unit construction cost (excavation and temporary support) versus GSI index values. Paraskevopoulou and Benardos (2013)

7.2 Method of Calculation of Critical Tunnel Length in Selecting Drill and Blast or TBM Tunneling Methods Based on the Construction Cost for the Same Geology and Same Tunnel Diameter

In this comparative study, the model published by Brockway (1982) from Robbins Company will be used using actual data. He made the following assumptions:

(a) Gripper type TBM of 4.8 m in diameter and 18.1 m^2 of cross-section, in D&B 18.1 m^2 horseshoe-shaped in cross-section
(b) Tunnel support varies in actual situations according to rock conditions encountered, so no support cost (rock bolts, ring beams, etc.) was included in the analysis.
(c) Each method was considered for excavation in Granite and Limestone, both rock types having an unconfined compressive of 137 MPa
(d) Production was based on three shifts per day, five days per week. Also included one shift per week (Saturday) of two men to perform general maintenance.
(e) The costs presented are for personnel, machinery, operating, and interest costs only directly involved in the actual tunnel excavation. Personnel and equipment outside the portal were not included due to the wide variation observed with various contractors and job sites.
(f) Energy costs for the TBM and D&B operations were not included.
(g) Nine men per shift are assumed to work in TBM excavation, and ten men are considered to work in Drill and Blast tunneling.
(h) TBM utilization was 50 per cent, and with a cutterhead rotation of 8 rpm, actual machine penetration rates of 1.8 m/h and 3.7 m/h were assumed to be for Limestone and Granite, respectively.
(i) In drill and blast drilling, 40 blast holes with drill lengths of 3.65 m, 1.2 m/min in Granite, and 2 m/min in Limestone were considered, and other parameters affecting advance per day were added.
(j) A total of 28 and 14 weeks for delivery and set up time for the TBM and Jumbo drills were considered.
(k) The following formula for the interest cost is used.

$$\text{Interest cost} = \text{n}\left[(\text{total cost} + \text{value after depreciation})/2\right]\text{i} \tag{7.1}$$

where; n = number of total project time, years, i = annual interest time, years

As a result, Figs. 7.4 and 7.5 were drawn,

Brockway (1982) concluded that for Granite with a 4 km tunnel length, the construction cost of TBM tunnelling is equal to D&B tunneling, and thereafter, it decreases. This critical value is calculated as 8 km for Limestone and 4 m for Granite.

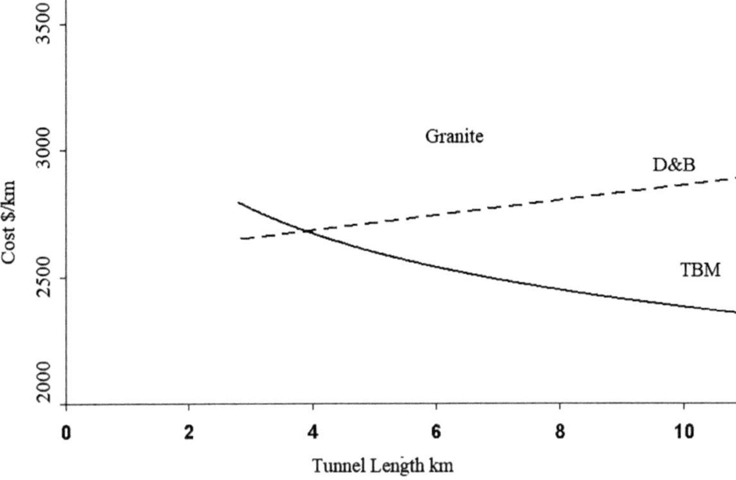

Fig. 7.4 The variation of tunnel construction cost with tunnel length in granite by Brockway (1982)

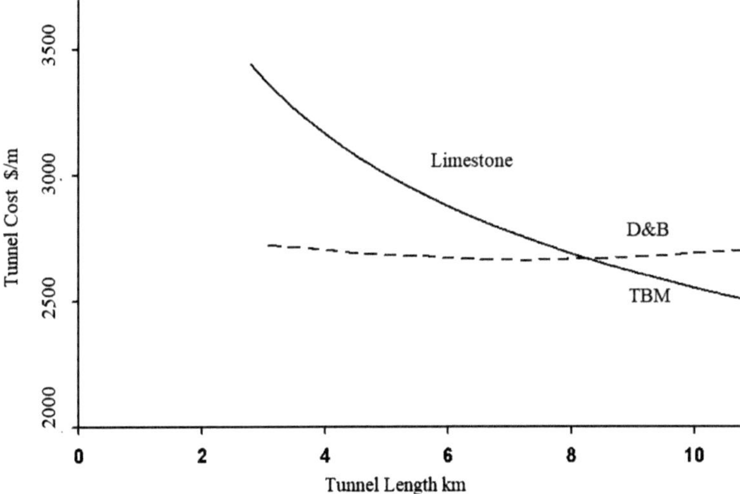

Fig. 7.5 The variation of tunnel construction cost with tunnel length in limestone, by Brockway (1982)

7.3 The Variation of the Tunnel Construction Cost with Tunnel Length in TBM Tunneling for a Given Geology and Tunnel Diameter

For an acceptable project cost calculation, it is necessary to use a valid performance prediction model and accurate assumptions to calculate daily advance rates. In this section, it is preferred to use the actual daily advance rates of 10.4 m with an EPB-TBM, obtained in the Kaynarca-Kartal-Kadıköy/Turkey Metro tunnels mentioned in Chap. 11. The other assumptions made are as follows:

(a) The rock formation of 55 MPa compressive strength is excavated.
(b) The equipment selected and their prices are given in Chap. 6
(c) Initial equipment cost as given in Chap. 6 as, 20,902,200 $
(d) EPB-TBM of 6.6 m diameter will work mainly in open mode.
(e) Equipment depreciation is 50%.
(f) Number of crew 154 is men/day (inside and outside); this is the mean number starting from the order of the equipment, finishing by the end of the project.
(g) It is accepted that the material for tunnel support is the same in both tunneling methods.
(h) Daily wage for one man is 200 $,

Maintenance cost (1 $/m^3$) Cutter cost is (2 $/m^3$)

(i) Project time = Mining time + TBM delivery 10 months + TBM set up time 2 months
(j) For the calculation of the interest cost, Eq. 7.1 in Sect. 7.2 is used

TBM equipment calculated cost is given in Table 7.1.
Project cost for different tunnel length are given in Table 7.2.

Table 7.1 TBM equipment cost for different tunnel length

Tunnel length km	Equipment depreciation, $	Excavated material, m^3	Cutter cost, $	Maintenance cost, $	Equipment cost, $
2	10,451,000	68,400	136,800	68,400	10,656,200
4	10,451,000	136,800	273,600	136,800	10,861,400
6	10,451,000	205,200	410,400	205,200	11,066,600
8	10,451,000	273,600	547,200	273,600	11,271,800
10	10,451,000	342,200	684,000	342,000	11,477,000

Equipment depreciation (%50) of the initial cost, maintenance cost 1 $/m^3$, cutter cost 2 $/m^3$

Table 7.2 TBM project cost/km

Tunnel length, km	Project time, days	Crew cost, $	Equipment cost, $	Interest cost, $	TBM project cost, $	Project cost $/km
2	192.3 + 360	17,010,840	10,656,200	2,726,628	30,393,668	15,196,834
4	384.6 + 360	22,933,680	10,861,400	3,887,840	37,682,920	9,420,730
6	576.9 + 360	28,856,520	11,066,600	4,929,558	44,852,678	6,280,487
8	769.2 + 360	34,779,360	11,271,800	6,170,329	46,051,160	5,756,395
10	961.5 + 360	32,386,200	11,477,000	7,052,146	50,915,346	5,091,535

Average daily advance 10.4 m/day, 154 men/day, 200$ day Annual interest 12%, Project time = Mining time + TBM delivery 10 months + TBM set up time 2 months. Personnel costs are the costs involved with machinery set up and mining

7.4 The Variation of the Tunnel Construction Cost with Tunnel Length in Drill and Blast Tunneling

In this section, it is preferred to use the actual daily advance rates of 2.7 m, obtained in the Kaynarca-Kartal-Kadıköy/Turkey Metro tunnels as mentioned in Chap. 11. The other assumptions made are as:

(a) The rock formation of 55 MPa compressive strength is excavated.
(b) The equipment selected is given in Chap. 6; initial equipment cost as provided in Chap. 6, is 2,074,000 $, and equipment depreciation is 50%.
(c) The number of crew is 116 men/day (inside and outside), this is a mean number starting from the order of the equipment, finishing by the end of the project.
(d) The daily wage for one crew is 200 $.
(e) It is not considered acceptable that the material for tunnel support is the same in both tunneling methods.
(f) Project time = Mining time + Drill Jumbo delivery 4 months + Jumbo set up time 1 month.
(g) For the calculation of the interest cost, Eq. 7.1 in Sect. 7.2 is used

The equipment cost calculated is given in Table 7.3.

Project cost for different tunnel length are given in Table 7.4.

Table 7.3 D&B equipment cost

Tunnel length, km	Equipment depreciation, $	Excavated material, m^3	Explosives and drill bit cost, $	Maintenance cost, $	Equipment cost, $
2	1,037,000	68,400	200,000	17,100	1,254,100
4	1,037,000	136,800	400,000	34,200	1,471,200
6	1,037,000	205,200	600,000	51,300	1,688,300
8	1,037,000	273,600	800,000	68,400	1,905,400
10	1,037,000	342,200	1,000,000	85,500	2,122,500

Equipment depreciation (%50) of the initial cost, maintenance cost 0.25 \$/m^3, explosives and drill bit 200,000 \$/km

Table 7.4 D&B project cost/km

Tunnel length, km	Mining time, days	Crew cost, $	Machinery cost, $	Interest cost, $	TBM project cost, $	Project cost, $/km
2	740.7 + 150	20,664,240	1,254,100	335,417	21,918,340	10,959,170
4	1481.4 + 150	37,848,480	1,471,200	671,194	39,990,874	9,997,719
6	2222.1 + 150	55,032,720	1,688,300	1,062,867	57,783,887	9,630,648
8	2962.8 + 150	72,216,960	1,905,400	1,505,920	75,628,280	9,453,535
10	3703.5 + 150	89,396,560	2,122,500	2,001,859	93,520,919	9,352,092

Average daily advance 2.7 m/day, 116 men/day, 200 $ day Annual interest 12%, Project time = Mining time + equipment delivery 4 months + equipment set up time 1 months

7.5 Critical Tunnel Length in Selecting TBM and D&Blast Tunneling Methods Based on Tunnel Construction

Figure 7.5 is drawn considering the calculations made in Sects. 7.3 and 7.4, which show that the construction cost in both tunneling methods equals at 4 km of tunnel length. This justifies Katuwall's arguments (2023) defining the equilibrium at 1.5–4.5 km of tunnel length depending on the geology, tunnel geometry, environment, and project timelines (Fig. 7.6).

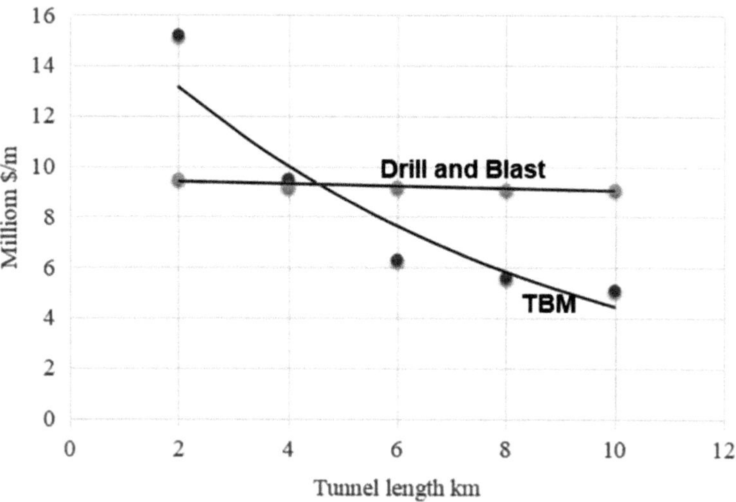

Fig. 7.6 The variation of tunnel construction cost in both TBM and Drill&Blast tunneling methods with tunnel length for a given geology and tunnel diameter, calculated by the authors of this book

7.6 Concluding Remarks

This chapter first discusses the work done in the past on the effect of tunnel length, tunnel diameter, geology, type of tunnel, and environment on the construction cost. Past studies showed that tunnel construction cost decreases with tunnel length, it increases with tunnel diameter, it decreases with increasing parameters like the geological strength index (GSI), tunnel types (wastewater, water, railway, highway, etc.) affect the cost of tunnel construction, the tunnel environment may lead to expensive protection arrangements for the mitigation of its impact. However, the relation between dependent and independent values is very scattered, which makes the question that at each level of tunnel length, the construction costs of TBM and Drill&Blast equal each other. For this, a comparative construction cost calculation is made in this Chapter considering the same tunnel diameter and tunnel geology. It is proved that at 4 km of tunnel length, as suggested by different authors, the construction costs of TBM and Drill&Blast equal each other at around 4 km of tunnel length.

References

Ahmed C (2021) Early cost estimation models based on multiple regression analysis for road and railway tunnel projects. Arab J Geosci 14:972. https://doi.org/10.1007/s12517-021-07359-x

Benardos A, Paraskevopoulou C, Diederichs M (2013) Assessing and benchmarking the construction cost of tunnels. Geo Montreal 2013:6

Brockway JE (1982) Tunnel boring machine or drill and blast. In: 14th Canadian rock mechanics symposium. Canadian Institute of Mining and Metallurgy, Vancouver, pp 11–16
Flyvbjerg B, Bruzelius N, Van Wee B (2008) Comparison of capital costs per route-kilometre in urban rail. Eur J Transp Infrastruct Res 8(1):17–30
HM Treasury (2010) Infrastructure cost review: technical report, obtainable in web. https://assets.publishing.service.gov.uk/government/uploads/system/uploads/attachment_data/file/192589/cost_study_technicalnote211210.pdf uploaded June 2022
Katuwall TB (2023) Selection of tunnel construction methods, Norhed II Project 70141 6: Conference 28 June 2023, Trondheim, Norway. https://doi.org/10.13140/RG.2.2.31837.59361/1
Kondrachova T, Grasselli G, Gaspari (2021) What about the cost of tunnels? A review of some recent Canadian tunnelling case histories. Tornto Underground, October 2021, p 9
Membah J, Asa E (2015) Estimating cost for transportation tunnel projects: a systematic literature review. Int J Constr Manage 15(3):196–218. https://doi.org/10.1080/15623599.2015.1067345
Paraskevopoulou C, Benardos A (2013) Assessing the construction cost of tunnel projects. Tunnell Undergr Space Technol 38(Sept 2013):497–505
Paraskevopoulou C, Boutsis G (2020) Cost overruns in tunnelling projects: investigating the impact of geological and geotechnical uncertainty using case studies. Infrastructures 5(73):1–35. https://doi.org/10.3390/infrastructures5090073
Petroutsatou C, Lambropoulos S, Pantouvakis J-P (2006) Road tunnel early cost estimates using multiple regression analysis. Oper Res 6:311–322
Rostami J, Sepehrmanesh M, Gharahbagh EA, Mojtabai N (2013) Planning level tunnel cost estimation based on statistical analysis of historical data. Tunn Undergr Space Technol 33:22–33
Sayadi AR, Hamidi JK, Monjezi M, Najafzadeh M (2015) A preliminary cost estimation for short tunnels construction using parametric method. Eng Geol SociTerritory 1:461–465

Chapter 8
Time for Mobilization Affecting the Selection of Tunneling Methods

Abstract Mobilization is a process that combines all the elements of a tunnel project and ensures they are ready to be executed. There is no doubt that the correct estimate of the time for mobilization for the tunneling project is one of the critical factors in the success of that project. It is generally accepted that the mobilization time of a TBM is higher than that of any equipment from a drill and blast tunneling project. Bearing in mind these realities, the main objective of this Chapter was to summarize the mobilization time of the equipment from TBM tunneling and drill and blast tunneling comparatively. The supply time of a TBM to be used in the TBM tunneling, which was subject to in Chap. 7, was reported as 10–12 months. However, for Kozyatağı-Kadıköy Metro tunnels having specific futures, the supply time of an EPB-TBM of 6.6 in diameter and 600 t of weight was around 19 months. Besides this, some manufacturers claim that Onsite First Time Assembly (OFTA) is used efficiently on several global projects to shorten overall delivery times to supply TBMs, saving several months in mobilization time.

8.1 Introduction

Mobilization is a process that combines all the elements of a tunnel project and ensures they are ready to be executed. A mobilization plan is vital to ensure the project runs smoothly, on the supply of the equipment, on budget, and on time. A detailed checklist will help provide visibility of key tasks required. There is no doubt that the correct estimate of the time for mobilization for the tunneling project is one of the critical factors in the success of that project. Chen et al. (2023) emphasize this respect, pointing out that:

(a) A technical scheme for transporting TBM components in advance should be prepared. This scheme should include an advanced investigation of the carrying capacity of the roads to be passed through, the selection and configuration of equipment for large-piece transportation, and safety measures for large-piece transportation.

N. Bilgin and C. Balci, *Critical Issues in Selecting Conventional and Mechanized Tunnelling Methods*, Springer Tracts in Civil Engineering,
https://doi.org/10.1007/978-3-031-89114-4_8

(b) Safety measures should be taken around the TBM lifting site, and all lifting equipment entering the site should be carefully inspected and accepted. At the same time, special operators should be checked for qualification.
(c) Traffic must be ensured during equipment transportation, and unique personnel should be assigned for safety and monitoring.

Mobilization time of a TBM depends on the project outlines. Mobilization time for the TBM in the Kozyatağı-Kadıköy Metro tunnel is given for each parameter of the project design as seen in the Table 8.1, for example for this project TBM is lowered down to the tunnel through a shaft which took 4.5 months for lowering the transported parts into the shaft and assembly of the machine, Akgül and Yurtaydın (2007)

It is generally accepted that the mobilization time of a TBM is higher than that of any equipment from a drill and blast tunneling project. Checking Tables 8. 2 and 8.3, one will see that the most time-consuming equipment supply for both tunneling methods is TBM, with a supply time of 10–12 months. For Kozyatağı-Kadıköy Metro tunnels for a specially designed EPB-TBM of 6.6 in diameter and 600 t of weight, the supply time was as given in the following Table 8.1 after Akgül and Yurtaydın (2007)

TBMs are usually assembled in factories, where the components are assembled and tested, then disassembled and shipped to the job site. Delivery of a machine can often be the critical path affecting project schedule, cost, manpower, etc. However, to shorten overall delivery times to supply TBMs, some companies have developed and efficiently used Onsite First Time Assembly (OFTA) on several global projects. Home (2010) has also cited that OFTA saved 3–4 months in Niagara and AMR water tunnels and 2–3 months in Mexico City Metro. One of the typical examples of OFTA from Turkey may be cited as the **Bakçenur** High-Speed Railway Tunnel Project. The tunnel is situated in Gaziantep Province, in Southeastern Turkey, an important center of agriculture and trade divided into nine districts. The Bahce-Nurdağı Railway Tunnel consists of two parallel 10.1 km tunnels. The Single Shield TBM of 8.0 m was assembled at the job site using Onsite First Time Assembly (OFTA). TBM assembly started on the job site on 15/07/2015 and terminated on 21/

Table 8.1 Mobilization time for a TBM in the Kozyatağı-Kadıköy Metro tunnel, after Akgül and Yurtaydın (2007)

Parameter	Months
Design phase	4
Manufacturing different parts of TBM	7
Assembly of the machine in the factory	1.5
Factor acceptance by the client and dismantling of TBM	1
Transport of the TBM machine to the job side including custom clearance	1.5
Lowering the transported parts into the shaft and assembly of the machine	4.5
Total supply time	19.5

09/2015, including gantry assembly and testing of all systems, which took 68 days. The job site location about 32 km from the Syrian border complicated the shipping of some parts and transporting TBM parts started in May 2015, Detlef and Bilgin (2018). Transport of the TBM parts in this mountain and the problematic area took 122 days, and the assembly of the TBM took only 68 days.

8.2 Time for Mobilization in Mechanized Tunneling

In general, the supply time of the equipment to be used in the TBM tunneling project, is given in Table 8.2.

Chapter 4 defined all equipment used in TBM tunneling; however, Chillers, segment molds, and segment plants were omitted. For segment plants and segment production, readers may consult the book by Bilgin and Acun (2024). We define these items briefly below. During the boring operation, a Chiller TBM (Tunnel Boring Machine) uses water to cool the shield of the TBM's cutterhead. The water used for cooling the cutterhead of the TBM works in a close circuit: by a pump, the water is

Table 8.2 Supply time of the equipment used in the TBM tunneling project

Equipment	Purpose of using	Specification	Supply* time months
EPB-TBM	Excavation and support	D = 6,6 m	10–12
MSV	Transportation of materials	50 Ton of capacity	9
Belt conveyor	Transportation of muck	With belt storage (315 + 110) kW	6–8
Ventilation fans and fan tubes	Ventilation	D1000 mm^{-2} * 75 kW motor	2–3
Segment moulds	Production of segments	Depending on segment geometry	6
Segment production plant	Production of segments	Depending on segment production capacity	8
Grout plant	Annulus grouting	20 m^3/h	4
Cooling system (Chiller)	Cooling for TBM hydraulics		3
Portal cranes forklift	Vertical transportation of materials	40 ton and 15 t capacity	3
	Handling of segments	15–20 ton lifting capacity	–
Elevator		30 m vertical capacity (10–12 person)	4

The information provided is from Dr. Ali Yüksel from Yapı Merkezi, Sinan Acun and Nadir Soyak from İÇTAŞ and Erhan Ünlü from ErTunnel

*Supply time, is the time from the order to the start of running the equipment in the tunnel

Fig. 8.1 Cooling system for TBM installed inside the container, courtesy of Cogede Italy

pumped inside the TBM, goes through a cooling circuit of the TBM cutterhead, and then the water comes out from the TBM and goes back inside a water storage tank. A cooling system for TBM is installed inside the container, as shown in Fig. 8.1. A segment production plant is one of the most important items of an entire mechanized tunneling system. A view of stationary segment plant from Eşme Tunnel is seen in Fig. 8.2, Bilgin and Acun (2024)

8.3 Time for Mobilization in Drill and Blast Tunneling

Supply time of the equipment used in the D&B tunneling project is important in selecting the tunnelling method, conventional tunneling or mechanized tunnelling. Table 8.3 is given in this chapter in order to give a clear idea to the readers on the subject.

The information provided is from Dr. Ali Yüksel from Yapı Merkezi, Sinan Acun and Nadir Soyak from İÇTAŞ and Erhan Ünlü from ErTunnel.

Chapter 3 summarized some of the D&B equipment with photographs. However, a few of them were not included in that Chapter. Now, in this chapter, we are providing below some Photos of the equipment outlined in Table 8.3 to help the readers visualize it. In Fig. 8.3, a Telehandler Manitou 1440 is shown. In Fig. 8.4, an Epiroc E2C two-boom Jumbo drill is illustrated. In Fig. 8.5, a typical tunnel inner lining form is given.

Fig. 8.2 A view of stationary segment plant from Eşme Tunnel, Bilgin and Acun (2024)

Table 8.3 Supply time of the equipment used in the D&B tunneling project

Equipment	Brand/Technical specification	Supply time
Excavator, with jack hammer	Hitachi 220, power 163 HP, backhoe 1 m^3	1 Week
Mini excavator	JCB, Cat 428	1 Week
Loader	Hidromek WL 640, 324 HP, 26, 3 t, 4.2 m^3	1 Week
Dump trucks	Ford Trucks 3542D DC Euro-6	1 Month
Telehandler	Manitou 1440,75 HP, engine power rating	1 Week
Wet shotcrete spraying machine	Tünelmak Adroit 450H/W, 100 HP, 11 t	3–4 Months
Dry shotcrete spraying machine	Meyco GM, 7 kW 9 round hole motor	1 Week
Driller (Jumbo)	Epiroc, E2C 125 kW, two boom rings, ~ 35 t	4–5 Months
Driller (Rock Drill)	Epiroc, FlexiROC T35/Engine 225 HP	3 Months
Grout pump	DSI MAI 400 NT	1 Week
Truck mixer	Ford Trucks 4142 M	1 Months
Ventilation fan and fan tube	Teknima, 75 kW, 1400 mm Diameter	15 Days
QC testing equipment	Depending on supplier	1 Month
Submersible pumps	Atlas Copco, 2–4 inches	1 Week
Generator	Atlas Copco, 500 kVA	1 Week
Track with crane	FASSI 450/VOLVO FH 480 6 * 4	3 Months
Tunnel inner lining form	–	3 Months
Stationery concrete pump	Tatmak	15 Days
Others	Hand tools	1 Week

Fig. 8.3 Telehandler Manitou 1440, courtesy of Manitou

Fig. 8.4 Epiroc E2C two booms Jumbo drill, with courtesy of Epiroc

Fig. 8.5 Tunnel inner lining form, with courtesy of Mesa

8.4 Concluding Remarks

Mobilization is a process that combines all the elements of a tunnel project and ensures they are ready to be executed. There is no doubt that the correct estimate of the time for mobilization for the tunneling project is one of the critical factors in the success of that project. It is generally accepted that the mobilization time of a TBM is higher than that of any equipment from a drill and blast tunneling project. Bearing in mind these realities, the main objective of this Chapter was to summarize the mobilization time of the equipment from TBM tunneling and drill and blast tunneling comparatively.

Checking Tables 8.2 and 8.3, one will see that the most time-consuming equipment supply for both tunneling methods is TBM, with a supply time of 10–12 months. For Kozyatağı-Kadıköy Metro tunnels for a specially designed EPB-TBM of 6.6 in diameter and 600 t of weight, the supply time was 19 months. However, some manufacturers claim that Onsite First Time Assembly (OFTA) is used efficiently on several global projects to shorten overall delivery times to supply TBMs, saving several months in mobilization time.

References

Akgül M, Yurtaydın Ö (2007) Construction, transport and assembly of a TBM selected for Kozyatağı-Kadıköy metro tunnels. In: Proceedings of the 2nd symposium on underground excavations for transportation, Istanbul, organized by ITU Mining Department and Turkish Chamber of Mines, Istanbul Branch, pp 73–82

Bilgin N, Acun S (2024) Practical management of tunneling with tunnel boring machines. CRC Press, Taylor and Francis Group. https://doi.org/10.1201/9781003558978

Chen K, Jiao S, Wang J (2023) TBM design and construction. Springer, p 867. ISBN 978-981-99-0058-9. https://doi.org/10.1007/978-981-99-0059-6

Detlef J, Bilgin N (2018) Turkey's hardest rock: the Bahce-Nurdag high-speed railway tunnel. 4:22–28

Home L (2010) To build a tunnel boring machine: why assembly on location. https://www.therobbinscompany.com›2010/09

Chapter 9
Benchmarking Metrics as an Aid for Rational Planning of Conventional and Mechanized Tunneling

Abstract Benchmarking metrics, if well investigated worldwide, may aid in rationally planning conventional and mechanized tunneling projects. This chapter uses benchmarking as a reference point to know how different tunneling projects have succeeded. For this purpose, data on conventional and mechanized tunneling projects from Turkey are analyzed and compared. The tunnel studies are from roadway, railway, utility, stormwater, irrigation, water conveyance, hydropower, and metro tunnel projects with information on geology, excavation method, mean daily advance rates, tunneling crew, and machine utilization time. Drilling and blasting pattern in Ovit roadway tunnel is given also as an example to one of most successfully opened tunnel. Later in Turkey. The authors investigated also several international projects for benchmarking purposes. These projects are: Gotthard Base Tunnel; sixteen UK tunnels from the transport and utilities sectors; tunnel Emisor Oriente, Mexico City; wastewater and water supply tunnels from the Boston area; USA, the Olmos Trans-Andean irrigation tunnel from Peru; the Niagara Tunnel Project from Canada; a survey of two authors from USA on 46 metro sized EPB-TBMs is also added to the analysis for benchmarking.

9.1 Introduction

The correct estimate of the efficient tunneling performance for both drill & blast and mechanized tunneling is vital for rational planning of all construction processes. Several theoretical models are used for this purpose. However, actual tunneling performance may be pretty different than the predicted one, which may be related to several unexpected factors, such as geological anomalies, a sudden change in the financial conditions of the client and the contractor, significant accidents affecting all the tunneling operations, usage age of the machines, etc. For this, benchmarking is used to leverage data to challenge conservative or optimistic estimates and set (and check) performance and cost targets. A benchmark may be defined as "a standard or a point of reference against which things may be compared and by which something

N. Bilgin and C. Balci, *Critical Issues in Selecting Conventional and Mechanized Tunnelling Methods*, Springer Tracts in Civil Engineering,
https://doi.org/10.1007/978-3-031-89114-4_9

can be measured or judged." Benchmarking is setting goals and measuring productivity based on best industry practices. It helps improve performance by learning from best practices. It involves regularly comparing different aspects of performance with best practices, identifying gaps, and improving situations. Benchmarking should not be treated as just a comparison. It is necessary to have a reference point to know how well one is doing, it helps organizations get ahead of the competition. When used appropriately, it will drive competition and innovation, securing better project outcomes. For this purpose, the data from Turkish and International Projects obtained from roadway, railway, utility, stormwater, irrigation, water conveyance, hydropower, and metro tunnel projects with information on geology, excavation method, machine design parameters, mean daily advance rates, tunneling crew, and machine utilization time are analyzed and compared with each other and with data obtained in different countries. The factors affecting the differences in the tunneling performances are outlined, resulting in benchmarking, and for each group of the tunnels, one or more are selected for detailed performance analysis. We hope that this will aid the rational planning of tunneling projects.

9.2 A Summary of Some Road and Railways Tunnels in Turkey

Table 9.1 shows the mean daily advance rates obtained when opening some Turkish road and railway tunnels. Tunnels with a length greater than 3000 km are deliberately chosen to provide a basis for comparative study. Detailed information on tunnels in Turkey may be found in Bilgin and Balcı (2016).

In mechanized tunneling, the best advance rate is obtained in Eşme- Salihli high-speed railway tunnel with an XRD EPB-TBM of 13.77 m diameter TBM. The tunnel is opened in a mélange consisting of gneiss, sandstone, claystone, mudstone, quartz, and silt, Jordan et al. (2023). Nurdağı railway tunnel is opened in the hardest rock formations in Turkey, with compressive strength changing from 70.6 to 327.4 MPa by a single shield TBM in a very hard meta-sandstone, meta-mudstone, and quartzite; the mean daily advance rate achieved in this tunnel was 9.5 m/day, Bilgin (2016), Jordan and Bilgin (2018). Eurasia road tunnel is opened in Trakya formation (sandstone, mudstone, limestone, andesite dykes), marine sediments (sand, silt, clay, gravel, and cobble), and transition zones by a Herrenknecht slurry TBM of 13.7 m diameter. The basic design parameters of TBMs used in these three tunnels are given in Table 9.2.

As Arıoğlu et al. (2016a; b) stated, "*Appropriate design of TBM parameters are crucial, since over design can cause higher costs whereas inadequate design can cause unexpected problems that can lead to delays in the project duration. In the Eurasia tunnel, TBM has a 247,300 kN thrust force capacity, and during the excavation utilized highest thrust force is 97% at the deepest point of the alignment, corresponding to the maximum face support pressure value of 10.8 bar, which is 90% of the design capacity*". Table 9.2 allowed us to understand better the effect of face

Table 9.1 Some Turkish road and railway tunnels opened by conventional and mechanized tunneling methods

Tunnels and case numbers	Length, km	Line	Start	Finish	m/day
1-Eşme-Salihli T-01 Railway Tunnel* Jordan et al. (2023)	3047	2	2021	2021	14.5
2-Nurdağı, Railway Tunnel*, Bilgin (2016), Jordan and Bilgin (2018)	9750	2	2017	2024	9.5
3-Eurasia Roadway Tunnel*, İstanbul, Arıoğlu et al. (2016a, 2016b)	3340	1 × 2	2011	2016	7.0
4-Zigana Gümüşane-Trabzon	14,481	2 × 2	2016	2023	6.4
5-Ovit, Rize Erzurum	14,346	2 × 2	2012	2018	7.4
6-Eğribel, Şebinkarahisar/Giresun	5905	2 × 2	2015	2021	3.1
7-Ilgaz Çankırı-Kastamonu	5486	2 × 2	2012	2016	4.0
8-Cankurtaran, Hopa-Borçka, Artvin	5228	2 × 2	2010	2018	2.1
9-Demirkapı, Antalya	5068	2 × 2	2017	2023	2.4
10-Sabuncabeli, Bornova, İzmir	4085	2 × 2	2012	2018	2.0
11-Troya, Ayvacık-Çanakkale	4007	2 × 2	2016	2022	2.3
13-Nefise Akçelik, Fatsa-ordu	3825	2 × 2	2000	2007	1.8
14-Orhangazi (Samanlı)	3591	3 × 2	2012	2016	2.5

*Eurasia road tunnel, Nurdağı and Eşme Railway tunnels are excavated with TBMs, the information on case number from 4 to 14 are obtained privately by the authors

pressure on penetration rate if the machine utilization factor is considered its effect on the daily advance rate. Referring to Table 9.2 for comparing the advance rates, we see the direct effect of the face pressure on the advance rate for similar machine utilization. For a similar TBM diameter, the thrust of the TBM in the Eurasia Tunnel is 14.9 times higher than that of the TBM in the Eşme -Salihli Tunnel for EPB face pressure around 1 bar and 10 bar for the Eurasia tunnel. For comparable machine utilization time and if the effect of ground conditions is excluded, we see that the daily advance rate in the Eurasia Tunnel is 2.3 less than the other tunnel. This supports the findings published by Bilgin and Yüksel (2023) that thrust and torque values almost increase to 2.2 and 1.7 times with EPB face pressure changing from 0.5 to 4 bar in all formations studied except for torque in silty-clayey sand. However, after 2 bar

Table 9.2 The basic TBM design parameters in Eşme-Salihli, Nurdağı and Eurrasia tunnels

Tunnel	Eşme-Salihli, railway	Nurdağı railway	Eurasia undersea roadway tunnel
TBM	Robbins XRE 451-379, working in EPB and hard rock mode, Jordan et al. (2023)	Robbins single shield, Jordan and Bilgin (2018)	Herrenknecht, slurry, Arıoğlu et al. (2016a; b)
Diameter, m	13.77	8.0	13.7
Total power kW	–	–	10,300
Cutter head power, kW	19 × 220 = 4180	10 × 330 = 3300	14 × 350 = 4800
Torque kNm	52,229/78344	4588 at 6.9 rpm. 14,453 at 0–3.3 rpm	23,389/34933
Thrust kN	16,577/282499	16,503/88543	247,300
Total weight, t	–	–	3300
Best m/day	32.4	18.2	18.0
Best m/week	178.2	112.5	
Best m/month	721.8	450.1	
Mean m/day	16.2	9.5	7.0

Stoppages are included for mean daily advance rates

of face pressure, the effect of EPB pressure on penetration and specific energy was more remarkable than the other parameters.

In conventional tunneling, the best daily advance values are obtained in Ovit and Zigana tunnels, with mean values of 7.4 and 6.4 m per day, respectively, both tunnels are opened from 4 faces, and the other tunnels from 2 faces, which makes the mean advance rate of 1.4 m per day for road tunnels having cross-section between 88.7 and 115.97 m^2. All road tunnels are tween tunnels opened in difficult ground conditions, with several faults and transition zones, and some tunnels have considerable water ingress.

9.2.1 *Selected Case Studies from Turkey for Railway and Roadway Tunnels*

9.2.1.1 Eşme-Salihli Railway Tunnel

The Eşme-Salihli Railway Tunnel is a short high-speed railway section measuring 3.05 km of the Ankara-İzmir High-Speed Railway Project for the Turkish State Railways (TCDD). A-13.77 m diameter mixed ground Rock/EPB TBM bored the tunnel at rates of up to 721.8 m in one month, making it the fastest TBM ever recorded

Fig. 9.1 Conceptional cross section of Eşme Salihli tunnel

over 13 m in diameter. The machine began excavating in altered gneiss, then passed through a mélange consisting of gneiss, sandstone, claystone, mudstone, quartz, and silt Jordan et al. (2023). The cross-section of the tunnel is seen in Fig. 9.1

A work study showed that 38% of the total time was spent on excavation, 27% on ring assembly, 21% on planned downtime such as maintenance, and 14% on unplanned downtime. When looking at planned downtime, most of the time was spent on cutterhead maintenance at 28%, followed by crew breaks at 22%. An average machine push took 60 min, while ring builds averaged 42 min, Jordan et al. (2023). Total number of the crew was 118.

9.2.1.2 Eurasia Tunnel

The 3340 km subsea Eurasia tunnel was excavated with a 13.7 diameter Herrenknecht Mixshield Slurry TBM. Tunnel alignment passes through Trakya formation (sandstone, mudstone, limestone, andesite dykes), marine sediments (sand, silt, clay, gravel, and cobble), and transition zones under a maximum hydrostatic pressure of ~ 11 bar and in a seismically active North Anatolian Fault zone (~ 16 km away). The mining operations took 476 calendar days with an average daily advance rate of 7.0 m/day for seven days/24 h with a TBM utilization of 31%. The maximum advance rate was realized at the marine sediment zone with 18.0 m/day. It should be noted that if the working days are considered and the major standstills due to seismic joint installation, religious holidays, and hyperbaric maintenance are excluded, the average rate and TBM utilization become 9.1 m/day and 41%, respectively (Arioglu et al. 2016a, b). TBM crew is approximately 40 persons, and each has an average of 2.8 TBM projects/person and 15.0 km TBM tunnel/person experience. The cost of this project with the 3-storey Great Istanbul Tunnel Project is about 3.5 billion dollars. The total project length is 14.14 km, with roads on the European side of 5.0 km, roads on the Asian side of 5.4 km in length, and a Bosphorus Crossing Tunnel of 3.34 km.

9.2.1.3 Ovit Tunnel

The Ovit Mountain Tunnel is a highway tunnel between Rize and Ispir Erzurum in NE, Turkey. The project's tender was first released to use a TBM. During the preparation of the portals and pilot trials with the drill and blast method, the job owner and the contractor, seeing some unexpected geological facts, decided to go to an independent arbitrator to reevaluate the process of the excavation, the geology of the tunnel line, and the risks of using TBM. Academicians from Istanbul Technical University were charged for this study, Tüysüz and Genç (2012) and Bilgin (2012). Their study showed that pretender documents were insufficient, and there were several uncertainties in recommending a TBM for the tunnel excavation. The region is influenced by the eastern Black Sea tectonics, which developed in the northeast, northwest, and east–west directions. The tunnel is situated within three primary geological formations. The first is Kackar granitoids, which are massive and jointed. The second formation, Kackar porphyrites, is more fractured than the granitoids. Dikes, sills, and hydrothermal weathering zones are the main structures of the porphyrites and granitoids, which are cut off by this formation in many locations. The third is Çatak Formation, located on the tunnel's SE section, consisting of basaltic lava and pyroclasts. The formation is cut by sediment rock layers like claystone, marl, and siltstone rocks. The geological cross-section of the tunnel is given in Fig. 9.2, as summarized by Guner et al. (2014).

The study carried out by Tüysüz and Genç (2012) and Bilgin (2012) showed that the most critical factors affecting the excavation methods are the geological discontinuities, which would affect the performance of a TBM in this tunnel, as face collapses, squeezing of TBM, etc. These geological discontinuities reported are cooling fractures with apertures of 30–40 cm in Kaçkar Porphyrites as seen in Fig. 9.3, hydrothermal alterations within Kaçkar Porphrites seen in Fig. 9.4, lavas and sediments in Çatak Formation seen in Fig. 9.5, and discontinuities as shown in Fig. 9.6

The overburden of the tunnel is about 850–900 m maximum and 400–500 m on average. Crosscuts connect the twin tunnels at intervals of 250 m. The tunnel length is 12.5 km, with a pilot tunnel of 1.7 km. The cost of the project is 1.114 billion Turkish Liras. (3.4472 TL = 1 USD on August 30) 324 million USD. Shotcrete, reinforced with wire mesh and rock bolts for rock reinforcement, is used as initial support. In difficult ground conditions, additional support with foreboding piles varying between 50.80 mm and 88.90 mm. The tunnels were driven from four faces. The drill and blast method was used in complex rock formations, and the impact hammer was used in softer ground. Tunnel driving performance recorded during the side visit of the

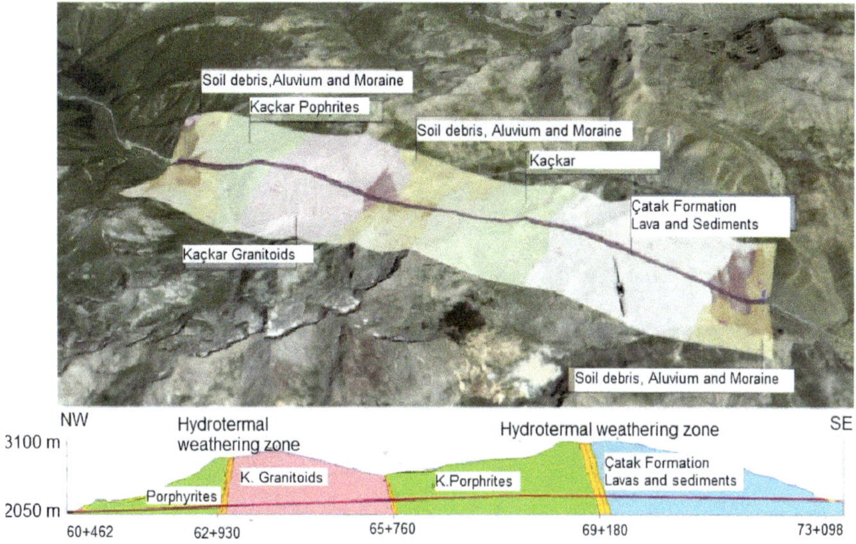

Fig. 9.2 Geological cross-section of Ovit Tunnel, Guner et al. (2014)

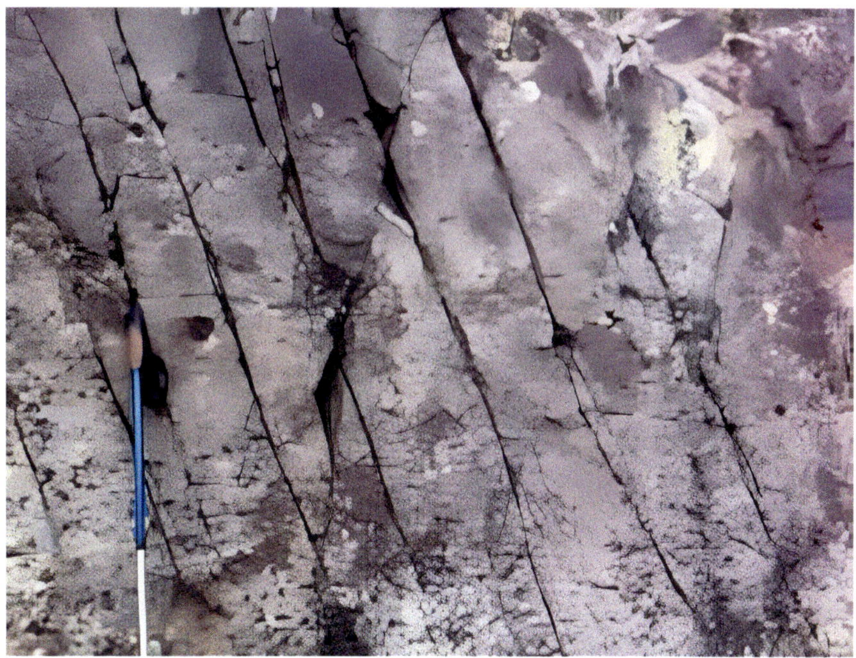

Fig. 9.3 Cooling fractures with apertures of 30–40 cm in Kaçkar Porphyrites, from ITU report after Tüysüz and Genç (2012)

Fig. 9.4 Hydrothermal alterations within Kaçkar Porphrites, from ITU report after Tüysüz and Genç (2012)

Fig. 9.5 Lavas and sediments in Çatak Formation, from ITU report after Tüysüz and Genç (2012)

Fig. 9.6 Tunnel face with several discontinuities in west tube, round 16. After Bilgin, the author 2012 ITU report

author between 12 and 17 December 2012 is given in Table 9.3, Bilgin (2012). Several side visits and reports submitted by academicians from the University verified the difficulties that might be raised during TBM tunneling. After that, the job owner and the contractor continued with the drill and blast tunneling method.

After that, in general, the blasting pattern and blasting parameters applied were as defined in Fig. 9.7.

The number of experienced crew, including the equipment used, is one of the main factors determining the success and efficiency of tunneling operations. The Ovit tunnel is given in this chapter as a typical example of a well-managed and completed project in Turkey carried out by the conventional tunneling method. Thus, a list of tunnel crew and the equipment used in the project is given in Tables 9.4 and 9.5 for comparative issues. Total number of crew is 101. Within the total pieces of equipment 37, the number of Lorries to transport the muck is 12. This is reflected in Figs. 9.8 and 9.9, given for the upper and lower parts of Ovit Tunnel of 53.6 m^2 and 35.1 m^2 in cross-section areas, Ozcelik (2016). Time to spend for muck removal in the upper part is % 23.1 of the shift time and 38.5 for the lower part of the tunnel, and the share of each working item is compared in Table 6 as a reference with those given in the work of Zare and Bruland (2006). The NTNU advance rate model for conventional tunneling given by these authors is based on state-of-the-art technology and equipment with Norwegian tunneling experience using round cycle time consumption and comprises drilling, charging, blasting, ventilation, loading, hauling, scaling, and rock support. In this model, weekly advance rate as a function of tunnel cross-section and equipment combination is presented by applying the model for 5 m drill hole length, 48 mm drill hole diameter, and parallel hole cut in a medium strength rock with medium strength characteristics, 101 working hours

Table 9.3 Tunnel driving performance with drill & blast method in Ovit Tunnel in December 2012, recorded by the author

Round No.	Start (km)	Finish (km)	(m)	Rock class	Date
R139	72 + 893.20	72 + 891.45	1.75	B2	12.12.2012
R140	72 + 891.45	72 + 889.70	1.75	B2	12.12.2012
R141	72 + 889.70	72 + 887.95	1.75	B2	12.12.2012
R142	72 + 887.95	72 + 886.20	1.75	B2	13.12.2012
R143	72 + 886.20	72 + 884.45	1.75	B2	13.12.2012
R144	72 + 884.45	72 + 882.70	1.75	B2	13.12.2012
R145	72 + 882.70	72 + 880.95	1.75	B2	14.12.2012
R146	72 + 880.95	72 + 879.20	1.75	B2	14.12.2012
R147	72 + 879.20	72 + 877.45	1.75	B2	14.12.2012
R148	72 + 877.45	72 + 875.70	1.75	B2	15.12.2012
R149	72 + 875.70	72 + 873.95	1.75	B2	15.12.2012
R150	72 + 873.95	72 + 872.20	1.75	B2	15.12.2012
R151	72 + 872.20	72 + 870.45	1.75	B2	16.12.2012
R152	72 + 870.45	72 + 868.70	1.75	B2	16.12.2012
R153	72 + 868.70	72 + 866.95	1.75	B2	17.12.2012
R154	72 + 866.95	72 + 86,520	1.75	B2	17.12.2012

per week are considered. The following assumptions for a road tunnel are made: the tunnel length is 3 km with a cross-section of 60 m^2, 92 charged holes of 48 mm and three large holes of 102 mm diameters are considered, and 15 polyester anchored bolts are taken per round of 5 m. The standard and gross round cycle times are estimated to be 371 and 431 min, which results in a 74.3 m average weekly advance rate and a 64 m gross weekly advance rate. For quick and easy comparison, the readers are advised to refer to Table 9.6 to see how the figures obtained in the Ovit Tunnel in the upper section, and those predicted by Zare and Bruland are close.

9.2.1.4 Zigana Tunnel

The new Zigana Tunnel, with its length of 14,481 km, is the longest in Turkey and ranks among the longest tunnels worldwide. Its cost is US $1 billion. It's a double-tube highway tunnel (2 × 2) constructed using conventional tunneling methods. The finished cross-section of the tunnel is 64.4 m^2. The maximum and minimum excavation areas are 127 m^2 and 99 m^2, respectively. The new tunnel crosses the mountain at an approximately 560 m lower elevation than the existing tunnel and shortens the distance by 8 km. There are six ventilation shafts in the tunnel alignment. The shafts are excavated with a Raise Boring Machine for safety reasons and to shorten tunneling operations' time. In the layout of Zigana Tunnel, Kızılkaya formation consisting of clayey limestone, sandy limestone, and tuff inter-bedded dacite and pyroclastic and

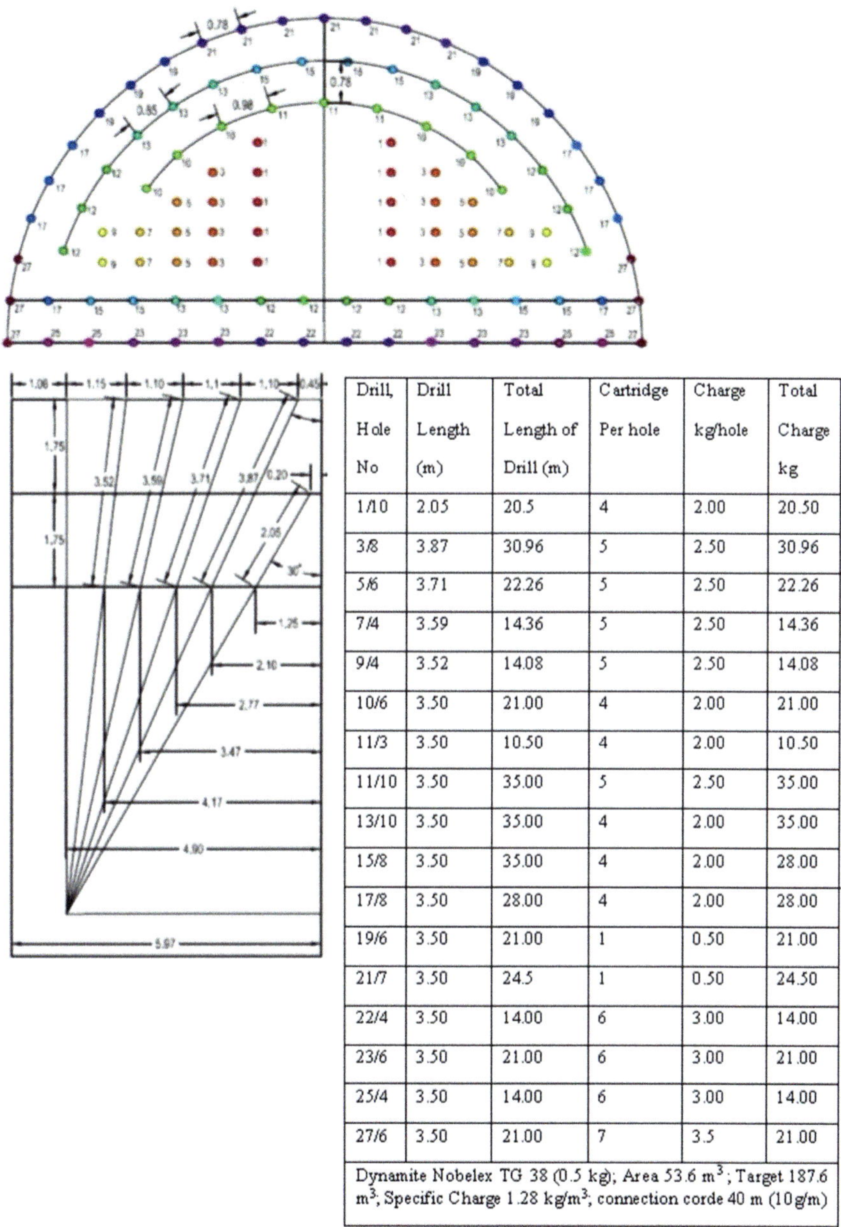

Drill, Hole No	Drill Length (m)	Total Length of Drill (m)	Cartridge Per hole	Charge kg/hole	Total Charge kg
1/10	2.05	20.5	4	2.00	20.50
3/8	3.87	30.96	5	2.50	30.96
5/6	3.71	22.26	5	2.50	22.26
7/4	3.59	14.36	5	2.50	14.36
9/4	3.52	14.08	5	2.50	14.08
10/6	3.50	21.00	4	2.00	21.00
11/3	3.50	10.50	4	2.00	10.50
11/10	3.50	35.00	5	2.50	35.00
13/10	3.50	35.00	4	2.00	35.00
15/8	3.50	35.00	4	2.00	28.00
17/8	3.50	28.00	4	2.00	28.00
19/6	3.50	21.00	1	0.50	21.00
21/7	3.50	24.5	1	0.50	24.50
22/4	3.50	14.00	6	3.00	14.00
23/6	3.50	21.00	6	3.00	21.00
25/4	3.50	14.00	6	3.00	14.00
27/6	3.50	21.00	7	3.5	21.00
Dynamite Nobelex TG 38 (0.5 kg); Area 53.6 m^3; Target 187.6 m^3; Specific Charge 1.28 kg/m^3; connection corde 40 m (10 g/m)					

Fig. 9.7 Blasting pattern in the upper part of Ovit tunnel

Table 9.4 The crew in Ovit Tunnel for two faces in one tunnel

The crew	The number
Tunnel chief	1
Shift engineers	2
Geo-technical engineer	1
F 101 chief foreman	1
Electrical foreman	1
Surveying	1
Foreman	4
Tunnel worker	20
Electrician	4
Office work	4
Loader operator	6
Excavator operator	6
Becoloader	2
Telehandler operator	2
Jumbo operator	5
Shotcrete operator	5
Concrete plant operator	2
Lorry driver for (2661 MT)	24
Mixer driver for (2661 MT)	10
Total	101

Table 9.5 Equipment used in Ovit Tunnel

Equipment	Number
Shotcrete robot	3
Jumbo	3
Loader	3
Excavator	3
Telehandler	1
Bekoloader	1
Lorry (2661 MT)	12
Mixer for (2661 MT)	5
Concrete plant	1
2 × 250 kW jet fan	1
Transformer, outside 1000 kW	1
Transformer, inside 630 kW	1
Generator 500 kV	2

Note In the 4th and 5th km of the tunnel, extra 200 kW fans were added. Tunnel ventilation is realized with 2800 mm diameter fan tubes. Lighting is realized with 400 kW projectors installed every 300 m

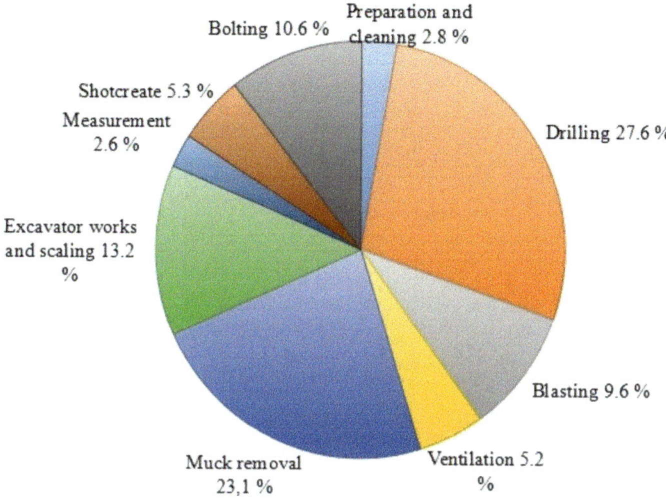

Fig. 9.8 Job distribution in drill and blast operations in the upper part of Ovit Tunnel of 53.6 m^2

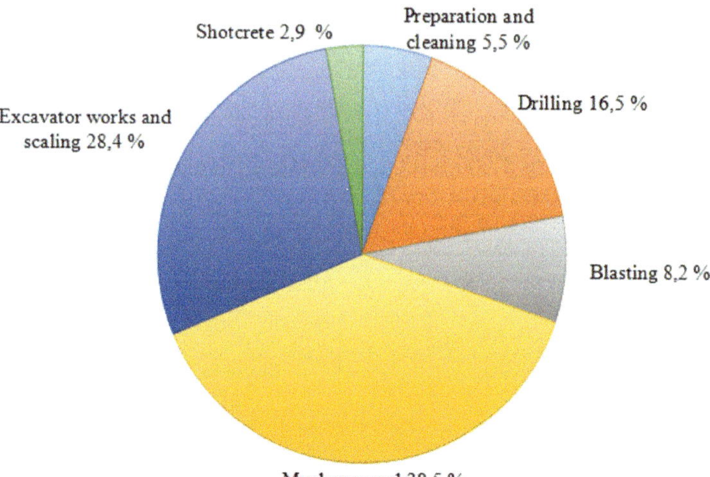

Fig. 9.9 Job distribution in drill and blast operations in the lower part of the Ovit Tunnel having 35.1 m^2 of excavation area

Çağlayan formation composed of limestone, sandstone inter-bedded andesite and basalts are found.

Table 9.6 The percentage of the works done in Ovit Tunnel and the the figures predicted by Zare and Bruland (2006)

Item	Ovit tunnel Upper part	Ovit tunnel Lower part	Conceptional after Zare and Bruland (2006)
Preparation and cleaning %	2.8	5.5	–
Drilling %	27.6	16.5	33.0
Blasting %	9.6	8.2	14.0
Ventilation %	5.2	–	3.0
Muck removal %	23.1	38.5	24.0
Excavator works and scaling %	13.2	28.4	12.0
Measurement %	2.6	–	
Shotcrete %	5.3	2.9	
Bolting %	10.6	–	
Rock support %			14.0

9.2.2 *Comparative World While Records in Road and Railway Tunnels*

Tunneling performance for the Turkish tunnels under investigation is compared and benchmarked with other available data from international projects. This is summarized in Fig. 9.10.

With a route length of 57.09 km Gotthard Base Tunnel it is the world's longest railway and deepest traffic tunnel. In a recent study, Falanesca et al. (2023) mentioned

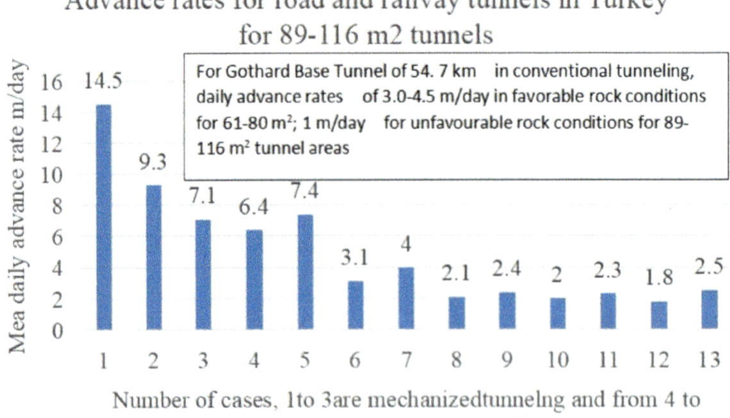

Fig. 9.10 Advance rates in road and railway drives in Turkey and comparison with some international projects

that for the Gotthard Base Tunnel, 54.7 km of the total 153 km of excavations were carried out by conventional methods. The daily advance rate with a conventional drive for single-track tubes with overburden up to 2400 m are:

- Mean advance rate in good rock conditions changes from 3.0 to 4.5 m/d for cross-section between 61 and 80 m^2
- Mean advance rate in difficult ground conditions is approximately 1.0 m/d for cross-section between 85 and 110 m^2
- Max. Advance rate recorder for the single-track tube Sedrun South East with 11.5 m/d.

Readers are advised to also read the paper by Falanesca et al. (2023) for benchmarking of gripper, single, and double shield TBMs advance rates.

A UK Infrastructure and Projects Authority study IPA (2018) obtained data for sixteen UK tunnels from the transport and utilities sectors in ground conditions ranging from soft rock to cohesive and non-cohesive soils. The construction methods included a Tunnel Boring Machine (TBM), backhoe shield, and excavator with precast concrete or shotcrete linings. A reduced number of projects resulted in higher quality and more reliable analysis. For both types of tunnels, mean daily advance rates of 9 m/day and 7.2 m/day were obtained for 50 m^2 and 79 m^2 cross-section tunnels. Compared with the others, these results reveal that tunneling methodology, cross-section of the tunnels, and ground conditions play an essential factor in a benchmarking study.

9.3 Utility Tunnels

More than half of the world's population lives in cities, and the trend is still rising. According to United Nations estimates, 68% of people will live in urban areas by 2050, https://www.un.org/development/desa/en/news/population/2018-revision-of-world-urbanization-prospects.html. The size of these cities creates challenges for urban planners. How can they deal with the huge volumes of wastewater produced and the need for clean water to cope with storm waters? In light of these facts, this chapter will focus on utility tunnels, which are used for various essential services such as water conveyance, irrigation, stormwater, wastewater, or sewage. In this chapter, we deal with utility tunnels separately from other tunnels because they have specific construction items that can prolong the termination of the tunnels. As seen in Fig. 9.11 in a wastewater tunnel as an example from Haliç tunnels, Istanbul, there is a secondary concrete lining after the precast segments and a rubber coating to protect the tunnel from acidic effects.

Environmental problems in big cities, problems of disposal of the muck, difficulties in opening the shafts, etc., are the factors that retard the tunnels to make in service. These tunnels can vary from small to large enough to accommodate human and vehicle traffic. The methods of utility tunnel construction, mainly realized by cut and cover method, excavation with roadheaders, jackhammer, excavation with

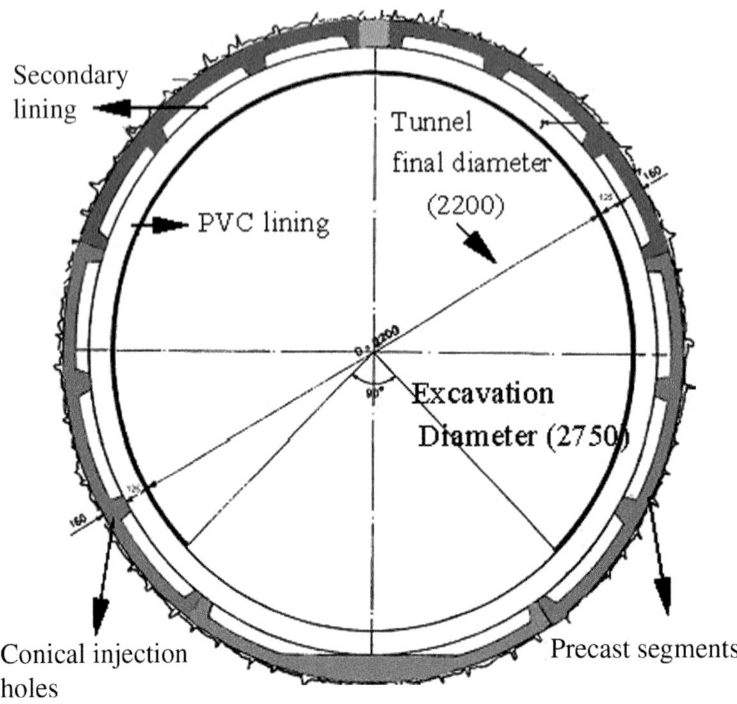

Fig. 9.11 A typical cross-section of a utility tunnel

TBM, micro tunneling or pipe jacking, etc., are practical and widely applicable. In this chapter, we will focus on excavating utility tunnels with roadheaders and TBMs, providing information that is directly relevant to the relevant work in the field. However, we should mention that in big cities, especially ones in China and Singapore, the trend is to integrate most utilities (electricity, wastewater, water, cables, etc.) in one multiple-purpose utility tunnel (MUT). The main advantages of MUT may be cited as a major reduction of construction costs, improved inspection and maintenance of utilities, more organized planning of underground space, etc. The main disadvantages may be security risks, coordination issues, etc. Luo et al. (2020).

9.3.1 *Some Utility Tunnels in Istanbul*

Many wastewater collectors, tunnels, and discharge lines exceeding 120 km in length were constructed during the efforts referred to as the Northern and Southern Golden Horn projects Haliç). The Golden Horn area and the whole of Istanbul were renovated in this process. Over the last 30 years, a 17,500 km wastewater collection system service was launched in Istanbul. In addition, 4500 km of rainwater collection services were commissioned, https://www.dailysabah.com/opinion/op-ed/istanbuls-golden-horn-from-past-to-present. Information gathered from the first sewer tunnels excavated by shielded roadheaders in Golden Horn since 1985 was and immense value in assessing the geology and geotechnical properties of the area and methods to be used in the future tunneling projects in Istanbul, Bilgin (1988). Thus a summary of tunneling performances by roadheaders and thereafter by TBMs will be given below.

Table 9.7 and Fig. 9.12 summarize some Turkish wastewater and stormwater tunnels, including length, excavation diameter, excavation method, daily advance rates, and rock formation.

Design parameters used in some of the wastewater tunnels are given in Table 9.8.

Table 9.7 Some utility tunnels in Turkey

Tunnel	Length, m	D, m	Excavation method	m/day	Rock formation
Collector (sewarage-wastewater) tunnels					
Eyüp Bilgin (1988)	1865	3.2	Shielded roadheader	4.7	Sandstone, siltstone, shale, diabaze, σc = 55–160 MPa
Küçüksu Bilgin et al. (2005a, b)	1037	3.2	Shielded roadheader	3.2	Mainly limestone and silt σc = 60–145 MPa, RQD 40–80%
Büyükçekmece, Özaydın et al. (2013)	4607	3.8	EPB-TBM	8.0	Limestone, mudstone with interbedded sanstone
Beykoz-Kavacık, Guclucan et al. (2008, 2009)	7190	3.2	Single shield TBM	6.2	Limestone, shale, carbonated shale, diabase
Ambarlı	6065	4.6	EPB-TBM	13.7	Sediments, Mudstone
Baltalimanı-Sarıyer-Tarabya Bilgin et al. (2005a, b)	13,270	2.9	Single shield TBM	11.5	Limestone, shale, sandstone σc = 32–81 MPa
Storm water tunnels					
Ayvalıdere, Özaydın (2024)	4662	4.0	EPB-TBM	8.9	Mudstone-sandstone, clay, sand
Unkapanı, Özaydın (2024)	1516	4.0	EPM-TBM	8.6	Mudstone-sandstone, clay, sand
Aksaray-Vatan, Özaydın (2024)	815	4.0	EPB-TBM	6.4	Mudstone-sandstone, clay, sand

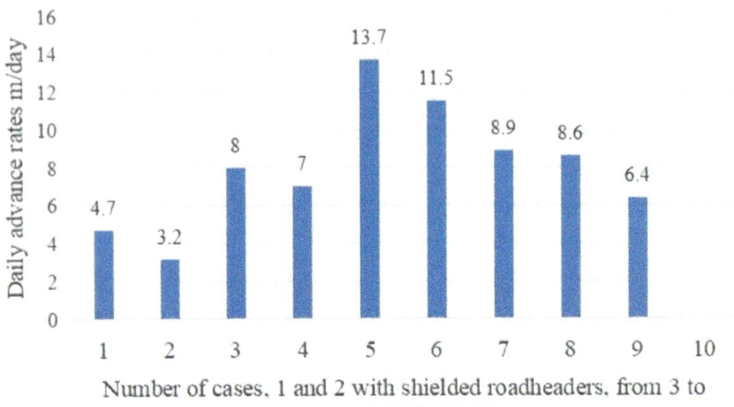

Fig. 9.12 Advance rates in wastewater and storm water tunnel in Turkey having 6 m–16.6 m^2 excavation area

Table 9.8 Design parameters of some TBMs used in utility tunnels in Istanbul

Tunnel	Beykoz-Kavack	Ambarlı	Baltalimanı-Sariyer
Geology	Gözdağ-Doloyaba-Kartal	Sediments-Mudstone	Doloyaba-Tuzla-Trakya
TBM	Single shield	EPB	Single shield
Diameter, m	3.2	4.6	2.9
Total power kW	400	600	620
Cutter head power, kW	250	400	400
Torque kNm	527 at 4.3 rpm 254 at 8.5 rpm	3117	725
Max, thrust kN	9950	16,625	3750
Best m/day	23	28	24.8
Best m/week	100	155	145
Best month	365	571	531
Mean m/day, including stoppages (m)	7	13.7 m	11.50

9.3.2 Cases for Shielded Roadheaders, Eyüp and Küçüksu Tunnels

9.3.2.1 Eyüp Tunnel

Eyüp Tunnel, with a length of 1865 m and an excavation diameter of 3.2, is a part of the combined interceptor sewerage system for the South Haliç (Golden Horne). A shield mounted roadheader with a 95 HP hydraulically driven boom with variable cutting speed up to 90 rpm, the cutting boom is built as an integral part of the shield body. The machine is illustrated in Fig. 9.13. Rock formations of the Carboniferous age are found in the area and consist of fine to coarse-grained strongly fractured mudstone, shales, sandstone, and conglomerates. Some diabase dykes have also been encountered while driving the tunnel, and these affected progress rates as they are significantly harder than the rock excavated along a major part of the route. The compressive strength of the rock formations changed from 70 to 150 MPa, and RQD values from 20 to 80%. Considering starting and finishing dates, a mean progress rate of 4.7 m/day was obtained. Tunneling activities were carried out on a three-shift/day working programme. Staffing for one shift consisted of 17 workers as follows: one foreman and one roadheader operator. Three man for segment erection, one erector operator, two locomotive drivers, three man for concrete injection, one man to supervise the belt conveyor, two men charged with the security in shaft top and bottom, 0ne surveyor and and one for miscellaneous works. The energy consumption for m^3 of excavated material was 42 kWh. Average pick consumption 0.12 pick cutter/m^3. Machine utilization was 54% of shift time. Machine downtime for band conveyor was 11.6% of shift time, for boom 10.7% of shift time, for erector 5.8 of shift time, for shield 4.4 of shift time, for replacement of the cutters 2% of the shift time for other cases 11.5% making a total down time of 46%.

9.3.2.2 Küçüksu Tunnel

Küçüksu tunnel is a part of sewage project which is situated between Küçüksu and Hekimbası in the Anatolian part of Istanbul. The same machine as used in Eyüp tunnel was used also for Küçüksu tunnel. The machine had a cutting power of 90 kW and total power of 224 kW. The cutting head is axial type having 36 conical cutters of 75° tip radius. The excavation of the tunnel started in 27th August 2002 and finished in 9th August 2003. The crew consisted of 5 civil engineers, three surveyors and 12 workers per shift. The excavated material was transported using Clayton locomotive of 21 HP and cars of 2.5 m^3 in volume. Two 7.5 kW ventilator and ventube of 500 mm in diameter realized the ventilation of tunnel. Precast segments prepared in tunnel side are used as initial tunnel support; each segment has a length of 0.75 m and thickness of 0.10 m. Each rig consists of 4 precast segments and an invert. In situ casted concrete lining was used as secondary tunnel support and PVC lining to protect concrete from harmful effect of the sewage water. The main rock

Fig. 9.13 Shielded Herrenknecht roadheader used in Eyüp and Küçüksu Tunnel

formation is limestone (72%) having compressive strength values changing from 60 to145.2 MPa and RQD values from 40 to 90%. 16%of the rock formations are andesite and diabase dykes with compressive strength from 74.1 to 163.8 MPa and RQD 80–90%. the mean cutter consumption is 0.33 cutter/m^3 varying from 0.077 cutter/m^3 in siltstone to 0.669 cutter/m^3 in limestone-diabase. The machine utilization time and daily advance rates in Küçüksu Tunnel is found less than the values obtained in Eyüp tunnel, this may be explained by the fact that rock formations in Eyüp Tunnel is more fractured than found in Küçüksu Tunnel. Tunneling performance with roadheader in Küçüksu Tunneli is illustrated in Fig. 9.14.

9.3.3 *Cases for TBMs*

9.3.3.1 Baltalimanı-Sarıyer-Tarabya Wastewater Tunnel

The tunnel is a part of sewerage project, which is planned to clean the sea pollution in Istinye—Tarabya and Büyükdere bays around Istanbul Bosphorus. The tunnel having a length of 13,270 m and final diameter of 2 m is excavated by a single shield TBM of 2.9 m in diameter and cutting power of 560 kW. The main rock formations are Paleozoic aged sedimentary rocks, limestone, sandstone–siltstone, andesite dykes and sediment fillings. Some geomechanical properties of the rock formations are given in Table 9.9. The mean daily advance rate of TBM was 11.5 m/

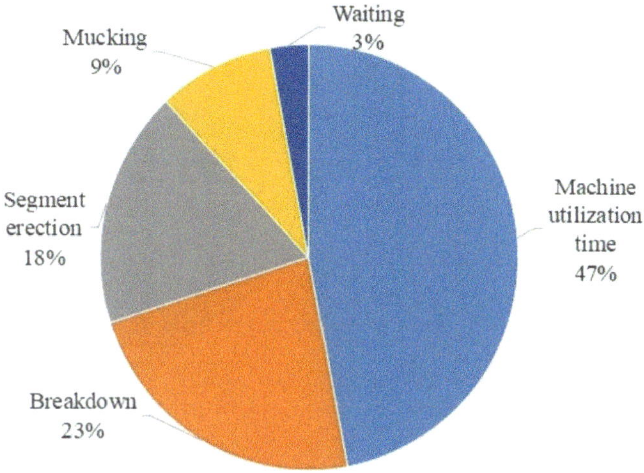

Fig. 9.14 Tunneling performance with roadheader in Küçüksu tunnel

Table 9.9 Summary of the rock properties encountered in tunnel route

Rock formation % of the total	Comppressive strength MPa	Tensile strength MPa	Elastic modulus MPa 10^3
Limestone 65%	44–81	4–5	9–15
Shale 17%	55–59	2–4	9–10
Sandstone–siltstone 12%	59	–	–
Dykes 1%	32–40	3–4	6–7
Sediments 5%	–		–

day with 35% machine utilization time. However, it dropped to 9.4 m/day if starting and finishing the tunneling activities are considered. Machine performance data is shown in Fig. 9.15. The summary of the rock properties encountered in tunnel route is given in Table 9.9.

9.3.4 Comparative World While Records in Utility Tunnels, International Cases

9.3.4.1 Tunnel Emisor Oriente, Mexico City

This is the longest wastewater tunnel in the world, with a length of 62.5 km and an inner diameter of 7 m. It was constructed between 2008 and 2019 using six EPB-TBMs of 8.89 m diameter. Three Robbins TBM were used for lots 3, 4, and 5. The estimated cost for the tunnel was 15 billion pesos. After completion, the tunnel was

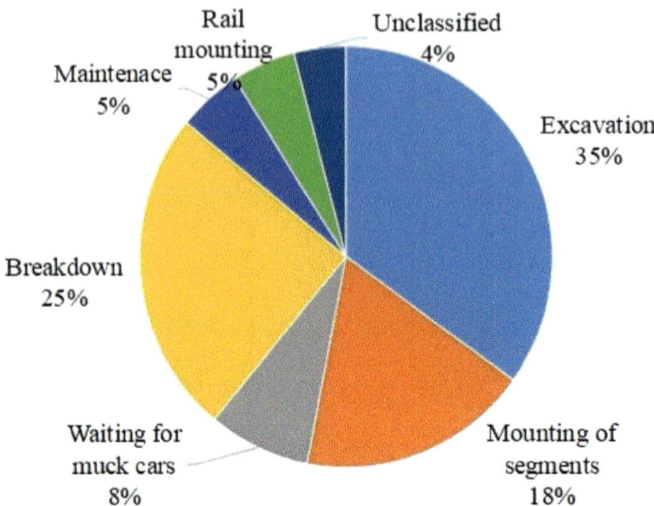

Fig. 9.15 TBM performance data in Baltalimanı-Sarıyer-Tarabya wastewater tunnel

realized to have exceeded budget and cost 30 billion pesos. The project is planned as 6 lots with very difficult geology consisting of Quaternary and Pliocene lacustrine deposits, basaltic ashes, clay, and volcanic deposits. The project has a total length of 62.5 km is completed with a mean daily advance rate of 15.8 m/day with 6 TBMs including 24 shafts up to 150 m deep, Gonzalez et al. (2014), https://en.wikipedia.org/wiki/Emisor_Oriente_Tunnel

9.3.4.2 Boston Area Some Wastewater and Water Supply Tunnels

The city of Boston, which has plenty of utility and water conveyance tunnels, is a good example of how the complex geology and structure of a town affect the utility tunneling performance. The mean daily advance rates when using drill and blast, TBM & drill and blast simultaneously, and TBM tunneling in tunnels having excavation diameters between 3.1 and 4.1 in different formations are summarized in Table 9.10. As seen from this table, tunneling methodology and geology have a considerable effect on the tunneling performance. A typical example to these tunnels is Inter-Island Tunnel, Boston tunnel summarized below.

Inter-Island Tunnel, Boston

The Boston Harbor Project's Inter-Island Tunnel extends from Deer Island south for 7.7 km to the Nut Island. The tunnel was driven with a TBM and completed at the end of the drive with a 4.3-m diameter cast-in-place lining. The excavation began beneath Deer Island in July 1992 and was completed in November 1995, and the contractor finished the liner in late 1997. The rock is chiefly sandy argillite tuff and

Table 9.10 Some wastewater and water tunnels in Boston area, Barosh and Woodhouse (2011, 2012)

Tunnel	Length,m	Lined diameter m	Method of excavation	Mean daily advance rate m/day	Rock formation
North Metropolitan	6300	3.1	Drill and blast	3.0	Argillite
Malden	1600	3.8	Drill and blast	4.4	Argillite, diabase
Dorchester	10,200	3.1	TBM and drill and blast	5.6	Argillite, sandstone, conglomerate, basalts
Braintree-Weymouth	4600	3.7	TBM	3.2	Argillite, dikes, volcanic rocks
Inter-Island	8000	4.3	TBM	7.4	Argillite

sandstone. Along the tunnel route, there are many faults with intervals of 3 to 24 m, and also the tunnel's alignment crosses some major faults, highly sheared diabase with higher water inflow. Problems in the tunnel construction resulted from the large fault zone, water inflow, and a fire. The broken rock in the fault zone collapsed onto the TBM and caused it to become stuck. This stoppage required hand mining the soft and broken rock for about 30 m in front of the machine to free it. The tunnel advanced only 270 m between May 1994 and June 1995 when crossing this fault zone. The rock fractures in the tunnel produced a larger quantity of groundwater than expected, with an inflow of 20 m^3/min. A fire, which resulted from igniting a conveyor belt used to remove the muck generated by the TBM, caused another delay. Still, all the miners escaped safely through an emergency shaft on Long Island. Hydrogen sulfide occurred in the rock excavated south of Deer Island and caused a minor problem during construction. In addition, argillite clogged the cutter head of TBM, causing delays throughout the tunnel, Barosh and Woodhouse (2011, 2012).

9.4 Irrigation and Water Conveyance Tunnels

Irrigation and water conveyance tunnels are the blood veins of the nations. Some irrigation and water conveyance tunnels in Turkey are given in Table 9.11, and typical detailed examples will be given thereafter, TBM performance and characteristics in selected irrigation and water conveyance tunnels in Turkey are given in Table 9.12. The emerging points from this table is that difficult geology, with extreme water ingress, faults and shear zones effected tremendously daily advance rates in Belpınar and Gerede Tunnels.

Table 9.11 Some irrigation and water conveyance tunnels in Turkey

Tunnel	Length, m	De = m	TBM	m/day	Rock formation
Irrigation tunnels					
Suruç, Ilci et al. (2003), Koçbay and Erdoğan (2015)	17,185	7.8	Double shield TBM	14.3	Marl
Belpınar, Sakallı et al. (2019)	5450	6.8	EPB-TBM	7.8	Periodotits, sepentinites
Water conveyance tunnels					
Gerede, Koçbay and Erdoğan (2015), Harding and Alpagut (2020)	31,600	5.2	3 Double shield TBM, 1 XRD	15	Volcanic units
Şanlı Urfa	26,400		Roadheader	4.2	Marl, mudstone
Melen, Koçbay and Erdoğan (2015)	3145	6.2	EPB-TBM	8.0	Limestone, sandstone, claystone

Table 9.12 TBM performance and characteristics in selected irrigation and water conveyance tunnels in Turkey

Tunnel	Suruç, İlci et al. (2013), Koçbay and Erdoğan (2015)	Belpınar, Sakallı et al. (2019)	Gerede, Harding and Alpagut (2020)
Geology	Eocene–Oligocene clayey limestone	Serpentinites and peridotites	Tuff, basalt, breccia, sandstone, shale, clay
TBM	Seli Double shield	Herrenknecht EPB	Robbins XRE TBM
Diameter, m	7.88	6.82	5.6
Total power kW	–	–	
Cutter head power, kW	4500	2785	8 × 210 = 1680
Torque kNm	4000	8307	13,896
Thrust kN	20,000	48,657	42,000
Total weight, t	1080	–	470
Best m/day	22.4	21	29.4
Best m/week	135	106	134.6
Best month	397	377	484
Mean m/day*	17.5/14.3	9.93/8.04	9.49/8.57

*Mean daily advance rates are given excluded stoppages and included stoppages separately

9.4.1 Suruç Tunnel

Suruç Tunnel, with a length of 17,185 m, is the longest irrigation tunnel excavated in Turkey. It is planned to irrigate the Eastern part of Turkey, Urfa. The expected water flow rate is 90 m^3/s and irrigate an area of 94,814 ha. A hard rock double shield TBM of 7.88 m diameter excavated the tunnel. The main rock formation is marl and clayey limestone with compressive strength changing between 10 and 44 MPa 15–70%. The excavation of the 7.88 m diameter tunnel started with double shield hard rock TBM in September 2010 and finished on 3rd March 2014, with a daily advance rate of 13.6 m/day. However, the clayey formation clogs the cutterhead and muck transport systems, decreasing the excavation efficiency. Observations made during the exaction showed that clogging of the cutterhead has an immense adverse effect on thrust force torque, depth of cut per rev, specific energy, and the excavation rate. In the majority, the thrust force to torque ratio decreased from 2.3 to 1.3. Specific energy increased from 4 to 24 kWh/m^3. In other words, clogging the cutterhead caused the energy to excavate a rock volume 6 times more than a clean head. Anti-clogging agents were used to prevent the cutterhead from clogging. Work distribution of the mechanical excavation in the Suruç tunnel is seen in Fig. 9.16. The most important point in this diagram is that the time spent on the excavation is 43% of the total shift time; this is due to both the homogeneous structure of the rock strata and the management of the experienced tunneling crew.

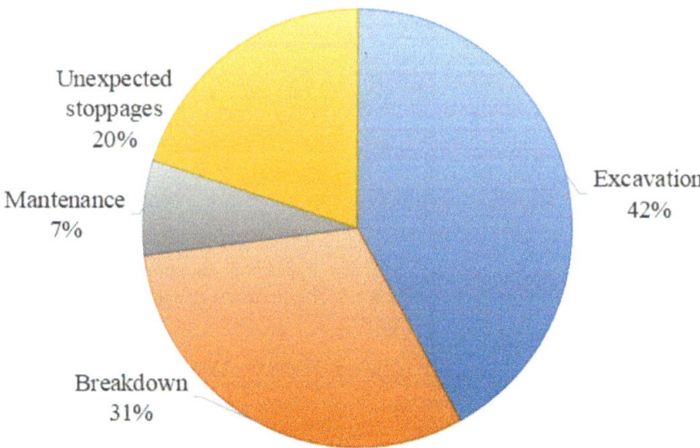

Fig. 9.16 Work distribution in double shield TBM in Suruç Tunnel, Suruç, Ilci et al. (2003)

9.4.2 Belpınar Tunnel

The Belpinar irrigation Tunnel, which is 5450 km in length, was opened with an EPB-TBM with a diameter of 6.82 m. The tunnel has a complex geology with serpentinites and peridotites, highly fractured in most cases with water ingress reaching up to 450 l/s. A mean daily advance rate of 8.04 m with stoppages and 9.93 m without stoppages was obtained with a maximum daily advance of 21 m. The excavation of the tunnel started on December 24th, 2014, by an EPB TBM and was completed on November 6th, 2016. The mean ratio of compressive strength to tensile strength ratio of the intact rock is extraordinarily low, changing between 3.8 and 4.1. The relationship between compressive strength to tensile strength ratio and optimum specific energy values in different tunnels in Turkey is seen in Fig. 9.17, Sakallı et al. (2019). Specific energy is defined as the energy spent per unit volume of the rock. This figure enabled us to compare the capability of the rocks according to the ratio of compressive strength to tensile strength ratios as defined as the brittleness of the rock. After Fig. 9.17 and the experiences gained from the past projects, as a first approach in non-fractured rock formations, the excavability may be classified as very hard to excavate if the brittleness of the rock is less than five, the rock is hard to excavate if the brittleness of the rock is between five to eight, the rock is hard to excavate if the brittleness is between eight to twelve, the rock is easy to excavate if it is higher than twelve.

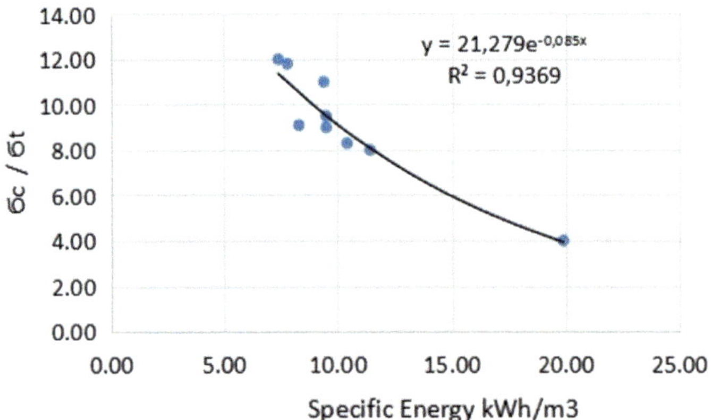

Fig. 9.17 The variation of specific energy with of brittleness (compressive strength to tensile strength ratio), Sakallı et al. (2019)

9.4.3 Gerede Water Transmission Tunnel, the 9 km Section of 31.6 km Long Tunnel

This project was executed in one of Turkey's most challenging geological areas. When two of the double shield TBMs became stuck and were unable to continue, the hybrid-type TBM solution was developed to complete the 9 km length of the final section of the tunnel. The machine is designed to fulfill the requirements of being sealable up to 20 bar hydrostatic pressure. 20 bar passes through 48 fault zones and high water pressures up to 26 bar. The crew succeeded in achieving a best monthly advance of 448 m and a mean value of 285 m per month Koçbay and Erdoğan (2015), Harding and Alpagut (2020)

9.4.4 International Projects

The Olmos Trans-Andean irrigation tunnel, Peru:

The first phase of this tunnel, 20 km long, was opened by drill and blast method and finished in 1950. This tunnel transfers water from the Huancabamba River on the Eastern side of the Andes to drought-ridden areas on the Pacific Ocean Watershed. Once the first phase of the tunnel project was operational, the scheme supplied more than 2 billion cubic meters of water annually for irrigation of 560 km^2 of farmland. In the final case, 13,900 km of the tunnel was opened with a 5.3 m diameter Robbins main beam TBM. Boring the tunnel started in late 2008, and after four years of extreme excavation through high cover and volcanic rock conditions, it was completed on December 20, 2011, with a mean production rate of 12.9 m/day. This is a remarkable success of a robust and well-designed TBM with an experienced contractor and crew. The complex geology consists of quartz porphyry, andesite, and tuff from 60 to 225 MPa UCS. Over 400 fault lines were present along the tunnel, including two major fault lines approximately 50 m wide. In excess, high in-tunnel temperatures exceeded 54 °C. The tunneling teams drove under a cover up to 2000 m, which caused tremendous rock stresses, resulting in more than 16,000 rock bursting events about 17% of which were classified as severe. McNally Support System Stabilizing Ground with Steel Slats. The system consists of steel slats anchored to the tunnel's roof by steel straps and rock bolts, effectively containing loose and unstable rock. These steel slats form an umbrella allowing the crew to work safely, https://www.robbinstbm.com/projects/olmos-trans-andean-tunnel/, uploaded July 2024.

9.5 Hydropower Tunnels

Some hydropower tunnels opened in Turkey are seen Table 9.13 with length of tunnel (l), excavation diameter (De), excavation method, daily advance rate and rock formations. TBMs used in these tunnels and the performance of the machines are tabulated in Table 9.14. What is important in Table 9.14, is that although, Ermenek Tunnel and Mavi Tunnel were driven in difficult ground, gripper TBMs and Double shield TBM achieved remarkable advance rates as 14.1 and 15.8 m/day. Remembering that jammed twice at chainages 13,800 and 8800 km spending 95 days for rescuing 95 days for salvage operations, Koçbay (2010) Koçbay and Erdoğan (2015)

9.5.1 *Uluabat Tunnel*

The project area is situated on the southern part of Uluabat–Bursa (Apolyont) Lake, The project consists of Cinarcik Dam at an elevation of 328 m which is constructed on Orhaneli River. A power tunnel of 11,465 km in length and a pipe conveyance of 1150 m delivers the water to the hydroelectric power plant. Tunnel excavation commenced in 2002 from the portal next to Uluabat Lake, using conventional excavation methods with roof bolts, shotcrete, wire mesh and steel arches as the primary

Table 9.13 Some hydropower tunnels in Turkey

Hydropower tunnels					
Tunnel	l = m	De = m	Ex. method	m/day	Rock formation
Uluabat, Bilgin and Algan (2012)	11,400	5.05	EPBTBM-	8.6	Meta-sandstone, meta-siltstone, graphytic schist
Kargı, Home (2014, 2015)	7800	9.84	Double shield TBM	9.0	Ophiolitic rocks, metapellites, graphitic schist
Kargı, Home (2015)	4000	9.84	NATM	5.8	Andsitic basalt, agglomerate
Doğançay, Bilgin et al. (2016)	6655		Double shield TBM	6.6	Limestone, sandstone, siltstone, quartzite
Ermenek, Koçbay, (2010)	8028	6.7	2 Gripper TBM	14.1	Ophiolitic melange, Limestone, sandstone,
Erik, Koçbay and Erdoğan (2015)	3856	3.9	Gripper TBM	6.8	Ophiolitic melange, Limestone, sandstone, claystone
Mavi, Koçbay and Erdoğan (2015)*	17,061	4.9	Double shield TBM	15.8	Clayey limestone, shale dolomitic limestone, sandstone

*Mavi tunnel delivers the water from Bağbaşı Dam for watering Konya Plains

Table 9.14 TBMs used in some hydropower tunnels and the performance of the machines

Tunnel	Uluabat, Bilgin and Algan (2012)	Kargı, Clark (2015)	Mavi, Koçbay and Erdoğan (2015)
Geology	Meta-sandstone, graphytic schist	Ophiolitic rocks, graphitic schist, metapellites	Dolomitic clayey limestone, shale, sandstone
TBM	EPB-TBM	Double shield TBM	Double shield TBM
Diameter, m	5.05	9.84	4.88
C. Head power, kW	2100	$13 \times 750 = 9750$	$6 \times 315 = 1890$
Torque kNm	2048 at 6.25 rpm	9864 at 3.66 rpm	2367 at 0–7.5 rpm
Max, thrust kN	29,000	20,904	25,538
Best m/day	28.8	39.0	26
Best m/week	198.4	–	245
Best month	583.2	723.2	823
Mean m/day, including stoppages (m)	8.6	9.0*	15.8

*For Kargı project, mean daily advance rate reached to 22.5 m/day, for fairly stable rock to 19 m/day, for non supporting rock to 8.5 m/day, for running /squezing rock to 1.8 m/day (Clark 2015)

tunnel support. Tunnelling operations were halted in November 2003 due to extreme roof deformations and floor heaves. However, a private investor took over the project from DSI (State Water Authority) and decided to continue the project with a 5.05 m diameter EPB-TBM. Contractor started the tunnel excavation by TBM on June 2006, and terminated on March 2010.

The Uluabat Tunnel case study and the detailed TBM performance analysis presented clearly showed how the squeezing characteristic of the ground can impact the progress rate of a EPB-TBM. 18 rescue galleries were constructed to free the trapped TBM. 192 days were spent on TBM rescue operations, Bilgin and Algan (2012). An average daily advance rate of 8.6 m/d was achieved including all stoppages such as TBM standstills and hand mining. The best daily and weekly advance rates were found to be 28.8 m and 198.4 m, respectively. The best monthly advance rate was 583.2 m in February 2007. The learning points from this case study emerge from Figs. 9.17 and 9.18. In the squeezing ground, the time spent on the installation of rings may go up to 43% which is expected to be 25% in non-squeezing ground, and machine utilization time is highly related to Q values. Machine is trapped with Q values going up to 0.20 and machine utilization time goes up steadily to a level of 25% and levels off steadily (Fig. 9.19).

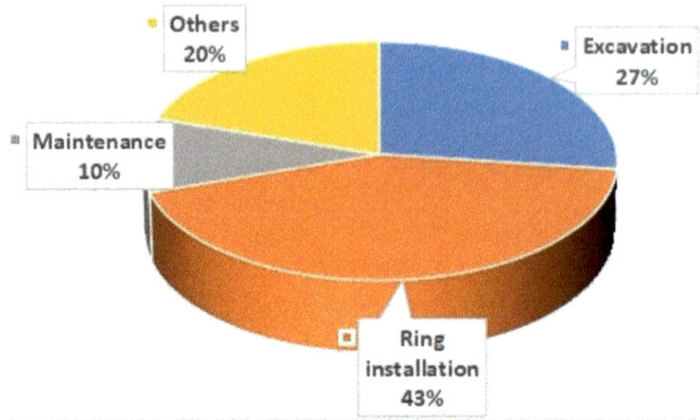

Fig. 9.18 Pie chart from work study in Uluabat Tunnel, from the archive of Bilgin, N

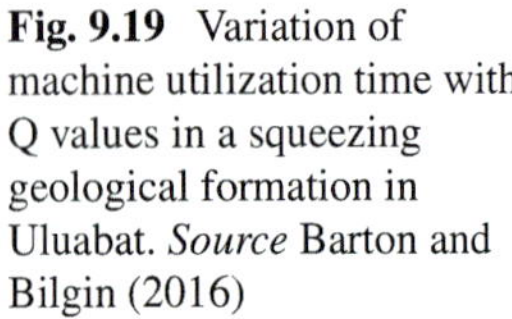

Fig. 9.19 Variation of machine utilization time with Q values in a squeezing geological formation in Uluabat. *Source* Barton and Bilgin (2016)

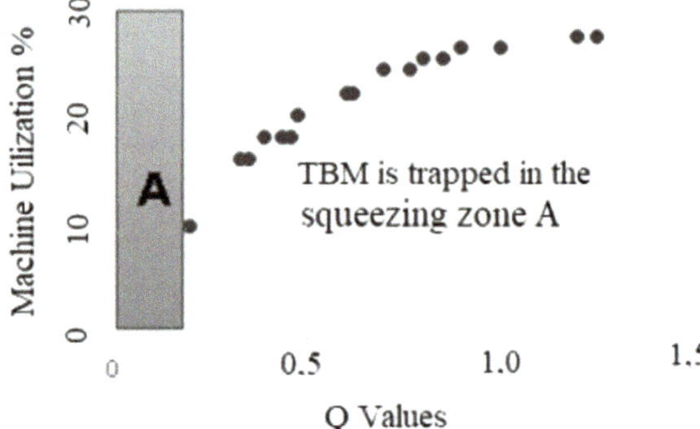

9.5.2 *The Niagara Tunnel Project*

The main beam Robbins TBM of 14.4 m diameter was selected to excavate the 10.2 km Niagara Hydropower Tunnel. TBM was assembled at the job site using Onsite First Time Assembly (OFTA) in less than 12 months. The TBM began boring in September 2006. TBM's drive was completed on May 13, 2011, with a mean daily advance rate of 6.4 m. While the tunneling part of the Project was completed, another two years were needed for concrete lining, etc. This is a typical example showing the need to differentiate the mean daily advance rate and the mean daily production rate. The mean daily advance rate of 6.4 m drops to 4.4 m per day, which will determine the tunneling budget. To prevent leakage, the finished 12.8 m diameter tunnel is fully lined with 600 mm thick cast-in-place concrete and a polyolefin waterproof membrane. The tunnel is located in Queenston shale, and it has some limestone, sandstone, and mudstone up to 200 MPa UCS. The rock along the tunnel bore path is known to have high in-situ stress, and there is potential for squeezing ground. An initial rock support lining of wire mesh, steel ribs, rock bolts, and shotcrete was installed as the TBM advanced. After about 793 m of excavation, the TBM entered the Queenston shale formation, where large rock blocks started to fall from the crown before rock support could be placed. Significant over-break up to 3 m above the cutterhead support was reported in some cases. Consequently, a ground support system to cope with the geology was designed, which consisted of 9 m long pipe spiels in an umbrella pattern at the crown of the tunnel. Using the new method, over-break was limited to about 0.9 m above the normal tunnel diameter. Nearly 500 m of very difficult ground was excavated using this method at average rates of about 3 m per day.

The tunneling crew consisting of 30 members operated the TBM and equipment 24 h a day, seven days a week. The crew has faced winter conditions reaching below − 20 °C at the jobsite, which caused the conveyor systems to freeze over with ice. Antifreeze was sprayed on the affected conveyors and the ice was chipped off in order to keep them running. The typical day is divided into two production shifts and a maintenance shift. The average pay for each worker is $150,000–$200,000 per annum.

https://www.robbinstbm.com/projects/niagara-tunnel-project/#Geology

https://niagarafrontier.com/tunneltechnical.html

The readers are advised to consult the book written by Brox (2021) for case studies and lessons learned on hydropower tunnels.

9.6 Metro Tunnels

The oldest section of the metro in Istanbul is the M1 line, which opened in 1989. As of March 2024, the system now includes 158 stations in service, with 36 more under construction. With 243.3 km, Istanbul has the 22nd longest metro line in the world and the 5th in Europe. Ankara and Izmir follow Istanbul with several km of Metro Lines. The summary of TBM design parameters and tunneling performance in some metro tunnels are given in Table 9.15.

One should be careful using performance values of TBMs. In EPB-TBM opening ratio and in any TBM the number of usages play important role in the performance values. This is clearly seen in Table 9.16 showing the affect of TBM design and the number of usages in Mahmutbey-Mecidiyeköy Metro Project Bilgin et al. (2017). As seen from this table the performance values of a new TBM with higher opening ration has better performance values compared to other EPB-TBMs.

9.6.1 Otogar Kirazlı Metro Tunnels

This metro line is an example of Turkey's most problematic metro tunnels. Otogar and Kirazlı Metro line is 5.77 km long. The excavation of this section began in May 2006 and was completed in June 2008. During tunnel excavations, 28 buildings were

Table 9.15 The summary of TBM design parameters and tunneling performance in some metro tunnels

Project	Kozyatağı-Kadiköy	Otogar-Kirazlı	Üsküdar-Ümraniye-Sancaktepe[a]	Marmaray Aynlık Cesme-Usküdar
Geology, Bilgin et al (2016)	Kartal Formation[b]	Güngören[c] Formation	Gözdağ[d] Formation	Trakya[e] Formation
TBM	2 Herrenknecht Open Mode + EPB	Herrenknecht + Lovat EPB	Herrenknecht EPB	2 Hitachi-Zozem Slurry
Diameter, m	6.6	6.5	6.6	7.85
Total power, kW	2000	963; 1622	2100	
Cutterhead power, kW	1260	630; 900	1260	2000-
Torque kNm	5200 at 1.6 rpm 1515 at 5.5 rpm	4350; 4450 2.5 rpm;1.95 rpm	2912; 5200 at 5.5 rpm and 3.1 rpm	4450 at 1.9 rpm
Max, Thrust kN	42,575	35,000/54000	20,000	75,000
Best m/day	30.8	24	25.5	15
Best m/week	152.6	119	137.5	57
Best m/month	438.6	419	–	230
Mean m/day[f]	10.4; 8.2	11.3; 5.5	11.0; 10.5	4.6; 3.5

[a]The values are given for the length between Çarşı and Ümraniye; [b]Kartal Formation: fossilized shale and limestone, interbedded sitstone and sandstone; [c]Güngören Formation: Quartenary, Plyocene and Miocene aged, sand, sandy clay, silty clay sometimes stiff clay; Gözdağ Formation; siltstone-claystone, sandstone sometimes dykes; [e]Trakya Formation.: Carboniferous interbedded, siltstone, mudstone and sandstone; [f]First value is without stoppages, the second value is with stoppages

Table 9.16 The affect of TBM design and the number of usages in Mahmutbey-Mecidiyeköy metro project Bilgin et al. (2017)

Geology; laminated frequently fractured interbedded siltstone, mudstone and sandstone with [a]RQD 0–80; [b]UCS = 45–120 MPa			
TBM	[c]EPB Herrenknecht	[d]EPB Lovat	[e]EPB Terrateck
Diameter, m	6.55	6.57	6.55
Opening ratio %	20	29	37
Rotational speed rpm	0–3.8	0–4.0	0–3.0
Torque kNm	4400	4500	5440
Max, Thrust kN	32,000	55,000	40,000
Best m/day	32.2	28.0	30.8
Best m/week	147.0	112.0	152.6
Best m/month	456.4	288.4	438.6
[f]Mean m/day	6.8; 5.8	5.3, 3.8	10.4; 8.2

[a]Rock Quality Designation, [b]Uniaxial compressive strength, [c]Second usage; [d]Third usage; [e]First usage; [f]first value is without stoppages; second value is with stoppages

demolished because of the severe damage. At these buildings, a total of 214 flats existed. In addition, rent and removal expenditures were also paid off in total for the 364 flats, along with the flats that were demolished and evacuated for precaution during the tunnel excavation. Demolished buildings, repaired buildings, rent payments, and other expenses totaled 35.6 million dollars. The total project cost was 225 million dollars. Consequently, the project cost increased by 15.8% because of difficult ground conditions Ocak and Bilgin (2009), Ocak (2009).

Waiting time due to excessive ground deformations in this project is around 28%, the mean daily advance (including waiting due to excessive deformations) is 5.5 m/day, and the mean daily advance (excluded waiting) is 11.2 m /day. The geological cross-section of the tunnel is given in Fig. 9.20, and the geotechnical properties of the formations are given in Table 9.17

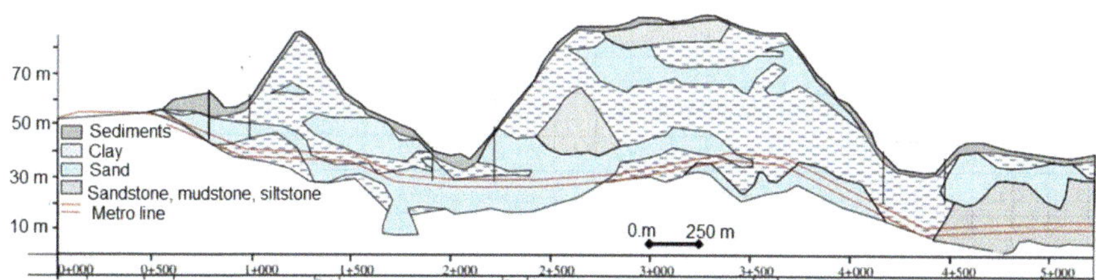

Fig. 9.20 The geological cross-section of the Otogar Kirazlı metro tunnel

Table 9.17 Geotechnical properties of the formations in Otogar-Kirazlı metro tunnels

Geological formation	Unit weight (kN/m^3)	Modulus of elasticity (kN/m^2)	Cohesion (kN/m^2)	Poisson ratio	Angle of friction degrees
Fill	18.0	5000	1	0.30	10
Sand	18.3	25,000	1	0.25	35
Very dense sand	18.5	30,000	1	0.30	30
Clay (Güngören fr)	16.5	20,000	20	0.35	14
Hard clay (Güngören fr.)	17.2	28,000	25	0.40	20

9.6.2 Benchmarking of EPB-TBM Performance in International Metro Projects

For benchmarking EPB-TBM performance in international Metro projects, a survey by Roby and Willis (2014) based on the average weekly advance rate will be used to measure performance. The survey is based on the data collected from 46 EPB projects worldwide. The 46 TBMs revived were located in 11 different countries and worked on 22 projects. 85% (39) of the TBMs were working on Metro tunnels, 2% (1) on a high-speed rail tunnel, 11% (5) on sewer tunnels, and 2% (1) on a gas pipeline tunnel. The geology on which the machines operated varies wildly from sedimentary rock and weathered rock through glacial till, gravel, sands, soils, and clays. Ground pressure averaged approximately 3.8 bar. Three different manufacturers manufactured the TBMs. The maximum diameter of machines reviewed is 6.95 m, and the minimum is 5.90 m. With an average diameter of 6.34, these machines can be considered "typical" metro TBMs. With 100% of the projects reporting their average weekly production based on 2 × 12-h shifts, seven days per week. Following is the summary of the average TBM weekly advance: maximum: 184.8 m/week average, minimum: 33.7 m/week average, average: 85.5 m/week average, which makes 12.2 m/day.

9.7 Concluding Remarks

This chapter uses benchmarking as a reference point to know how different tunneling projects have succeeded. For this purpose, data from national and international roadway, railway, utility, stormwater, irrigation, water conveyance, hydropower, and metro tunnel projects with information on geology, excavation method, mean daily advance rates, tunneling crew, and machine utilization time are analyzed and compared with each other. For a comparative study, the most important parameter to consider is geology. The classification made by Clark in highly changeable geology in the Kargı project for TBM excavation and conventional tunneling method, as good stable rock, fairly stable rock, and non-self-supporting rock is taken as a basis, and later this classification is modified by the author using experience gained in Uluabat Tunnel by Barton and Bilgin (2016). This modified classification is given in Table 9.18.

The Ovit tunnel is given in this chapter as a typical example of a well-managed and completed project in Turkey carried out by the conventional tunneling method. Thus, a list of tunnel crew and the equipment used in the project is given within the chapter as comparative issues. Total number of crew is 101. Within the total pieces of equipment, 37. It is important to note that the percentage of the works done in Ovit Tunnel and the figures predicted by Zare and Bruland (2006) using the NTNU advance rate model are in the same order (Table 9.19).

Mean advance rate in good rock conditions changes from 3.0 to 4.5 m/d for cross-section between 61 and 80 m^2. Mean advance rate in difficult ground conditions is approximately 1.0 m/d for cross-section between 85 and 110 m^2. Maximum advance rate recorded for the single-track tube is from Sedrun South East with 11.5 m/d.

A summary of the case studies done for benchmarking in mechanized tunneling for roadway, railway, utility, irrigation, water conveyance, hydropower, and metro tunnels are separately given in Table 9.20. The points emerging from this table are given below.

Eurasia tunnel of a 13.7 m diameter is an undersea tunnel opened by a slurry TBM, face pressure going up to 11 bar, and geological difficulties were the main reason for the low advance rate and machine utilization time. Eyüp Utility tunnel is an excellent example of the application of a shielded roadheader in high-strength

Table 9.18 Modified rock classification of given by Clark (2015) for Kargı tunnel and for Uluabat tunnel given by Barton and Bilgin (2016), Turkey

Rock mass condition	RMR	Q	Rock class	TBM. mean daily advance m/day	Drill & Blast, mean daily advance m/day
Good stable rock	71–100	> 40	A	22.5	8
Fairly stable rock	41–70	4.-40	B	19	5.5
Non-self-supporting formation	21–40	0.2–4.0	C	8.5	1.5
Running/squeezing formation	< 20	< 0.2	D	1.8	N/A

Table 9.19 Benchmarking for conventional tunneling

Tunnel	Excavation area m^2	Working face	Rock class	Mean m/day	Workers per shift for one face
Zigana	99–127	4	B-C	6.4	52
Ovit	99–127	4	B-C	7.4	51
Ilgaz Çankırı	99–127	2	C	4.0	48
Nefise Akçelik	99–127	2	C-D	1.8	60

Table 9.20 Benchmarking for mechanized tunneling

Tunnel	D, m	Rock class	TBM	M.U %	Mean m/day	Workers per shift
Roadway, Undersea Eurasia	13.7	B-D	Slurry	31	7.0	40, excluding slurry plant
Railway, Eşme	13.77	A-B	EPB, XRE	38	14.5	36
Railway, Nurdağı	8.0	A-B	Single shield	45	9.5	32
Utility, Eyüp	3.2	A-B	Shielded roadheader	54	4.7	17
Utility, Ambarlı	4.6	B-C	EPB-TBM	43	13.7	24
Utility, Sarıyer-Tarabya	2.9	B-C	Single shield	35	9.4	25
Irrigation, Suruç	7.8	B, C	Double shield	42	14.3	28
Irrigation, Belpınar	6.8	C-D	EPB	35	7.8	30
Water conveyance undersea, Melen	6.2	A-B	EPB	36	8.0[a]	27
Hyropower Ermenek	6.7	B	Gripper	37	14.1	33
Hydropower Uluabat	5.05	D	EPB	27	8.6[b]	27
Metro, Mahmutbey Mecidiyeköy	6.6	B	EPB	45	8.17[c]	30
Metro, Otogar Kirazlı	6.6	D	EPB	22	5.5/11.2[d]	22

D is TBM diameter, M.U is machine utilization time is defined as % of machine working time per shift time, a) the value is for a new TBM b) in this tunnel, the TBM was trapped 18 times c) this value is for a new TBM d) in the first value, waiting time due to the excessive deformations are is included, in the second number waiting time is excluded

rocks. Melen is a tunnel undersea, with continuous probe drilling, in a few cases core drilling was applied, which decreased machine utilization time and daily advance rates. In Uluabat Tunnel TBM jammed 18 times and rescue galleries were opened decreasing the machine performance. Mahmutbey Mecidiyeköy Metro Tunnel is a typical example showing the effect of opening ratio and the impact of using age of the EPB-TBM, this is explained within the chapter. In Otogar Kirazlı Metro tunnel the performance of EPB-TBM was low due to excessive ground deformations, note that when time spent to cope with deformation is excluded, mean daily advance rate becomes 11.2 m/day. The best values obtained for railway, utility, irrigation, and hydropower tunnels are 14.5 m, 13.7 m, 14.3, and 14.1 m, respectively.

The performance values obtained in different international projects are as follows.

Tunnel Emisor Oriente, Mexico City, is the longest wastewater tunnel in the world, with a length of 62.5 km and an inner diameter of 7 m. It was constructed between 2008 and 2019 using six EPB-TBMs of 8.89 m diameter. The project, which has a total length of 62.5 km, was completed with a mean daily advance rate of 15.8 m/day, including 24 shafts up to 150 m deep, Gonzallez (2020)

In Boston, USA, an Inter-Island wastewater tunnel with a length of 8000 km and diameter of 4.3 m was opened with an EPB-TBM. In a very complex geology, a mean daily advance of 7.4 m was obtained.

The Olmos Trans-Andean irrigation tunnel in Peru, which has a length of 13,900 km, was opened with a 5.3 m diameter main beam TBM. Boring the tunnel started in late 2008, and after four years of extreme excavation through high cover and volcanic rock conditions, it was completed on December 20, 2011, with a mean production rate of 12.9 m/day.

Niagara Hydropower Tunnel, 10.2 km long, was opened with a main beam TBM of 14.4 m diameter. The TBM began boring in September 2006. The TBM's drive was completed on May 13, 2011, with a mean daily advance rate of 6.4 m. While the tunneling part of the Project was completed, another two years were needed for concrete lining, etc.

A survey by Roby and Willis (2014) based on the average weekly advance rate is used to benchmark the EPB-TBM performance in international projects. This survey obtained an average advance rate of 12.2 m/day. It is interesting to note that this number is higher than the values of metro projects in Turkey. This may be explained by the fact that the geology in Turkey is highly complex compared to that in other countries. The study on benchmarking is believed to be a good basis for planning TBM projects.

References

Arioglu B, Gokce HB, Malcioglu F, Arioglu E (2016) TBM tunnel under the bosphorus for the Istanbul strait road crossing project. Geomech Tunnell 9(4):303–309

Arioglu B, Gökce H B, Ergin A (2016b) A unique project under variable geological and working conditions: Eurosia Tunnel. In: 2nd International conference on tunnel boring

Barosh PJ Woodhouse, D (2011/2012) Boston area water supply and wastewater tunnels, a surway. Civil Engineering Practice, pp 349–409

Barton N, Bilgin N, (2016) Fast or slow progress with TBM in ideal or faulted condition, Eurock 2016 Cappadocia, Turkey

Bilgin, N (1988) Roadheader performance in Istanbul, Golden Horn clean-up contribute valuable data.Tunnel and Tunnelling, June, p 41–44

Bilgin N (2012) A study on the possibility of using a TBM in Ovit Tunnel, report written for the Makyol, Istanbul Technical University, p 26

Bilgin N (2016) An appraisal of TBM performance in Turkey in difficult ground conditions and some recommendations. Tunn Undergr Space Technol 57:265–267

Bilgin N, Algan M (2012) The performance of a TBM in a squeezing ground at Uluabat, Turkey. Tunnell Undergr Space Technol 32(2012):58–65

Bilgin N, Balci N (2016) Tunnels and tunneling in Turkey. In: Feng X-T (ed) Rock mechanics and engineering, Volume 5: surface and underground projects. CRC Press, Francisand Taylor aBalkema book
Bilgin A, Yüksel A (2023) The effect of EPB face pressure on TBM performance parameters in different geological formations of Istanbul. Tunn Undergr Space Technol 138(2023):105184
Bilgin N, Feridunoglu C, Tumac D, CInar M, Palakci P, Gunduz O, Ozyol L (2005) The performance of a full face tunnel boring machine (TBM) in Tarabya (Istanbul). In: Erdem and Solak (eds) Underground space use: analysis of the past and lessons for the future. Taylor & Francis Group, London, ISBN 04 1537 452 9
Bilgin, N, Tumac, D, Feridunoglu, C, Karakas, AR, Akgul M (2005) The performance of a roadheader in high strength rock formations in Küçüksu tunnel. In: Erdem and Solak (eds) Underground space use: analysis of the past and lessons for the future. Taylor & Francis Group, London, ISBN 04 1537 452 9
Bilgin N, Copur H, Balci C (2016) TBM excavation in difficult ground conditions, case studies from Turkey, Earns and sohn, Germany, p 346
Bilgin N, Acun S, Ates, U, Murtaza M, Çelik Y (2017) The factors affecting the performance of three different TBMs in a complex geology in Istanbul. In: ITA-IATES world tunnel congress, WTC 2017, Bergen, Norway
Brox D (2021) Practical guide to rock tunneling, Vancover, Canada, p 368
Clark J (2015) Extreme excavation in fault zones and squeezing ground at the Kargi HEPP in Turkey SEE tunnel: promoting tunneling in SEE region. In: ITAWTC 2015 congress and 41st general assembly May 22–28, 2015, Dubrovnik, Croatia
Falanesca M, Merlini D Schürch R. Marclay R (2023) Benchmarking tunnelling production rates: challenging case histories of mechanized and conventional tunnelling in different geological conditions. In: Anagnostou, Benardos, Marinos (eds) Expanding underground. Knowledge and passion to make a positive impact on the world. ISBN 978-1-003-34803-0. Open Access: www.taylorfrancis.com
Gonzallez R (2020) Completing Mexico city's mixed ground mega tunnel: Emisor Oriente. In: North American tunneling conference, p 16
Gonzalez R, Olivares A , Telle MA (2014) Learned from the construction of the 62 km tunnel Emsor Oriente in Mexico's challenging and varied ground. In: Proceedings of the world tunnel congress 2014—tunnels for a better life. Foz do Iguaçu, Brazil
Guclucan Z, Meric S, Algan M, Palakci Y, Bilgin N., Bilgin AR, Balci C, Tumac D (2008) The use of a TBM in difficult ground conditions in Beykoz—Kavacik sewerage tunnel. In: Proceedings of the world tunnelling congress, underground facilities for better environment and safety, Agra, India
Guclucan Z, Meric S, Palakci, Y, Bilgin, N, Balci, C, Copur, H, Namli, M, Bilgin AR, Kandemir E (2009) The use of theoretical rock cutting concepts in explaining the cutting performance of a TBM using different cutter types in different rock formations and some recommendations. In: Proceedings of World Tunnelling Congress, Safe Tunnelling for the City and for the Environment, Budapest, Hungary
Guner P, Eryigit, HA, Sayin A, Kocak B (2014) An overview of the construction of Turkey's longest road tunnel focusing on ground support using macro synthetic fibres as shotcrete reinforcement. In: Negro, Cecílio, Bilfinger (eds) Proceedings of the world tunnel congress 2014, Iguassu Falls—Brazil. São Paulo: CBT/ABMS, 225 p
Harding D, Alpagut Y (2020) Tunneling through 48 fault zones and high water pressures on Turkey's Gerede water transmission tunnel, ITA-AıtesWorld tunnel congress12–15 May Kuala Lumpur.
Home L (2014) The Kargi challenge, a study of TBM boring vs. conventional excavation. TunnelExpo Turkey organized by Turkish Tunneling Society
Home L (2015) Hard rock TBM tunneling in challenging ground: developments and lessons learned from the field, symposium, TBM DiGs in difficult conditions, Singapore
https://en.wikipedia.org/wiki/Emisor_Oriente_Tunnel

https://www.un.org/development/desa/en/news/population/2018-revision-of-world-urbanization-prospects.html
Ilci N, Temel M, Sezgin S, Akpınar T, Guarasio S, Polat C, Bilgin N (2003) Clogging and squeezing effect of marl-clayey limestone on the performance of a hard rock TBM in Suruc Tunnel,Turkey. In: Anagnostou G, Ehrbar H (eds) World Tunnel Congress 2013 Geneva underground—the way to the future. Taylor & Francis Group, London. ISBN 978-1-138-00094-0
IPA (2018) Case study: benchmarking tunnelling costs and production rates in the UK. Infrastructure and Projects Authority, p 14
Jordan D, Bilgin N (2018) Turkey's hardest rock: the Bahçe-Nurdağ. High Speed Railway Tunnel 4:22–28
Jordan D, Alpagut Y, Kiliç Ş (2023) Record-setting large diameter mixed ground tunneling in Turkey. In: Anagnostou, Benardos, Marinos (eds) WTC 2023, Expanding underground. knowledge and passion to make a positive ımpact on the world
Koçbay A (2010) TBM performance analysis and a typical example, Ermenek (Karaman) energy tunnel, DSI (State Hydrolic Works) Technical bulletin, Number 108, July
Koçbay A, Erdoğan D (2015) TBMs and some xamples for prformance analysis. In: 3th National symposium on ırrigation systems. Ministry of Forest and Irrigation, Ankara, pp 11–25
Luo Y, Alaghbandrad A, Genger TK, Hammad A (2020) History and recent development of multi-purpose utility tunnels. Tunn Undergr Space Technol 103:32
Ocak I (2009) Environmental effects of tunnel excavation in soft and shallow ground with EPBM: the case of Istanbul. Environ Earth Sci 59:347–352. https://doi.org/10.1007/s12665-009-0032-6
Ocak I, Bilgin N (2009) The performance of two EPB mac.hines in Istanbul in metro tunnel drivagesin soft and shallow ground. In: ITA-AITES WTC 2009, Budapest, Hungary
Ozcelik M (2016) Criteria for the selection of construction method at the Ovit Mountain Tunnel (Turkey), KSCE (Korean Society for Civil Engineers). J Civ Eng 20(4):1323–1328
Özaydın YT (2024) Stormwater tunnels in Istanbul. In: International tunneling conference organized by Turkish tunneling society, 23–24 November, Istanbul
Özaydın YT, Avunduk E, Çopur H (2013) EPB-TBM performance in excavation of Büyükçekmece wastewater tunnel. In: 3rd International symposium on on underground for transportation, symposium organized by Turkish Tunneling Society and Turkish of Mines, 29–30 October, Istanbul
Roby J, Willis D (2014) Raising EPB performance in metro-sized machines. In: Proceedings of the world tunnel congress 2014—tunnels for a better Life. Foz do Iguaçu, Brazil
Sakalli M, Talu D, Bilgin N, Aksoy IH (2019) Problems associated with an EPB-TBM in a complex geology with serpentinites and peridotites in Turkey. In: Viggiani P, Celestino (eds) WTC 2019. Taylor & Francis Group, London, ISBN 978-1-138-38865-9
Tüysüz O, Genç C (2012) The geological study of the outline of the Ovit Tunnel, Istanbul Technical University
Zare S, Bruland A (2006) Estimation model for advance rate in drill and blast tunneling. In: International symposium on utilization of underground space in urban areas, 6–7 November Sharm El-Sheikh, Egypt

Chapter 10
Some Examples of Changing Tunnelling Method from Conventional to Mechanized Tunneling or Vice Versa

Abstract In this chapter, two case studies from Turkey, the Uluabat hydropower tunnel and T 26 high-speed railway tunnel; one from Taiwan, the Hsuehshan road tunnel, and one from the Venezuela Yacambu Quibor water conveyance tunnel were given to clarify when and why the tunneling method is changed during the progress of the job operations. Uluabat hydropower project started with drill and blast and changed to TBM tunneling due to the metadetritics and graphitic schist leading to high deformations in the tunnel. 11,465 m long tunnel finished in 8 years. T26 high-speed tunnel encountered rock formations with the same geotechnical characteristics as the Uluabat tunnel. The tunnel started with drill and blast and later switched to TBM tunneling. After the collapse of the tunnel, TBM was buried, and tunneling was carried out with drill and blast. This tunnel of 6100 m in length was completed in 13 years. The excavation of the Hsuehshan tunnel, 12.9 km in length with two-lane tunnels and a pilot tunnel, was started with 3 TBMs. After a complex geology with several water-bearing faults, the tunnel was carried out with drill and blast. The project began in 1974, but it took 34 years to complete for technical, political, and financial reasons. Over those 34 years, the project was stopped and started again eight times using TBMs, roadheaders, and conventional tunneling methods. Given the selected case studies, the need for detailed geological studies and proper selection of the Tunneling method are emphasized.

10.1 Introduction

A detailed geological investigation and its correct interpretation are necessary to select a tunneling method properly. Otherwise, tunneling operations may become a nightmare for the job owner and the contractor. Typical examples are the Uluabat and T26 tunnels in Turkey, the Hsuehshan tunnel in Taiwan, and the Yacambu Quibor tunnel in Venezuela. Uluabat hydropower project started with drill and blast and changed to TBM tunneling due to the metadetritics and graphitic schist leading to high deformations in the tunnel. Deformations up to 1 m were encountered in the tunnel, with the closure of the tunnel section area by up to 10%, Bilgin and Algan

N. Bilgin and C. Balci, *Critical Issues in Selecting Conventional and Mechanized Tunnelling Methods*, Springer Tracts in Civil Engineering,
https://doi.org/10.1007/978-3-031-89114-4_10

(2012), Hasanpour et al. (2017). T26 high-speed tunnel encountered rock formations with the same geotechnical characteristics as the Uluabat tunnel. The tunnel started with drill and blast and was later moved to TBM tunneling. After the collapse of the tunnel, TBM was buried, and tunneling was carried out with drill and blast; 6100 m was completed in 13 years, Bilgin (2016), Fugro Sial (2016), İncecik and Poşluk (2018), Aydan and Hasanpour (2019), Gökçeoğlu et al. (2022).

The excavation of the Hsuehshan tunnel, which is 12.9 km in size, has two-lane tunnels, and a pilot tunnel was started with 3 TBMs. One TBM was buried under heavy water inrush and excessive mud coming through a significant fault, Lin and Yu (2005), Lee and Leng (2005). Tsai et al. (2015). After encountered a complex geology with several water-bearing faults, the tunnel was carried out with drill and blast. The Yacambu Quibor tunnel project began in 1974, but it took 34 years to complete for technical, political, and financial reasons. Over those 34 years, the project was stopped and started up again eight times using TBMs, roadheaders, and conventional tunneling methods, Hoek (2001), Ramonet (2008), Hoek and Guevera (2009), Jimenez and Senent (2012). The need for detailed geological studies and proper selection of the Tunneling method are emphasized in this chapter, given the selected case studies.

10.2 Uluabat Hydropower Tunnel, Turkey

The project area is situated in the southern part of Turkey's Uluabat–Bursa (Apolyont) Lake. The project consists of the Çınarcık Dam at an elevation of 328 m constructed on Orhaneli River. A power tunnel of 11,465 km in length and a pipe conveyance of 1150 m was planned to deliver to the hydroelectric power plant, Bilgin and Algan (2012).

10.2.1 The Geology and Some Properties of the Rock Formations

The geologic map of the area is given in Fig. 10.1, and a general view of the Karakaya formation is seen in Fig. 10.2. The tunnel route within chainages 11 + 465 m/7 + 750 m and 6 + 000 m/1 + 792 m consisted of Karakaya formation of Triassic aged metadetritic rocks like fine-grained meta-claystone, meta-siltstone, meta-sandstone, and graphitic schists. It is important to note that the primary properties of sandstone, claystone, and siltstone changed entirely due to tectonic deformations, causing fragmented metadetritic formations. The physical and mechanical properties of metadetritic rocks are given in Table 10.1, Bilgin and Algan (2012)

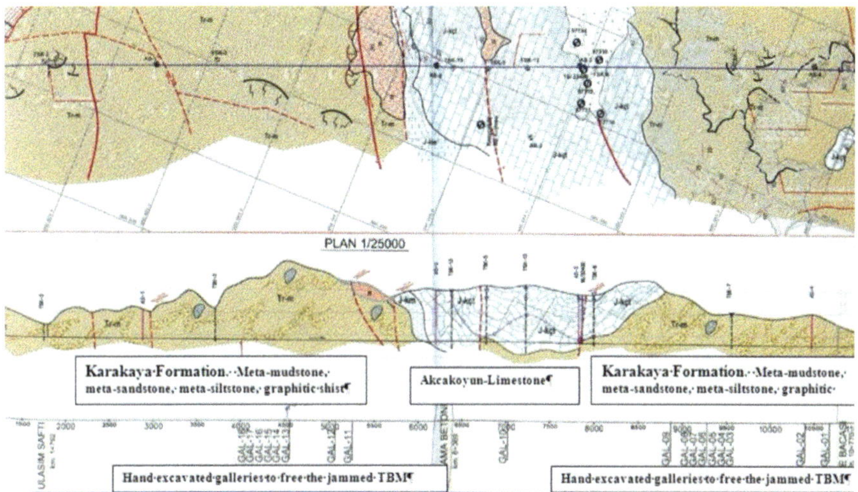

Fig. 10.1 The geological profile of the Uluabat tunnel, Bilgin and Algan (2012)

Fig. 10.2 Typical view of Karakaya formation through TBM opening, Bilgin and Algan (2012)

Table 10.1 Some mechanical and physical properties if the rocks in Karakaya and Akçakoyun formation, Bilgin and Algan (2012)

Formation	UCS (MPa)	BTS (MPa)	E (GPa)	ɣ (g/cm)^3	RMR
Karakoyun	1–96	2.5–4	1.53–39.5	2.4–2.7	III-V
Akçakoyun	24–125	3.5–5.4	9.1–126.6	2.4–2.8	III

UCS (Compressive strength, BTS (Brazilian tensile strength), E (Elastic modulus), ɣ (Bulk density), RMR (Rock mass rating)

10.2.2 Starting the Project with Conventional Tunneling Method

Tunnel excavation commenced in 2002 from the portal next to Uluabat Lake, using conventional excavation methods with roof bolts, shotcrete, wire mesh, and steel arches as the primary tunnel support. Tunneling operations were stopped in November 2003 due to extreme roof deformations and floor heaves. A typical deformation in the steel arches and deformation curve on the left wall of the tunnel is seen in Figs. 10.3 and 10.4. The deformation increases gradually after a specific time. It accelerates and levels off after 1 year, showing typical creep behavior. The figure shows the final deformations and closures of the cross section in the tunnel after a specific time for different stations. 5. As seen from this figure, deformations up to 1 m were encountered in the tunnel with the closure of the tunnel section area by up to 10% (Fig. 10.5).

It was decided that conventional tunneling was not feasible for the project due to the slow advance rates, resulting in the stoppage of tunneling operations for about two and half years. However, a private investor took over the project from DSI (State Water Authority) and decided to continue the project with a 5.05 m diameter EPB-TBM. The contractor started the tunnel excavation by TBM from chainage 11 + 465 m in June 2006 and terminated on March 2010 in chainage 1 + 792 m.

10.2.3 Switching the Project to Mechanized Tunneling

A single-shielded Herrenknecht EPB-TBM was selected for the project with specifications summarized in Table 10.2.

An average daily advance rate of 8.6 m/d was achieved, including all stoppages, such as TBM standstills and hand mining. The best daily and weekly advance rates were 28.8 m and 198.4 m, respectively. The best monthly advance rate was 583.2 m in February 2007. The TBM was stuck 18 times at different tunnel locations during tunnel excavation, as shown in Fig. 10.1. A general view of the squeezing ground around the TBM shield is given in Fig. 10.6. Rescue galleries were driven next to the TBM to free the shield, and 192 days were spent on these operations.

Fig. 10.3 Deformations in the steel arches on the left wall of the tunnel, Bilgin and Algan (2012)

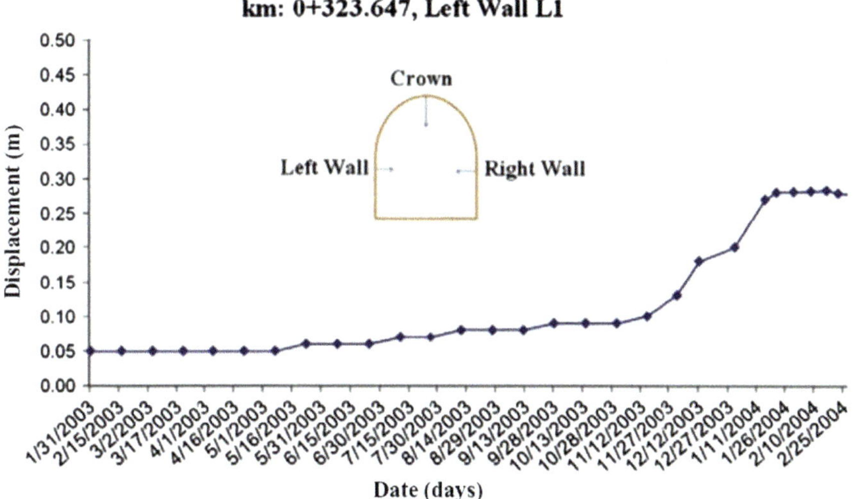

Fig. 10.4 Deformation against time on the left wall of the tunnel at km. 0 + 323, Bilgin and Algan (2012)

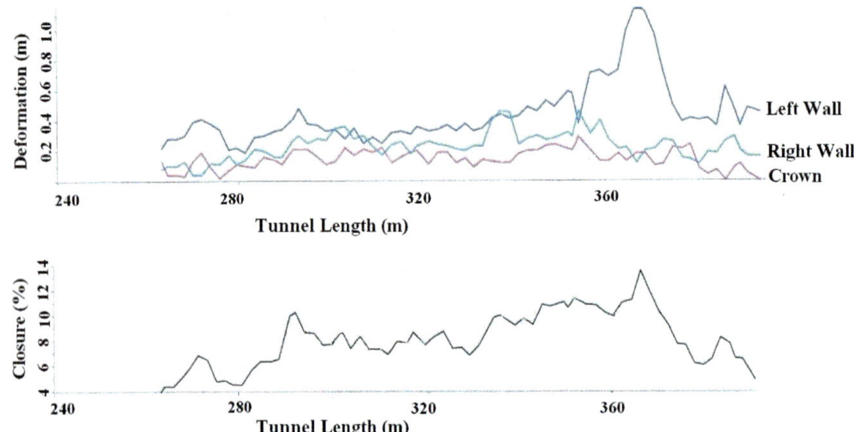

Fig. 10.5 Final deformation after leveling off and closure in the tunnel after a specific time, Bilgin and Algan (2012)

Table 10.2 Characteristics of Herrenknecht EPB-TBM, Bilgin and Algan (2012)

TBM parameter	Value
TBM diameter	5.05 m
Total power	2100 kW
Maximum thrust force	29,000 kN
Nominal torque	2048 kNm at 6.25 rpm
Number of discs	34
Disc diameter	433 mm
Number of scrapers	48
Shield diameter	4990 mm
Shield weight	69 t
Shield length	12 m
TBM weight	335 t
TBM length	108 m
Length of the band conveyor	65 m
Width of band the conveyor	650 mm
Power of the band conveyor	37 kW
Speed of the band conveyor	0–2.5 m/s
Capacity of segment erector	1800 kg

Fig. 10.6 General view of the squeezing ground around the TBM shield, Bilgin and Algan (2012)

10.3 T26 High-Speed Tunnel, Turkey

The T26 tunnel is part of the Ankara-Istanbul high-speed railway project. It is located in the Northeast of Turkey, 10 km Southwest of the town of Bilecik. The overburden changes between 30 and 236 m. The tunnel passes through a steep mountain range, where the old railway line and the existing highway are located right next. This project enables the trains to run up to 250 km/h, reducing the journey from 7 to 3 h.

10.3.1 The Geology and Some Properties of the Rock Formations

Karakaya formation is the primary formation in the tunneling area with Pazarcik melange. Karakaya Formation contains fault zones in several places, which is similar to the Karakaya Formation found in the Uluabat Energy Tunnel. The tunnel is planned to pass through the predominated graphitic schist, which is characterized as dark gray to black colored and frequently includes chlorite schist with the size of some tens of meters and rarely marble sheer bodies with dimensions of some few tens of meters. Graphitic schist is moderately weathered to fresh, weak to weak, and highly sheared

Table 10.3 Some physical and mechanical properties of graphitic schist and chlorite schist, from the archive of the author Bilgin

Rock	Compressive strength MPa	GSI	Density kN/m^3	Cohesion kPa	Int. friction angle ()°
Graphitic schist	2–7	20	22–23	100	15°
Chlorite schist	7	30	24	100	28

along the foliation surface. Therefore, the schistose surface is polished and coated with clay (Table 10.3)@@.

10.3.2 *Starting the Project with Conventional Tunneling Method*

Construction of Tunnel T26 started in January 2009 at the north portal. The planned tunnel length was 4950 m, and the chosen construction method was according to the principles of the NATM. The tunnel cross-section was divided into three sections: top heading, bench, and invert. The top heading has a height of 5.9 m, the bench 2.75 m, and the invert 3.2 m. The horseshoe-shaped cross-section has a total height of 11.85 m and a total width of 13.75 m. The excavated section area is 136 m^2. The works were already interrupted at the start by a landslide, which partially buried the portal area, as seen in Fig. 10.7. After a detailed investigation, the tunnel route moved 174 m to the North, which increased the tunnel length from 4950 to 6100 m.

In September 2009, the conventional tunneling started in the new alignment. During the tunneling operations, heavy displacements in the crown were observed. The deformation increased steadily. At chainage, + 0.298 km under an overburden of 30 m, displacements of up to 80 cm were measured in the crown. Due to geological problems, shear zones, fault zones, and low RQD values, the daily advance rates were very slow. Hence, a decision was made not to continue to excavate the tunnel conventionally. Hence, the job owner and the contractor decided to continue with a tunnel boring machine (TBM) with a diameter of 13.77 m.

10.3.3 *Switching the Project to Mechanized Tunneling and Collapse of the Tunnel*

At the end of 2011, the tunnel excavation method was changed to mechanical excavation using a 13.7 m diameter single shield TBM. TBM entering the portal is seen in Fig. 10.8.

The characteristics of the TBM are given in Table 10.4.

Fig. 10.7 The portal affected by the landslide, Gökçeoğlu et al. (2022)

Fig. 10.8 TBM entering the portal, from the archive of the author Bilgin

Table 10.4 Characteristics of single shield TBM used in the project

Machine diameter	13770 mm
Installed power	9700 kVa
TBM length (inc. backup)	80 m
Shield length	10.45 m
Weight (inc. backup)	2170 t
Number of thrust cylinders	2 × 15
Stroke	2800 mm
Installed thrust force	84,464 kN at 350 bars
Muck removal	by Screw Conveyor and Belt Conveyor
Cutterhead power	16 × 350 kW (5600 kW)
Rotation speed	0–4/min
Nominal torque	16,056 kNm
Breakaway torque	24,083 kNm
Overload torque	25,689 kNm

Face collapses and jamming of the cutterhead started at chainage 216 + 300 m and continued thereafter. TBM manufacturer decided to close some of the openings within the cutterhead. After several problems occurred during TBM excavation, the adverse situation thrust of the TBM was increased from 84,464 to 170,000 kN (breakout), and the torque was increased from 24,083 kN m (overload) to 36,000 kNm. Face collapsing continued frequently depending on the weak zones developed within shear zones, slowing down tremendously advance rates. After several discussions between the consultants, the machine manufacturer, and the contractor, bentonite and foam were decided to be used to stop face collapses. With several modifications to TBM, 2 bars of mean face pressure could be obtained. Segments started cracking in May 2012, and tunnels collapsed gradually, damaging the TBM. As a safety concern for steel arches, shotcrete of 30 cm thickness with steel fibers of 25 kg/m^3 of concrete was used. Vibrating strain gauges and crack meters were installed to observe the deformation. The failure of the segments and mitigation works to strengthen the tunnel support with steel arches are seen in Figs. 10.9 and 10.10. For the sake of safety, all tunneling activities were stopped thereafter.

10.3.4 Rehabilitation of the Tunnel and Finishing the Tunnel with Conventional Tunneling

In 2013, the excavation was interrupted in the T26 tunnel after the collapse of the tunnel, and the TBM was stuck and destroyed to some extent. Subsequently, no site work occurred inside the tunnel for about 4 years. Nevertheless, redesign studies of the T26 tunnel were initiated in 2016, and then Fugro Sial, 2016 conducted relevant

Fig. 10.9 The failure of segments, from the archive of the author, Bilgin

studies for the design phase according to NATM. The tunnel excavation was finally re-started in 2017, and NATM has successfully proceeded. At the end of 2017, excavations were initiated again with conventional tunneling, which was completed in 2024.

10.4 Hsuehshan Tunnel, Taiwan

The 12.9 km Hsuehshan tunnel, part of the Taipei-Ilan expressway, includes two separate two-lane tunnels and a pilot tunnel. TBMs initially planned to bore all three tunnels from the eastern portal. At that time, this tunnel was the longest road tunnel in Southeast Asia and the fifth longest in the world. The tunnel decreases driving time between Taipei and Ilan by 90 min. The Hsuehshan tunnel is among the most difficult TBM projects in the world regarding adverse geology, water inrushes, tunnel size, and tunnel length. Construction began in 1991 with a pilot tunnel to better understand the tunnel route's geological features. Drilling on the two main tunnels started in July 1993. Cross cut section of the main and pilot tunnels is seen in Fig. 10.11. The project faced 19 floods and 29 collapses of the tunnel face during construction. The tunnel system consisted of 57 tunnels, including two main tunnels for road traffic between Ilan and Taipei and 28 pedestrian connecting passages located every 350 m.

Fig. 10.10 Steel arches used to strengthen the tunnel support, from the archive of the author Bilgin

Maximum overburden is 515 m. Tunnel construction took 15 years to complete and cost US $2.83 billion. Tunnel construction used 370,000 m^3 of concrete, 2000 km of cables, and 2000 lighting units, Lee and Leng (2005), Lin and Yu (2005), Tsai et al. (2015), https://tunnelbuilder.com/News/Hsuehshan-Tunnel-Breaks-Through, uploaded July 2024.

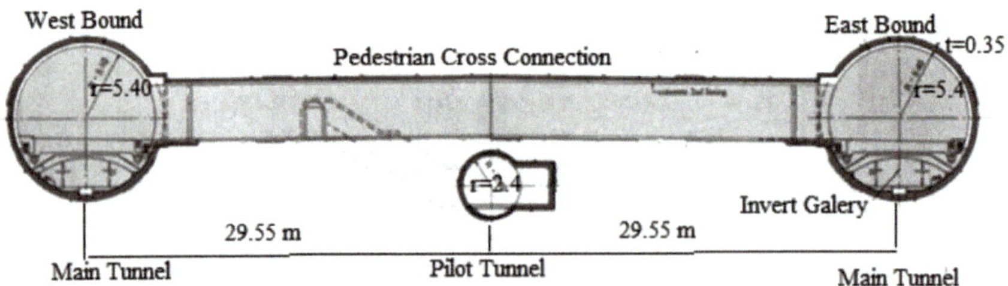

Fig. 10.11 Cross-cut section of two primary and pilot tunnels, Lee and Leng (2005)

10.4.1 The Geology

This tunnel's geological formations are of Eocene, Oligocene, and Miocene-aged sedimentary rocks. Major geologic structures encountered along the tunnel alignment are mainly fold structures with six major and several minor faults. The entire zone is heavily affected by tectonic movements and bears significant water. In addition to six identified water-bearing faults, many problematic and unanticipated shear zones exist. Lin and Yu (2005), Lee and Leng (2005), Tsai et al. (2015).

10.4.2 Driving Hsuehshan Tunnel with TBM and Drill and Blat Tunneling Method

For the main tunnels, two double shield TBMs of 11.7 m diameter and one double shield TBM of 4.8 m diameter for the pilot tunnel were initially imported for the excavation of the Hsuehshan tunnel; the main characteristics of the TBMs are given in Table 10.5.

Work started in July 1991 with a drill and blast pilot tunnel to assess geological conditions. When the pilot tunnel had gone 522 m, a 4.8 m diameter TBM was launched in December 1992. The TBM was abandoned in February 1996, and only 1 km of the pilot tunnel was completed due to the challenging ground conditions. The TBM in the northbound tunnel was buried in a significant collapse due to a large water inrush of 750 l/s, which caused a severe delay. To increase the tunneling advances, some additional working faces were created through shaft No.2. 4.8 m diameter TBM commenced its first boring of the pilot tunnel in December 1992; one month later, it was trapped. Westbound main tunnel TBM started excavating in May

Table 10.5 General specifications of TBMs used in Hsuehshan tunnel, Lee and Leng (2005)

Parameter	Pilot tunnel	Main tunnel
Manufacturer	Robbins	Wirth
Diameter, m	4.8	11.7
Cutter numbers	34	83
Cutter disc diameter	432	432
Cutter head speed, rpm	4.5–4.9/8.0–9.8	0–4
Cutter head torque, kNm	4.5 rpm/2010; 9.8/1005	0.95 rpm/30,000; 4 rpm/7200
Cutter head power, kW	960, 6 electric motors	4000/18 hydraulic motors
Thrust kN	7562	50,600/78700
Stroke	1.2	1.5
Machine length, m	177	250
Machine weigh, t	360	2300

1996; unfortunately, after excavating and advancing 456 m only, TBM was crushed and buried on December 15th 1997 due to unexpected huge water inrushes of the amount of 750 ~ 800 l/s and 18 bar water pressure. Eastbound TBM commenced its boring in December 1996. However, it securely passed the first major fault under the protection of the canopy (umbrella arch) method, a mixed excavation method with the conventional D&B and TBM. However, when it encountered extremely poor ground in July 1997, the TBM mining operation was called to stop and decided to adopt a long distance Canopy method, Lee and Leng (2005). 12.9 km long Hsuehshan Tunnel was finally broken through in 2005, regardless of TBM, Canopy, or D&B methods, Tsai et al. (2015). The project was completed in 1995, but the job has lasted 12 years. Ten engineers and workers have been killed, and an 11.7 m diameter TBM worth US $8.84 million was destroyed on the westbound drive of the main tunnel. Water inflows were a significant problem throughout construction and resulted in project retards.

10.5 Yacambú-Quíbor Tunnel, Venezuela

The Yacambú-Quibor tunnel is an average 4 m internal diameter 23.3 km long water transmission tunnel excavated through the Andes near Barquisemeto in Venezuela. The maximum overburden of the tunnel is 1270 m, and the construction of this tunnel was accepted by many as one of the most challenging tunnels in the world. It commenced in 1974 and it took 34 years to be completed. Initial attempts to use an open-face TBM in 1976 failed, as did attempts to use heavy support to resist squeezing. Only after the introduction of yielding support in about 1991 was reasonable progress made, Hoek and Guevera (2009). Besides the geological difficulties, political and financial problems also played an immense role in the retardation of the project.

10.5.1 The Geology

The phyllitic rock mass that dominates the mountain range in the Yacambu´-Quibor area ranges from strong and reasonably massive silicified phyllites in the dam area to severely tectonically deformed graphitic phyllite along most of the tunnel alignment. The graphitic phyllite has a squeezing characteristic and compressive strength change between 15 and 50 MPa, and the estimated GSI value is about 24, tunnelbuilder.com32-33/08 uploaded May 2024. The original site investigations, dating back to the 1970s, involved side surveys, core drilling, and the construction of several exploration adits. However, limited drilling was done at the tunnel portals, and only three vertical boreholes were attempted along the tunnel alignment. The deepest had to be abandoned at 300 m depth because of technical difficulties. In conclusion, the geology wasn't predictable, and the geological problems of the tunnel alignment weren't revealed correctly. It was only after the introduction of yielding support in about 1991 that reasonable progress was made.

10.5.2 Method of Construction and Problem Encountered

Table 10.6 summarizes the method of construction and the difficulties encountered. As seen from this table, the inadequate study of the geology and weak and squeezing characteristics of the graphitic schist played an essential role in the improper selection of TBMs. Open-type gripper TBM was unsuitable in the weak foliated rock. High squeezing characteristic of the graphytic schist trapped and buried one of the TBMs. In one stage, the contractor also used a roadheader. Only after yielding support was introduced in 1991 that reasonable progress was made with drill and blast. From 1995, an Atlas Copco 282 two-boom jumbo has been used on each of the two drives, with 2″ diameter 3.70 m R32/R38 drill bits. Rock reinforcement has been secured by Atlas Copco's Swellex bolts (1.50 m, 2 m, 3 m, and 4 m), 7 mm welded mesh made in Venezuela, and H200 steel arches. Shotcreting was performed with a Putzmeister pump, an Aliva 260, and a Shotcrete Technologies shotcreting robot at each face. The concrete admixtures were supplied by Sika and MB.

Table 10.6 Summary of the method of construction and problems encountered, compiled from Hoek and Guevera (2009)

Contract number, starting and ending	Method of construction and problems
1. 1975/1977	Two 4.8 m diameter Robbins gripper type TBMs, they could only progressed 700 m in the intake and 1000 m in the outlet drives. Open type TBMs were considered unsuitable in the weak foliated rock
2. 1978/1979	Graphtic phyllite was a serious problem. The TBM in the outlet drive was removed from the tunnel. The TBM in inlet drive was trapped in the squeezing rock and several years later it was removed piece by piece
3. 1981/1984	Drill and blast was performed in the outlet which was advanced 4350 m, inclined adit advance 1900 m. The intake drive was blocked by the TBM
4. 1984/1988	It was concentrated on on the inclined adit and headings. The intake adit was excavated by TBM
5. 1991/1997	Drill and blast in the outlet drive and a Dosco roadheader in the intake drive
6. 1977/2002	Sixth, seventh and eighth contracts were all carried out by the same Venezuelan contractor using conventional drill and blast methods
7. 2002/2005	
8. 2005/2008	

10.6 Concluding Remarks

A detailed geological investigation and its correct interpretation are necessary to select a tunneling method properly. Faults, squeezing rock formations, water-bearing strata, transition zones, and geotechnical characteristics of the rock formations will determine the method of tunneling, TBM, or drill and blat. Otherwise, tunneling operations may become a nightmare for the job owner and the contractor. An extensive literature survey will help to avoid repeating mistakes that have already been made in the past. Karakaya formation with the graphitic schist as a main rock formation caused extreme deformations in the Uluabat tunnel, deformations up to 1 m were encountered in the tunnel with the closure of tunnel section area by up to 10%, and the conventional tunneling method switched to mechanized tunneling with EPB-TBM, tunnel of 6100 m in length was completed in 13 years. The same story repeated in the T26 high-speed railway tunnel. Although the results of the Uluabat tunnel were discussed at the international level in journals and conferences, the conventional tunneling method was selected for the T26 Tunnel, which was planned to be executed in the Karakaya formation, having similar geotechnical characteristics as in the Uluabat tunnel. Due to high deformations in the tunnel, the tunneling method was switched to mechanized excavation with a single shield TBM. This choice wasn't correct. Squeezing characteristics of graphitic schist and over-excavation were the main reasons for the collapse of the tunnel, TBM was crushed and lost thereafter. After redesigning the tunnel support, the conventional tunneling method was used. Thereafter, the tunnel of 6100 m in length was completed in 15 years. If a proper literature survey could be done before the start of the project, the years and financial losses could be saved. The contractor should benefit from the lessons learned from the past.

Water inrush is always a big problem in the tunnel. One example is the Hsuehshan tunnel in Taiwan, which is among the most difficult TBM projects in the world in terms of adverse geology, water inrushes, tunnel size, and tunnel length. The TBM in the northbound tunnel of the project was buried in a significant collapse due to a large water inrush of 750 l/s, which caused a severe delay. The contractor carried on with the drill and blast with an umbrella arch. It is strictly advised that hydrogeology should be studied in detail in long tunnels.

The Yacambú-Quibor tunnel in Venezuela is an average 4 m internal diameter 23.3 km long water transmission tunnel excavated through the Andes in Venezuela. The maximum overburden of the tunnel is 1270 m, and the construction of this tunnel was accepted also by many as one of the most challenging tunnels in the world. It commenced in 1974 and it took 34 years to be completed. Initial attempts to use an open-face TBM in 1976 failed, as did attempts to use heavy support to resist squeezing. It was only after the introduction of drill and blast tunneling methods and yielding support in 1991 that reasonable progress was made. This project tells us that open-type TBMs are unsuitable in such difficult geology, and the feasibility of EPB-TBM should be considered.

References

Aydan Ö, Hasanpour R (2019) Estimation of ground pressures on a shielded TBM in tunneling through squeezing ground and its possibility of jamming. Bull Eng Geol Env 78:5237–5251. https://doi.org/10.1007/s10064-019-01477-3

Bilgin N (2016) An appraisal of TBM performances in Turkey in difficult ground conditions and some recommendations. Tunn Undergr Space Technology 57:265–276

Bilgin N, Algan M (2012) The performance of a TBM in a squeezing ground at Uluabat, Turkey. Tunn Undergr Space Technol 32:58–65

Fugro Sial Geosciences Consulting and Engineering Ltd (2016) Ankara Istanbul high-speed railway project Vezirhan Kosekoy Section, Geological-geotechnical excavation and support design project of T26 tunnel, report of exit portal of t26 tunnel. Unpublished work

Gökçeoğlu C, Aygar EB, Nefeslioğlu HA, Karahan S, Güllü S (2022) A geotechnical perspective on a complex geological environment in a high-speed railway tunnel excavation, a Case Study from Türkiye. Infrastructures 7:155. https://doi.org/10.3390/infrastructures7110155

Hasanpour R, Schmitt J, Ozcelik Y, Rostami J (2017) Examining the effect of adverse geological conditions on jamming of a single shielded TBM in Uluabat tunnel using numerical modeling. J Rock Mech Geotech Eng 9:1112–1122

Hoek E (2001) Big tunnels in bad rock. ASCE J Geotech Geoenviron Eng 127(9):726–740.1

Hoek E, Guevara R (2009) Overcoming squeezing in the Yacambu´-Quibor tunnel, Venezuela. Rock Mech Rock Eng 42:389–418 https://doi.org/10.1007/s00603-009-0175-5

https://tunnelbuilder.com/News/Hsuehshan-Tunnel-Breaks-Through, uploaded July 2024

Incecik M, Poşluk E (2018) T26 on the Ankara–Istanbul high-speed rail route, tunneling under challenging conditions. Geomech Tunnell 11(5). https://doi.org/10.1002/geot.201800026

Jimenez R, Senent S (2012) Teaching the importance of engineering geology using case histories. In: McCabe, Pantazidou, Phillips (eds) Shaking the foundations of geo-engineering education. & Francis Group, London, ISBN 978-0-415-62127-4

Lee WC, Leng, YT (2005) Management of TBM construction in Hsuehshan main tunnels and pilot tunnel. In: Proceedings of the international symposium on design, construction, and operation of long tunnels, Taipei, Taiwan, January 2005, pp 223–235

Lin CC, Yu CW (2005) Discussion and Solution of the TBM Trapped in the Westbound Hsuehshan tunnel. In: Proceedings of the international symposium on design, construction, and operation of long tunnels, Taipei, Taiwan, January 2005, pp 383–393

Ramonet D (2008) Venezuela completes Yacambú-Quíbor water transfer tunnel international EIR August 15, 2008, p 32

Tsai MS Lu, TB, Lee YH, Lu FS (2015) Planning and design of the Hsuehshan tunnel. In: Proceedings of the international symposium on design, construction, and operation of long tunnels, Taipei, Taiwan, January 2005, pp 7–22

Chapter 11
Using TBM and Conventional Tunneling Methods in the Same Project–The Hybrid Solution

Abstract This chapter discusses the hybrid approach to tunneling using drill and blast and TBM in detail, answering why and when this hybrid method is used. The First Kaynarca-Kartal-Kadıköy Metro Line in Istanbul, with twin tunnels of 26.5 × 2 km, is given as an example. Almost 2/3 of this line is excavated with drill and blast; the rest is excavated with 2 EPB-TBM. The methodology, advantages, and results of using two different tunneling methods in this highly populated area of Istanbul are discussed. Another example of a hybrid tunneling project is the Kargı hydropower project in Turkey. The 7.8 km of the total length of 11.8 km of the tunnel is excavated with a double shield TBM of 9.84 m diameter, and 4 km of the tunnel from the inlet part is opened with NATM due to the challenging ground conditions faced by the TBM. The Bahçe-Nurdağ railway tunnel in Turkey consists of two parallel 9.75 km tunnels excavated by drill and blast and by 8 m diameter single shield TBM. Risk analysis showed that the first km of this tunnel was risky for TBM since the tunnel's starting point was highly disturbed by the tectonic stresses created by the East Anatolian fault, and thereafter, this short length of the tunnel was opened by drill and blast. The final example is the Mount Macdonald tunnel, the longest railway tunnel of 14 km in Canada. The contractor for the East Tunnel of 8478 m used a 6.8 m diameter gripper-type TBM for the top heading. The bench was then excavated by drill and blast. The west tunnel of 6066 m was excavated by drill and blast with another contractor.

11.1 Introduction

Factors such as the tunnel length, the geology, the safety concerns, the initial investment cost, time scheduling of the project, etc., determine the methodology to be applied for tunneling, mainly drill and blast or TBM. The deliberate selection of both TBM and drill-and-blast in the same project may often be a simple matter of common sense, giving schedule advantages and cost savings, as noted by Barton (2012a, b). As Barton and Bilgin (2016) emphasized, tunnel lengths of 5 km, perhaps 10 km, and even a world record equaling 15 km can be driven in one year by a TBM in

N. Bilgin and C. Balci, *Critical Issues in Selecting Conventional and Mechanized Tunnelling Methods*, Springer Tracts in Civil Engineering,
https://doi.org/10.1007/978-3-031-89114-4_11

favorable conditions. In contrast, the very best drill-and-blast record so far is 5.8 km in 54 weeks, and usually, it is closer to 50 m/week. A factor with TBM tunneling that has not been widely acknowledged or used in planning is a general deceleration as the tunnel gets longer. This is seen in open-gripper case records and double shield TBM, in each case following speed-up in the 'learning curve' period. The same deceleration trends are seen in the current world record TBM performances, with diameters from 3 m and beyond 12 m. It is found that the steepest deceleration occurs when the rock mass quality Q is low. Specific case records from Turkey confirm this reality. Sometimes, hybrid TBM and D&B tunneling methods are best. Although the hybrid method reduces construction time and cost, sometimes using both tunneling techniques is challenging. Therefore, choosing the proper excavation method and equipment is essential to keep the project on schedule. This important topic will be treated in this chapter by giving three examples from Turkey and one from Canada, with the primary objective of comparing two different tunneling methods and showing the merit of the hybrid method.

Kaynarca-Kartal-Kadıköy Metro Line, having twin tunnels of 26.5 × 2 km in length, is the first example of a hybrid tunneling method given in this chapter. It was realized with great success in a highly populated area of Istanbul. Almost 2/3 of this line is excavated with drill and blast and the rest with 2 EPB-TBM, Özcan (2011), Şenol (2014), Şenol et al. (2014). A cautious and controlled blasting model applied during tunneling minimized vibrations and environmental problems created by the conventional tunneling method. Kahriman et al. (2007, 2010), Özer 2008) and Alan (2018) discussed this model in detail in national and international symposiums.

Another example of hybrid tunneling given is the Kargı hydropower project in Turkey. The 7.8 km of the total length of 11.8 km of the tunnel is excavated with a double shield TBM of 9.84 m diameter, and 4 km of the tunnel from the inlet part is opened with NATM due to the challenging ground conditions faced by the TBM, Yurt et al. (2014), Clark and Chorley (2014), Clark (2015), Home (2015), Bilgin (2016a, b), Home (2016). This project is one of the most discussed and published on hybrid tunneling in Turkey. The last example from Turkey is the Bahçe-Nurdağı high-speed Railway tunnel. It is in Southeastern Turkey's Gaziantep province, characterized by a complex fractured rock within the Eastern Anatolian Fault. With a population of nearly 1.7 million, the province is overhauling its public transportation with a rail line between the towns of Bahçe and Nurdağı. The tunnel consists of two parallel 9.75 km tunnels excavated by both NATM and TBM of 8 m diameter single shield TBM, Bilgin et al. (2016, 2017), Jordan and Bilgin (2018a, b).

One of the earliest publications on the hybrid tunneling method is on the Mount MacDonald railway tunnel in Canada, as reported by Kuasel and Hansmire (1987). This tunnel will be a final example of the hybrid tunneling method. The top heading of the East Tunnel, 8428 m in length, was excavated with a 6.8-m diameter Robbins gripper-type TBM, and the bench was then excavated by drill and blast. The West Tunnel of 6066 m was excavated by drill and blast by a second contractor. This remarkable achievement in complex geology was completed in two years.

11.2 Kaynarca-Kartal-Kadıköy Metro Project

The construction of the Kozyatağı-Kadıköy metro project was awarded to Yapı Merkezi-Doğuş-Yüksel -Yedigün-Belen Jv. in 2005, the project's second and third parts, from Kaynarca to Kozyatağı 17.7 × 2 km in length, were awarded to Astaldi-Makyol-Gülermak Jv in 2008.

The general layout of the Kaynarca Kadıköy Metro line of 26.5 × 2 km in two lines is given in Fig. 11.1

The first part of the project from Kozyatağı to Kadıköy 8800 m in length with a double line, was finished with 2 EPB-TBMs with a mean daily advance rate of 10.5 m/day. Around 3.4 km of this line was opened with drill and blast, and the rest was opened with 2 EPB TBMs. The platform tunnels of this project were opened with drill and blast, impact hammers, and roadheaders. This topic was discussed in detail in Chap. 5. This project is unique in that around 2/3 of the line was excavated with drill and blast in a highly populated area of Istanbul. A planned synchronized hybrid tunneling method enabled the contractor to finish the project from Kaynarca to Kozyatağı before the scheduled time. Therefore, each part of this project will be discussed separately to help the readers benefit from the results of similar projects.

Fig. 11.1 The general layout of the Kaynarca Kadıköy metro line

11.2.1 Kartal-Kozyatağı Metro Line 13.2 km in Length Excavated by Drill and Blast

The geologic cross-section between Kartal and Kozyatağı is given in Fig. 11.2. Sedimentary formations of Paleozoic ages with complex geology are found in the area. The main geologic formations excavated are the Kartal Formation of the Devonian age, consisting of fossilized shale and limestone; the Doloyaba Formation of the Devonian age, composed of laminated mudstone-shale-limestone; and the Kurtköy formation of the Ordovician age, consisting of conglomerate, sandstone, and mudstone. Mean values of physical and mechanical characteristics of the studied areas are given in Table 11.1.

Eight station shafts existed between Kartal and Kozyatağı, enabling four faces from each shaft for drill and blast operations. It was decided to open the 13,100 m tunnel with conventional tunneling methods, Özcan (2011), Şenol (2014), Şenol et al. (2014). The tunnel area was highly populated, and a unique blasting methodology was used to minimize the blasting vibrations, as given by Özer (2008). From one tunnel face, a daily advance rate of 2.7 m/day was possible with the drill and blast method. A pie chart for the work executed for the drill and blast method is given in Fig. 11.3. This pie chart shows that the most time-consuming items are mucking at 27.1%, waiting at 13.8%, bolting at 12.8, and shotcrete at 12.6%. There is room to

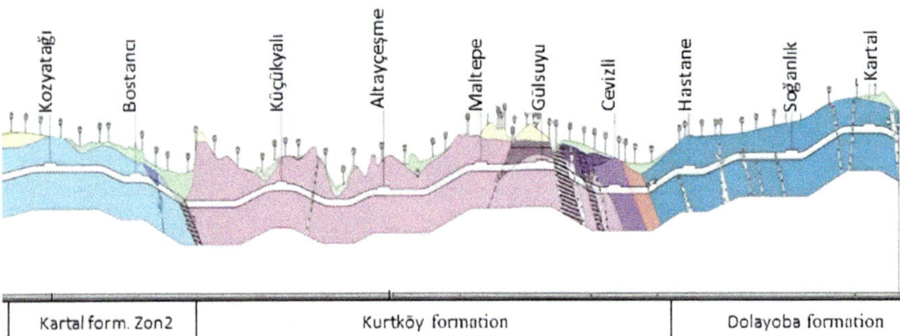

Fig. 11.2 Geologic cross section between Kartal and Kozyatağı, Kahriman (2008)

Table 11.1 Mean values of some physical and mechanical characteristics of the rocks in the studied areas Şenol (2014)

Formation	γ kN/m^3	σt MPa ($\pm$ sd)	σ_c MPa ($\pm$ sd)	RQD $\pm$ sd
Kartal	26.4	4.8 ($\pm$ 2.2)	34.3 ($\pm$ 21.5)	48.5 ($\pm$ 11)
Doloyoba	27.0	6.4 ($\pm$ 1.5)	45.2 ($\pm$ 17.5)	41.3 ($\pm$ 24)
Kurtköy	26.9	7.7 ($\pm$ 4.3)	52.7 ($\pm$ 29.3)	17.7 ($\pm$ 16.5)

γ is density, σt tensile strength, σ_c is the compressive strength, RQD is rock quality designation

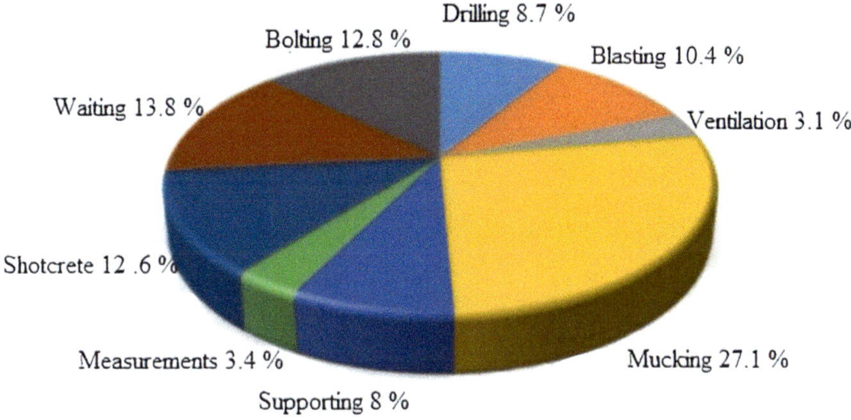

Fig. 11.3 Work distribution in drill and blast, 2.7 m/day per round, Şenol (2014)

increase the tunneling efficiency in drill and blast by improving the time spent on mucking and lost time.

Smooth blasting design principles developed by Kahriman (2008) were applied to the tunnels, considering the tunnel dimensions, hole diameter, length, and mechanical properties of the rock units. For this, drilling and blasting are planned in two stages: the tunnels' upper and lower half. Parallel cutting and V-cutting blasting designs with a single empty hole of 76 mm diameter were used for the cutting holes. All other holes in the designs were 39–42 mm in diameter. To minimize the vibration and environmental problems, the charge per delay was kept at a minimum. Usually, Powergel Magnum 365—Trimex is used as an explosive, and nonelectric detonators are used for the initiation system, Kahriman (2008).

11.2.1.1 Design Model for Single Empty Hole Parallel Cutting

In this model, drill and blast are realized in the upper and lower half stages. For smooth blasting, 56 holes were drilled in hard formations for the upper half, including one empty hole in the middle of the cutting holes. In the design created for the upper half excavation, the hole pattern and lag sequence hole pattern are shown in Fig. 11.4. Nonelectric delayed detonators are used for ignition Kahriman (2008). The amount of explosive material to be used in the hard formation pattern for the tunnel's upper and lower half excavation with a cross-sectional area of 32 m^2 is given in Table 11.2 and Table 11.3. At the same time, the amount of explosive material (specific charge) to be used per cubic meter is also calculated and given in Tables 11.2 and 11.3, Kahriman (2008).

In the lower half of the tunnel, 22 holes are drilled, with a cross-sectional area of 14 m^2; for this design, the hole pattern and lag sequence hole pattern are shown in

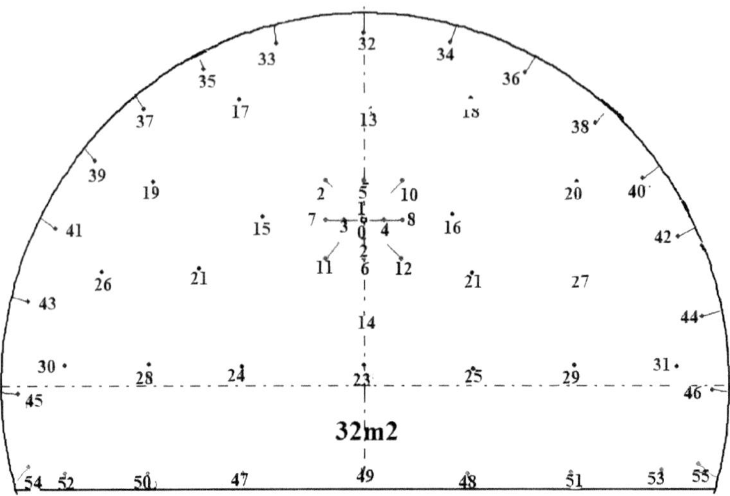

Fig. 11.4 For the upper half excavation, the hole pattern and lag sequence hole pattern, Kahriman (2008)

Table 11.2 Design parameters for the upper part of the tunnel for the single empty hole parallel cutting design, Kahriman (2008)

Type of holes	Number of drill hole	Drill hole length, m	Explosive type	Charge per drill hole (kg)	Total charge (kg)
Stopping holes	19	3.0	Emulsion	2.509	47.671
Cut holes	12	3.0	Emulsion	2.509	30.108
Bottom hole	9	3.0	Emulsion	2.509	22.581
Surrounding hole	15	3.0	Emulsion	0.660	9.9
Empty hole	1	3.0	Emulsion		
Total holes	56		Emulsion		110.26
Specific charge (kg/m^3) 1.23					

Table 11.3 Design parameters for the lower part of the tunnel for the single empty hole parallel cutting design, Kahriman (2008)

Type of drill holes	Number of holes	Drill hole length m	Type of explosive	Charge per drill hole (kg)	Total charge (kg)
Stopping holes	11	3.0	Emulsion	2.509	27.599
Surrounding holes	11	3.0	Emulsion	0.660	7.26
Total holes	22		Emulsion		34.859
Specific charge 0.89 kg/m^3					

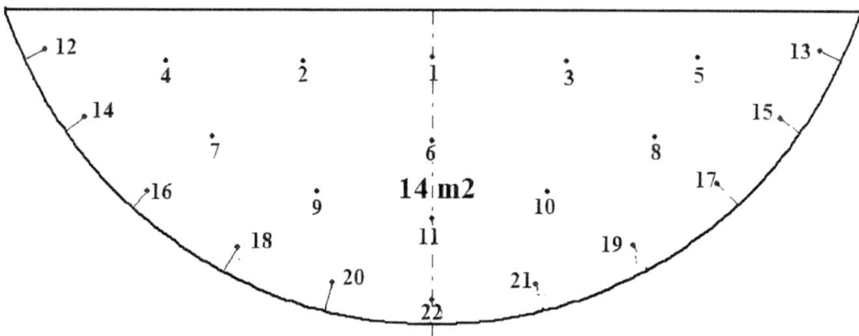

Fig. 11.5 Drill pattern in the lower part of the tunnel for the single empty hole parallel cutting design, Kahriman (2008)

Fig. 11.5. At the same time, the amount of explosive material to be used per cubic meter (specific charge) has also been calculated in Table 11.3 Kahriman (2008).

11.2.1.2 Design Model for V-cut

In this design, drill and blast are also carried out in two stages, the upper and lower half. In the upper half excavation, 70 holes are drilled for smooth blasting; this design's hole pattern and lag sequence hole pattern are shown in Fig. 11.6. Nonelectric delayed detonators are used for ignition. The amount of explosive material to be used in the hard formation pattern for the upper half excavation of the tunnel with a cross-sectional area of 32 m^2 is given in Table 11.4 Kahriman (2008).

The following method was used in the lower half excavation.

A total of 22 holes are drilled in the lower half excavation with the hole pattern shown in Fig. 11.7. Nonelectric delayed detonators for ignition are used, and the amount of explosive (specific charge) to be used per cubic meter for the lower half excavation of the tunnel with a cross-sectional area of 14 m^2 is given in Table 11.5. Kahriman (2008).

11.2.1.3 Kartal-Kaynarca Metro Line Excavated by 2 Identical EPB-TBMs

Due to the limited time to complete the project, the contractor decided to open the twin tunnels of 4.6 km in length from Kartal to Kaynarca by two identical Herrenknecht EPB-TBM, 505 and 506 of 6.57 m diameter. TBM used is given in Fig. 11.8, and the details of TBM are tabulated in Table 11.6. In this section, the performance of EPB-TBMs will be given in detail. The accumulated data will help us understand the comparative study of drill and blast and TBM tunneling. Table 11.7 shows the grouped

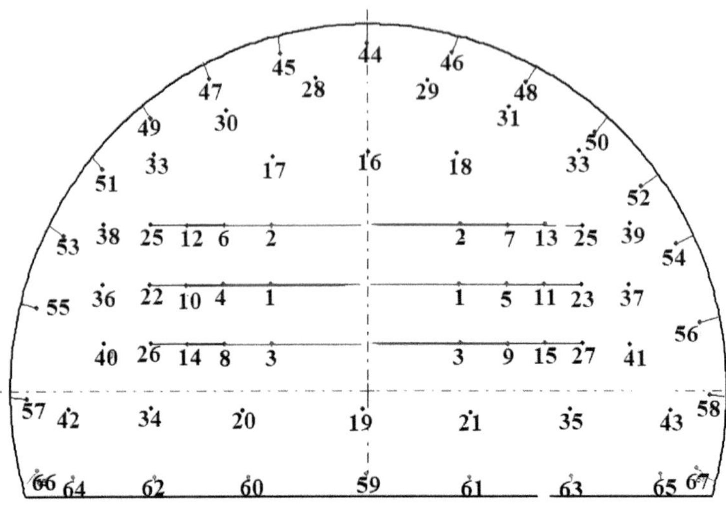

Fig. 11.6 Drill pattern in the upper part of the tunnel for V cut design model, Kahriman (2008)

Table 11.4 Design parameters for the upper part of the tunnel for V cut design model, Kahriman (2008)

Hole type	Number of holes	Length of drill hole	Explosion type	Charge per drill hole kg	Total charge kg
Stopping holes	28	3.0	Emulsion	2.509	70.252
1. Row cut holes	6	1.4	Emulsion	1.158	6.948
2. Row cut holes	6	2.5	Emulsion	1.93	11.58
3. Row cut holes	6	3.5	Emulsion	2.702	16.212
Bottom hole	9	3.0	Emulsion	2.509	22.581
Surrounding holes	15	3.0	Emulsion	0.660	9.9
Total kg	70				137.473
Specific charge 1.53 kg/m^3					

data based on penetration mm/rev. This table is used to analyze the performance of EPB-TBM number 505, Özcan (2011)

TBMs finished the tunnels, which are 4600 m in length, with a main daily advance rate of 11.6 m/day, the best daily advance of 40.5 m, and the best weakly advanced of 224 m.

Material consumption is one of the most critical factors in the TBM tunneling economy. The consumption of some materials is summarized in Table 11.8. Work distribution in TBM excavation, for the mean daily advance of 11.6 m/day, is given

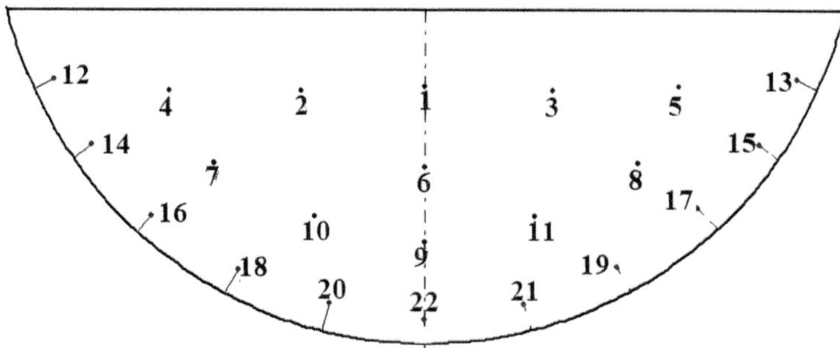

Fig. 11.7 Drill pattern in the lower part of the tunnel for V cut design model, Kahriman (2008)

Table 11.5 Design parameters for the lower part of the tunnel for the V-cut design model, Kahriman (2008)

Type of drill holes	Number of holes	Drill hole length m	Type of explosive	Charge per drill hole (kg)	Total charge (kg)
Stopping holes	11	3.0	Emulsion	2.509	27.599
Surrounding holes	11	3.0	Emulsion	0.660	7.26
Total holes	22		Emulsion		34.859
Specific charge 0.89 kg/m^3					

in Fig. 11.9. The time spent on excavation is 39%, time spent on segment mounting is 28%, and time spent on repairs and waiting is 33%. This figure is in good agreement with the general trend of a well-managed TBM operation.

Figure 11.10 shows the variation of production rate in m^3/h and the variation of specific energy in kWh/m^3 with respect to penetration in mm/rev. This figure clearly shows the merit of deeper penetration, production rate increases, and specific energy levels off after an optimum value of 14 mm/rev.

The variation in the percentage of geologic formations is seen in Fig. 11.11, with Some physical and mechanical properties of the geologic formations encountered in the Kaynarca-Kartal Metro line. Some physical and mechanical properties of the geologic formations encountered in the Kaynarca-Kartal Metro line are given in Table 11.9

Figure 11.12 shows the variation of thrust force with penetration. Thrust force decreases with penetration according to the strength of the rock, starting at 110 MPa and ending at 46 MPa.

Fig. 11.8 TBM used in Kartal Kaynarca Tunnel, from the archive of the author Bilgin

Table 11.6 Main characteristics of EPB-TBM used in the project, Bilgin

Machine diameter	6.57 m
Rotational speed	0.0–5.33 rpm
Number of cutters	32 single, 4 double
Number of rippers	72
Cutterhead power	8 × 160 kW
Cutterhead torque 1	3663 kNm nominal
Cutterhead torque 2	5018 kNm maximum
Maximum thrust	42,575 kN at 350 bar
Opening ratio	%35

11.3 Kargı Project

Another example of hybrid tunneling is the Kargi hydropower project. As Barton mentions, "*There are significant numbers of TBM projects that end up with difficult decisions to be made, namely to complete the projects by drill-and-blast from the other end of the tunnel. The deliberate selection of TBM and drill-and-blast may often be a simple matter of common sense, giving schedule advantages and cost*

Table 11.7 Grouped performance data of EPB-TBM number 505, based on penetration, (Özcan 2011), Şenol (2014)

Data	Penet mm/rev	rpm mm/rev	Torque MNm	Thrust MN	Cutting rate m^3/h	Power kW	Specific energy kWh/m^3
10	3.0	4.6	1.0	12.5	28.2	475.9	16.9
13	4.0	4.3	1.1	11.7	34.4	479.2	13.7
52	5.0	4.5	1.4	11.9	45.4	635.4	14.0
92	6.0	4.5	1.5	11.4	55.4	725.5	12.2
174	7.0	4.4	1.7	11.0	62.4	760.8	12.1
323	8.0	4.6	1.7	10.5	74.7	835.1	11.2
411	9.0	4.5	1.7	9.5	83.2	825.1	9.9
546	10.0	4.5	1.7	8.9	91.5	821.0	9.0
400	11.0	4.4	1.7	8.3	99.0	790.7	8.0
303	12.0	4.2	1.6	7.9	101.4	707.3	6.9
118	13.0	4.0	1.7	7.7	104.9	693.8	6.6
62	14.0	3.7	1.7	7.6	104.1	662.1	6.3
33	15.0	3.4	1.7	7.7	102.4	604.7	5.9
10	16.0	3.4	1.7	8.1	109.0	613.4	5.6
Mean	9.5	4.2	1.6	9.6	78.3	687.6	9.9
Minimum	3.0	3.4	1.0	7.6	28.2	475.9	5.6
Maximum	16.0	4.6	1.7	12.5	109.0	835.1	16.9
s.d ±	4.2	0.4	0.3	1.8	28.2	117.8	3.6

Table 11.8 Material consumption in EPB-TBM no 505, Özcan (2011)

Material	Consumption
Black gres, tail gres	217 barrel, 217 × 250 = 54,250 L
Wr 89 white gres	130 barrel, 130 × 250 = 32,500 L
Gres for main bearing	250 L
Monthly water consumption	11,000 m^3
Daily foam consumption	1250 L
Grout per ring	4.5 m^3

savings. This action is the preliminary level of hybrid tunneling. Very often, time is lost while waiting for TBM delivery and assembly, and great advantages could be gained by selective use of drill-and blast for more than just the standard TBM assembly chamber and starter tunnel", Barton (2012a, b)

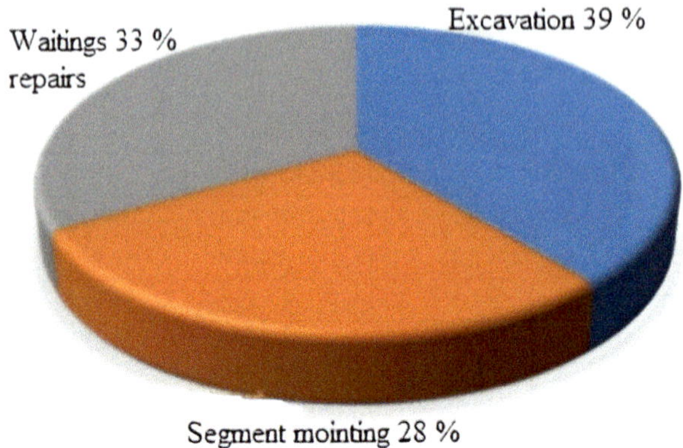

Fig. 11.9 Pie chart of the work distribution in TBM excavation for the mean daily advance of 11.6 m/day, Şenol (2014)

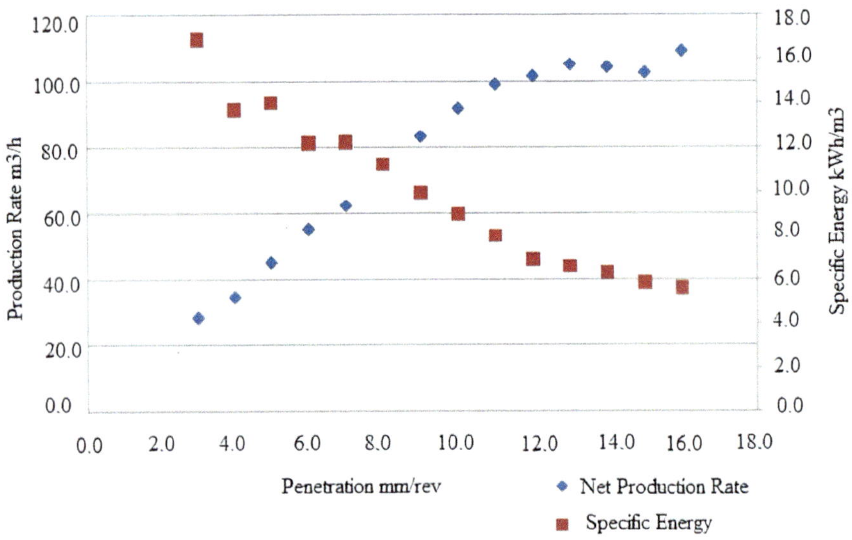

Fig. 11.10 The variation of production rate and specific energy with penetration rate, Şenol (2014)

11.3.1 *Project Description and the Geology*

The Kargi Hydropower Project is situated on the Kızılırmak River, downstream of Osmancık Town in Corum Province of Turkey. The project is developed by Kargı Kızılırmak Energy A.S, a subsidiary of Statkraft Energy A.S, which is owned 100% by Statkraft, Europe's largest producer of renewable energy. EMRA, Energy Market Board Authority, issued the license to build and operate for 49 years. Construction

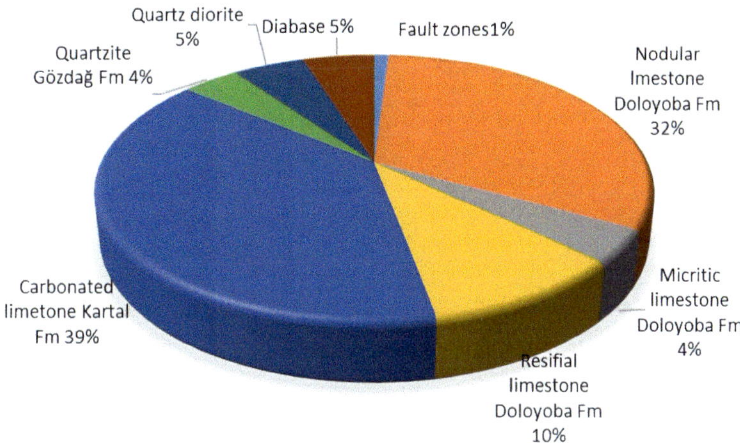

Fig. 11.11 The variation of the percentage of geologic formations, Şenol (2014)

Table 11.9 Some physical and mechanical properties of the geologic formations encountered in the Kaynarca-Kartal Metro line, Şenol (2014)

	γ kN/m³	UCS MPa	E_i	c kPa	ϕ	GSI
Carbonated shale Kartal formation	26	32	5300	280	47	45
Nodular limestone Doloyoba formation	26	34	5200	300	51	47
Resifal limestone Doloyoba formation	26	61	14,100	1000	56	61
Micritic limestone Doloyoba formation	26	65	15,300	400	56	45
Diabase	26	46	4550	750	61	64
Quarz Diorite	26.5	105	–			–
Quartzite Gözdağ formation	26.5	110	–	1300	62	–

started in January 2011. The estimated mean annual production is 470 GWh, approximately 1% of Turkey's hydropower capacity. The installed capacity is 102 MW, the gross head is 75 m, the maximum discharge is 167 m^2/sec, the type of dam is an earth-fill dam with a height of 13.5 m, inundated area is 4 km^2. The project area is located within the Northern Anatolian Fault Zone, primarily responsible for earthquakes in Turkey. The tunnel was driven through a mountainside with up to 600 m of overburden. The expected geology along the tunnel alignment consisted of Kirazbaşı complex Kargı ophiolites (including sandstone, siltstone, and marl and graphitic shist) for the initial 2300 m, followed by 1000 m of Kundaz metamorphites (including marble, and metapelite), and the remaining 8500 m consisted of Eocene aged Beynamaz Volcanites (including basalt, agglomerates and andesite). The anticipated strength of the rock was up to 140 Mpa. Multiple fault zones and transition zones added to the complexity of the geological conditions, Yurt et al. (2014). The

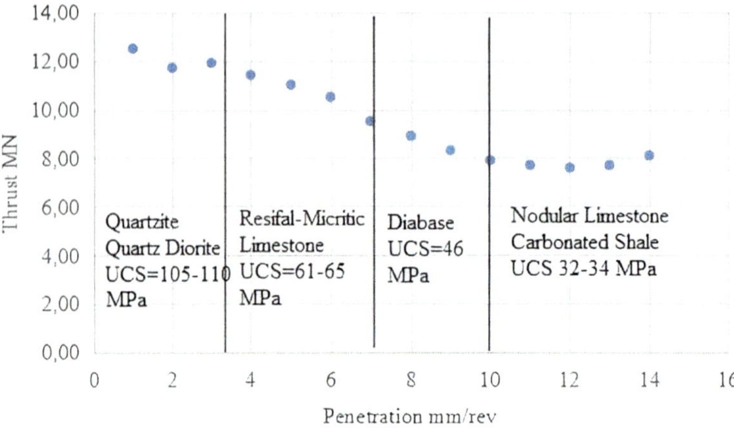

Fig. 11.12 The variation of thrust force with penetration, Şenol (2014)

project area's general layout and the tunnel route's geologic cross-section are seen in Figs. 11.13 and 11.14, Yurt et al. (2014), Bilgin (2016a, b).

Fig. 11.13 The general layout of the Kargı hydropower project area from the archive of the author Bilgin

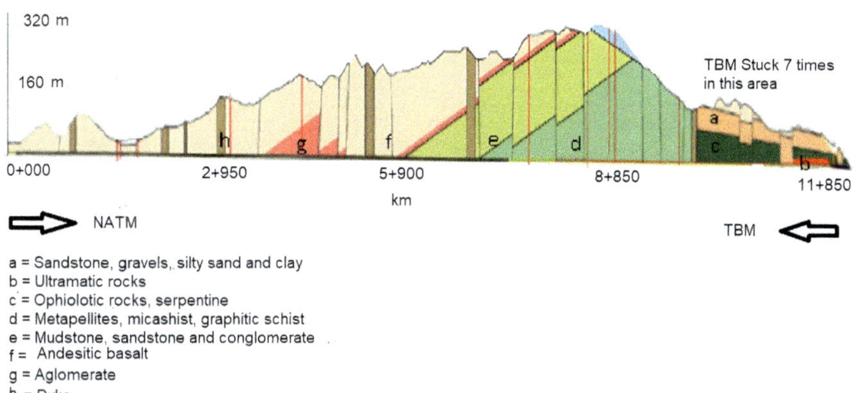

Fig. 11.14 The geologic cross section of Kargı hydropower tunnel Bilgin (2016a, b)

11.3.2 The Selection of TBM and the Performance

A Robbins crossover double shield TBM with a 9.84 m diameter, as seen in Fig. 11.15, was selected for the project. The general characteristics of the machine are as given by Clark (2015), 19-inch back-loading hard rock disc cutters, 13 × 370 kW variable frequency drive motors, 0–8.05 rpm cutterhead speed, 20,904 kN maximum operating cutterhead thrust, 9864 kNm cutterhead torque at 3.66 rpm and 1900 mm thrust cylinder stroke.

TBM was launched in April 2012. After only a boring of 80 m, TBM became stuck in a section of the collapsed mixed ground face of hard rock and running ground. In the first 2 km of the tunnel, the TBM jammed 7 times, and bypass tunnels were required to free the TBM. For remedial works, 281 days were spent on rescue galleries, setting umbrella arch, changing the seals of the main bearing, cleaning cutterhead and buckets, and setting up the belt conveyor. The distribution of the TBM stoppages is seen in Fig. 11.16.

As Clark mentioned in his article published in (2015), "*Custom-made gear reducers were ordered and retrofitted to the cutterhead motors to mitigate the effects of squeezing ground or collapse. They were installed between the drive motor and the primary two-stage planetary gearboxes. During standard boring operations, the gear reducers operate at a ratio of 1:1, offering no additional reduction and allowing the cutterhead to reach design speeds for hard rock boring. When the machine encounters loose or squeezing ground, the reducers are engaged, which reduces cutterhead speed, but the available torque is increased. After installing the canopy drill and increasing available cutterhead torque, the TBM traversed several sections of adverse geology, including stretches of severe convergence, without becoming trapped. The typical thrust force for standard boring operations using the main thrust cylinders on the Kargı machine is approximately 21,000 kN. Several times, a thrust force of up to 136,000 kN was applied through the auxiliary thrust system*

Fig. 11.15 Robbins double shield XRD TBM used in Kargı Project, from the archive of the Author Bilgin

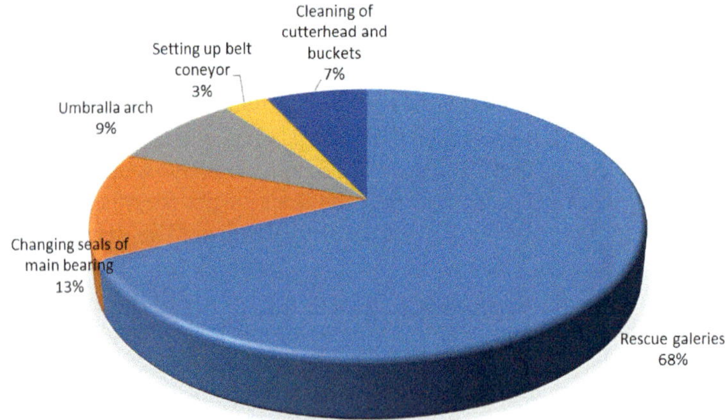

Fig. 11.16 Distribution of the TBM stoppages of 281 days due to face collapses and squeezing of TBM, Yurt et al. (2014)

before the machine could be freed from squeezing ground, Clark (2015). The new configuration allowed the carrying out of systematic probe drilling that enabled a continuous probe drilling program along 5 km along the tunnel, and an umbrella arch was applied in 9 different areas to stop face collapse excavation of the tunnel,

Bilgin et al. (2016). After a lengthy discussion between the contractor and project owner, it is agreed that the delays in TBM drives necessitated hybrid tunneling as an alternative to decrease the job termination time. Thereafter, it was decided to excavate the 7.8 km of the total length of 11.8 km of the tunnel with XRD TBM and 4 km of the tunnel from the inlet part with drill and blast tunneling method.

11.3.3 Using Drill and Blast Tunneling in Kargı

Time for mobilization took 4 months for drill and blast. The equipment listed and the support regime applied in this tunneling method are given in Tables 11.10 and 11.11. These tables may serve as a guide for similar applications. Full-face drill and blast were used, as seen in Fig. 11.17.

Table 11.10 The equipment used in drill and blast, from the archive of the author Bilgin

Equipment	Amount
Atlas Copco XE3C Jumbo Atlas	1
Atlas Copco E2C Jumbo Atlas	1
Meyco Potenza P. Shotcrete Rabot	1
Titan IS 26 P. Shotcrete Rabot	1
Manitou MTX-1840 Man Basket	1
Manitou MTX-1440 Man Basket	1
Volvo L110E Loader	1
Volvo EC290BLC Excavator	2
Mercedes Axor 3340 Haulage Truck	8
Komatsu WA 480 Loader	1
HMK-102S Hidromek HMK-102S	1

Table 11.11 Rock class and support regime, Clark (2015)

Rock class	Length (m)	Shotcrete tickness (cm)	Lattice girders m	Rock bolts/m
II	770	6	0	4 × 3
III	3100	15	0	6 × 4
IV	130	30	1	8 × 4

Fig. 11.17 Full-face drill and blast operation in Kargı Tunnel, from the archive of the author Bilgin

11.3.4 Comparison of TBM and Drill and Blast Tunneling Method in Kargı Tunnel

Due to the delays resulting from adverse geological conditions, it was decided to reduce the length of the TBM-driven tunnel to 7.8 km and drive the remaining 4 km from the inlet portal by utilizing drill and blast methodology. Drill and blast operations commenced from the inlet portal in July 2012, and the tunnel was driven utilizing drill, blast, and NATM support techniques. It was expected that advance rates in the drill and blast section would be relatively high due to the Beynamaz Volcanites' competent rock along this alignment section. From the launch of the TBM in March 2012 up until the end of July 2013, the TBM encountered 2228 m of extreme geological conditions, which required seven bypass tunnels and nine umbrella arches; hence, a direct comparison between the drill and blast operations and TBM operations is not possible for this period. As explained in this chapter, the geology of the Kargı hydropower tunnel is complex. A comparison between the performance of TBM and drill and blast should be made considering geological conditions to determine when each excavation method is the best. Although a direct comparison cannot be made between the TBM and drill and blast operations for the extreme conditions faced by the TBM during the first 2228 m of boring, a comparison can be made of the performance while boring through various geological conditions faced on the project. Table 11.12 shows the average comparative advance rates for the TBM and drill and blast operations in multiple ground conditions. It includes the performance of the TBM both before and after the modifications were carried out. It is clear from this table that the mean daily advance rate of TBM after modification

Table 11.12 Advance rates in different ground conditions, Clark (2015)

Ground condition	TBM before modification	TBM after modification	Drill and blast
Good stable rock	N.A m/day	22.5 m/day	8 m/day
Fairly table rock	9.7 m/day	19 m/day	5.5 m/day
Non self-supporting rock	4.7 m/day	8.5 m/day	1.5 m/day
Running/squeezing rock	1.5 m/day	1.8 m/day	N.A

Table 11.13 Overall comparisons of drill and blast and TBM, Clark (2015)

Parameter	TBM	Drill and blast
Boring length (km)	7.8	4.0
Months to complete	28.0	23.0
Best month (m)	723.2	281.5
Best day (m)	39.6	12
Monthly average (m)	271.4	173.8
Weeks out of advance	38	8
Time to mobilize, month	16	4

was double in different geological conditions. This point is very important in mechanized tunneling, so the contractor and the TBM manufacturer should jointly control the TBM performance and make necessary modifications to improve tunneling efficiency. The comparison in non-self-supporting rock again shows a double increase in the production rates of the TBM after the modifications were carried out. This increase can be attributed to the custom-built pipe roof drill, the increase in cuterhead torque, and the mucking system that enabled faster clearance of the buildup of loose material in the telescopic shield area of the TBM. The drill and blast operations achieved only 18% of the production rates of the modified TBM in non-self-supporting rock.

Table 11.13 makes overall comparisons of drill and blast and TBM tunneling. It clearly shows that drill and blast, in certain conditions, may not be a competitor to TBM tunneling. However, as Kargı's example shows, if they are used as hybrid tunneling methods in the same project, the profit is indisputable.

11.4 Bahçe Nurdağı Tunnel

This tunnel is a typical example of one affected by a central fault zone and tectonic stresses, making TBM drives a risky tunneling method. Due to this fact, a risk analysis was carried out for all the tunnel routes, and a drill and blast with probe drill was applied in the risky part of the tunnel.

11.4.1 Project Description and the Geology

The Bahçe Nurdağı tunnel is located at the borders of Osmaniye and Gaziantep in southeastern Turkey. The 17 km long double-track railway project includes the longest railway tunnels in Turkey. The railway bridges a gap in one of the busiest railway sections between Toprakkale and Malatya, where trains can only proceed with the support of an additional loco. With this project, the route between Bahçe and Nurdağı is 15 km shorter, with gentler gradients (dropping from 27 to 15%), wider curves (minimum radius will increase from 500 to 1500 m) and proceed at faster speeds (from 40 to 120 km/h). The tunnel is for railway transportation, and the project concerns two tubes, T1 and T2, 8 m in diameter and a length of 9750 m each. The excavation of T2 started in October 2014 with drill and blast tunneling method from chainage 13 + 450 m up to 12 + 600 m, including probe drilling and thereafter with single shield TBM. The excavation of T1 started with TBM in February 2021 and finished in August 2023.

Mixed ground conditions prevail in the project, ranging from abrasive, interbedded meta-sandstone and meta-mudstone with quartzite veins to highly weathered shale and dolomitic limestone. The chainage from 13 + 450 m to 12 + 400 concerns the Karadağ Limestone of the Mesozoic age, which is affected by the East Anatolian Fault (EAF).The geological formation from the chainage 12 + 400 m to 4 + 850 m concerns of middle Ordovician aged Kızlaç Formation of very massive interbedded meta-sandstone, meta-quartzite, and meta-mudstone having very high strength and abrasivity characteristics. However, the massive characteristic of the rock formation changes from chainage 4 + 850 m to 3 + 700 m, affected by local faults and shear zones; in this zone, RQD values are very low, and high water ingress is also expected. The meta-sandstone mainly comprises quartz with minor feldspar, opaque, and rock fragments. A thin chlorite cement holds the grains together. The meta sandstone can be classified as quartz arenite. The mean quartz content is 68% with grain sizes changing between 0.2 and 0.3 mm, and the mean value of Cerchar abrasivity is 4.2. The meta-mudstone consists of angular quartz grains embedded in a voluminous matrix of sericitic muscovite, opaque, chlorite, feldspar, and quartz. It shows a distinct cleavage defined by the parallel orientation of the mica grains. The mean quart content is around 48%, with grain sizes between 0.05 and 0.1 mm. The mean value of Cerchar abrasivitiy of this rock is 2.5.

The strength characteristics and brittleness index as the ratio of compressive strength to tensile strength play an important role in determining the penetration of disc cutters and, thus, net penetration rates of TBM s. The maximum compressive strength of rocks is 301.3 MPa for meta-sandstone and 327.4 MPa for meta-mudstone, with mean values of 213 MPa and 136.1 MPa consecutively. The ratio of mean compressive strength to tensile strength values for both rocks is around 8.4, indicating that maximum disc penetration will be low, considering the maximum disc bearing capacity, Bilgin et al. (2017), Jordan and Bilgin (2018a, 2018b).

Table 11.14 Recommended and manufacturer TBM design parameters of single shield TBM, Jordan, and Bilgin (2018b)

Parameter	Recommended value	TBM manufacturer value
TBM diameter	8 m	8 m
Disc number and diameter	53; 48.3 cm	53; 48.3 cm
Maximum load per disc	311 kN	311 kN
Disc spacing	80 mm	77.6 mm
Total power	3250 kW	10 × 330 kW
Cutterhead torque	4500 kN	4588 kNm at 6.9 rpm
Exceptional torque		14,453 kNm at 0–3.3 rpm
Recommended thrust based on disc bearing capacity	16,483 kN	16,503 kN
Maximum thrust		88,543 kN

11.4.2 The Selection of TBM

The job owner, TCDD, and the contractors İntekar Yapı and Biskon Yapı decided to charge the Department of Mining Engineering of Istanbul Technical University to carry out full-scale cutting tests in the laboratories for recommendation of TBM design parameters, estimation of the TBM performance, and later to do a risk analysis for the possibility of using drill and blast tunneling method in the critical parts of the tunnel. The recommended and manufactured TBM design parameters of single shield TBM are given in Table 11.14, and the selected TBM is shown in Fig. 11.18, Jordan and Bilgin (2018a).

11.4.3 Risk Analysis for TBM and Decision to Start with Drill and Blast

East Anatolian Fault (EAF), a large fault zone about 550 km long, is one of the primary sources of earthquakes, fracturing the rock mass greatly and applying tectonic stresses to the ground. High water ingress is expected in this area. Karadağ Limestone discharges the water at the toe of the Mountain at Nurdağı side. Several springs are present along the EAF. The tunnels excavated close to North and East Anatolian Faults led to the development of a risk classification system explained in detail by Bilgin (2016a, b). According to this classification system, the use of TBM in the Bahçe-Nurdağı tunnel between 13 + 450 m and 12 + 800 m is very risky, and from 12 + 800 m to 12 + 200 m is risky and favorable up to 4 + 850 m. First, excavating the first 1250 m up to 12 + 200 km was recommended by drilling and blasting the tunneling method, Bilgin et al. (2016, 2017). The recommendation is found reasonable by the job owner and the contractors since a predetermined time

Fig. 11.18 Single shield TBM used in Bahçe-Nurdağı railway tunnel, Gökçeoğlu 2022

was also necessary for the TBM manufacturer to provide the machine to get ready to run in the tunnel. When the tunnel reached 12 + 600 km, advancing 850 m, TBM was prepared to transport inside the tunnel without waiting for conventional tunneling to reach the chainage of 12 + 200 km. Readers will understand better the above paragraph consulting the geological cross-section given in Fig. 11.19 Bilgin (2016a, b).

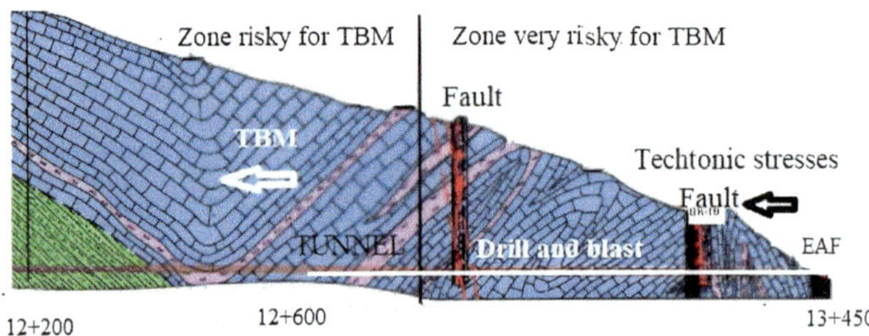

Fig. 11.19 The geological cross-section of the tunnel from 13 + 450 km to 12 + 200 km and the direction of tectonic tresses show risky zones for TBM, Bilgin (2016a, b)

11.4.4 The Performance of TBM

Surprisingly, the rock formation from 12 + 450 m up to 11 + 450 m was a very weak graphitic schist with a high water ingress. A water ingress of 400 l/s was encountered between chainages 12 + 420 m and 12 + 450 m in the transition zone between graphitic schist and weathered limestone (W3-W4). Within 1000 m of this weak zone, the TBM became stuck several times due to face collapses and the squeezing effect of the graphitic schist. Two bypass galleries were excavated at ring 1071, chainage 11 + 840 m, and ring 9250, chainage 12 + 075 m. Despite bad ground conditions, the TBM excavated the first 1050 m in 270 days, averaging 4 m daily. The TBM succeeded in hard and abrasive rock formations after chainage 11 + 400 m. Typical values for March 2015 reached 11.8 m per day in hard interbedded meta sandstone and meta mudstone as 11.8 m per day. Thereafter, mean daily advance rates changed from 10 to 16 m. Typical mean cutter consumption is 4.83 m/disc in quartzite, 4–9.4 m/disc in meta sandstone, and 12.7 m/disc in interbedded meta sandstone and meta mudstone.

11.4.5 Drill and Blast and Probe Drilling Within Chainages 13 + 450 and 12 + 600 km

Classical NATM techniques and probe drilling using Atlas Copco M2D and Mine Master Jumbo 1.7 were used between 13 + 450 and 12 + 600 km. The probe drilling was carried out at least at five different places on the tunnel face, and debris was collected carefully to identify the rock formation ahead of the face. Probe drilling rate is used in difficult geological conditions to indicate water ingress into a tunnel, gas potential ahead of a tunnel, fault, changing strata, and transition zones. However, the probe drilling rate is also related to the type of drill rig and drilling parameters such as feed, percussion, and rotation pressures, in other words, the thrust and the torque of the drill rig. Drilling rate measurements should be carried out under constant, percussive, and rotary pressures to compare the changing situations ahead of a tunnel. Before evaluating the measurements, drilling rate values should be normalized using thrust or rotary pressures for comparable values. However, it should also be emphasized that more than one probe drill hole will be necessary for difficult ground conditions and large-diameter tunnels. Bilgin and Ates (2016). A typical tunnel face and recording data are seen in Figs. 11.20 and 11.21. Probe drilling served to advance safely in a risky ground subjected to tectonic stresses and liable to high water ingress.

Figure 11.21 shows Probe drill data of the tunnel face, and Fig. 11.20 is from the author's archive, Bilgin.

Fig. 11.20 A typical tunnel face subjected to probe drilling, from the archive of the author Bilgin, N

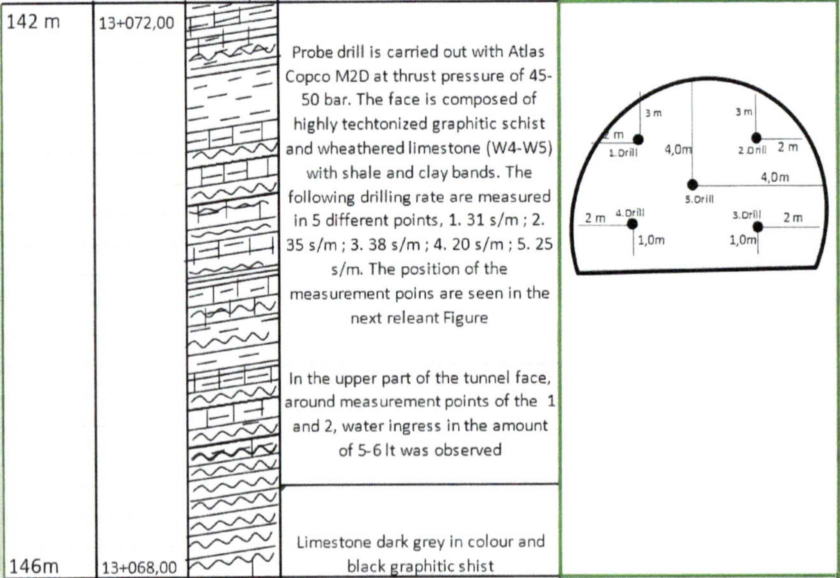

Fig. 11.21 Probe drill data of the tunnel face

11.5 Mount Macdonald Tunnel, Hybrid Tunneling, Drill, Blast and TBM

The Connaught Tunnel alone could not meet the increasing freight demands of coal, sulfur, and potash production in British Colombia, and the Mount Macdonald, 14.7 km in length, was built between 1984 and 1988 to increase the transport capacity of the Canadian Pacific Railway. Connaught Tunnel. Being 91 m lower than the Connaught Tunnel, the Mount Macdonald Tunnel climbs at a lower grade. This action allowed CPR to remove the pusher locomotives needed for the Connaught route. The new tunnel is 5.1 m wide and 7.8 m high and was the first to use the concrete "Pact-Track" floor system. This system eliminates the need for wooden ties and rock ballast, reducing maintenance costs. The tunnel also has a ventilation system with four 1.68-megawatt fans, a ventilation shaft, and a control and monitoring center. https://en.wikipedia.org/wiki/Mount_Macdonald_Tunnel, uploaded in May 2024) A view of the portal of the tunnel is seen in Fig. 11.22. The technical details below summarize an outstanding article published by Kuesel and Hansmire in (1987). All credit must be given to the authors.

Fig. 11.22 Eastbound freight entering the Eastern portal of CPR Mount Macdonald tunnel, photograph by David Davies. https://www.railwayage.com/mw/cp-plans-rogers-pass-tunnel-overhaul/?RAchannel=home, uploaded in May 2024

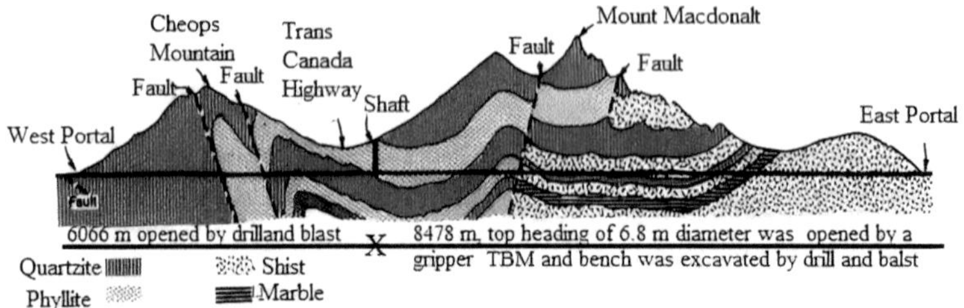

Fig. 11.23 Geological profile of the Mont Macdonald Tunnel, Kuesel and Hansmire in (1987)

11.5.1 The Geology

As seen in Fig. 11.23, the main rock formations consist of quartzite, phyllite, and schist in narrow bands. The West Tunnel was driven predominantly through high-strength quartzite, whereas the East Tunnel was opened mostly in softer schist with some quartzite and a small marble zone. Quartzite has a maximum compressive strength of 280 MPa. Fortunately, the quartzite occurred in bands and alternated with much softer phyllite and mica schist. Five major faults were distinguished during field mapping. Three joint sets correlated with the fold geometry were commonly associated with quartz veins or highly fractured rock zones. In general, the tunnel is pretty dry. The maximum overburden of the tunnel was 1700 m.

11.5.2 Method of Construction, Summary of Drill and Blast and TBM Tunneling

The west tunnel, 6066 m long, was opened by conventional tunneling method, primarily by drill and blast. The East tunnel had a length of 8478 m, and a top heading of 6.8 m diameter was opened with Robbins gripper TBM, and the bench was opened by drill and blast. The shaft and the west and east tunnels were constructed by three experienced contractors competent in their fields. The shaft was sunk through 90 m of saturated overburden with the aid of ground freezing. Conventional drill and blast methods sank the remaining 260 m in rock. Conventional drill-and-blast tunneling was used for the West Tunnel. The soil and mixed-face section at the portal was driven heading and bench. After encountering rock, the tunnel was excavated full-face using a Tamrock rail-mounted, six-boom jumbo. Typical progress was about 15 m per day with a three-shift operation or 75 m per five-day week. Concreting was done as a single arch and sidewall pour with a 24-m-long form. West Tunnel had safety rock reinforcement, but 30% was unsupported. The contractor for the East Tunnel used a 6.8-m diameter Robbins gripper TBM. Typical progress was about 15 m per day with a three-shift operation, or 75 m per five-day week, a North American record at that time for this size of tunnel. High rates were achieved only in weaker rocks having a

Table 11.15 Material consumption in ground support for East And West Tunnel, Kuesel and Hamsmire (1987)

Type of ground support	West		East	
	Estimated	Actual	Estimated	Actual
Atlas Copco Swellex, 2.4 m long	8100	13,200	15,500	10,400
Epoxy resin grouted bolts, 25 mm diameter	7100	2500	20,000	2500
Wire mesh, m^2	15,000	24,700	38,500	4900
Shotcrete for m^2	7500	600	22,500	3900

mean compressive strength of 70 MPa, where hourly penetration rates typically were 3–4 m/hr but as excellent as 5 m/hr on occasion. In the much stronger quartzites, the TBM penetration rate normally decreased by half to a low of about 1 m/hr. Cutter wear increased substantially, and machine availability decreased correspondingly. On one shift, all 52 cutters were replaced. Material consumption in ground support for the East And West Tunnel is given in Table 11.14.

It is reported that TBM excavation eliminated blast-induced fracturing, requiring less rock reinforcement than anticipated. Much of the tunnel required no support. Occasionally, when the rock bedding was oriented flat to the tunnel, slabs or wedges required support. Table 11.15 shows fewer than the estimated number of swelling bolts were used, and only 38% of the East Tunnel required support, Kuesel and Hamsmire (1987).

11.6 Concluding Remarks

This chapter discusses the hybrid approach to tunneling using drill and blast, and TBM in detail answering why and when this hybrid method is used. An example is the First Kaynarca-Kartal-Kadıköy Metro Line in Istanbul, with twin tunnels of 26.5 × 2 km. Almost 2/3 of this line is excavated with drill and blast; the rest is excavated with 4 EPB-TBM. The methodology, advantages, and results of using two different tunneling methods in this highly populated area of Istanbul are discussed. A cautious and controlled blasting model applied during tunneling minimized vibrations and environmental problems created by the conventional tunneling method. This project is unique in that most of the line was excavated with drill and blast in a highly populated area of Istanbul. A planned synchronized hybrid tunneling method enabled the contractor to finish the project from Kaynarca to Kozyatağı before the scheduled time. Therefore, each part of this project was discussed separately to help the readers benefit from the results of similar projects.

Another example of a hybrid tunneling project is the Kargı hydropower project in Turkey. From the launch of the TBM in March 2012 up until the end of July 2013, the TBM encountered 2228 m of extreme geological conditions, which required seven bypass tunnels and nine umbrella arches. TBM was modified to increase the machine's torque and thrust. The 7.8 km of the total length of 11.8 km of the tunnel is excavated with a double shield TBM of 9.84 m diameter, and 4 km of the tunnel from the inlet part is opened with NATM due to the challenging ground conditions faced by the TBM and shorten the construction time. In summary, the time for mobilization was 4 months for conventional tunneling and 16 months for TBM tunneling. In reasonably stable rock before modification of TBM, the daily advance rate was obtained as 9.7 m; after modification of TBM, this value went up to 19 m/day; for drill and blast, the mean advance rate was around 5.5 m/day.

Bahçe-Nurdağı tunnel is for railway transportation, and the project concerns two tubes, T1 and T2, 8 m in diameter and a length of 9750 m each. This tunnel is a typical example of one affected by a central fault zone/North Anatolian Fault and tectonic stresses, making TBM drives a risky tunneling method. Due to this fact, a risk analysis was carried out for all the tunnel routes, and a drill and blast with probe drill was applied in the risky part of the tunnel. The excavation of T2 started in October 2014 with the drill and blast tunneling method from chainage 13 + 450 m up to 12 + 200 m, including probe drilling, and thereafter, the excavation continued with single shield TBM. The excavation of T1 started with TBM in February 2021 and finished in August 2023.

The final example is the Mount Macdonald tunnel, the longest railway tunnel of 14 km in Canada. The geology is complex, with quartzite and phyllite schist in narrow bands, foliation zones, and faults. The maximum overburden of the tunnel was 1700 m. The contractor of the East Tunnel of 8478 m used a 6.8-m diameter gripper-type TBM for the top heading; the bench was then excavated by drill and blast. Typical progress was about 15 m per day with a three-shift operation or 75 m per five-day week. The tunnel is 5.1 m wide and 7.8 m high. The West Tunnel of 6066 m was excavated by drill and blast with another contractor. In West Tunnel, the typical progress was about 15 m per day with a three-shift operation or 75 m per five-day week. The shaft was sunk with a third contractor through 90 m of saturated overburden with the aid of ground freezing. Conventional drill-and-blast methods excavated the remaining 260 m of rock. The whole tunneling operation was terminated in two years. The authors of this book believe that this remarkable achievement is due to the hybrid tunneling method with good planning, efficient management, and correct equipment selection. Especially the choice of three different experienced contractors was one of the key factors in these problematic ground conditions.

References

Alan E (2018) Determining "k" and "β" coefficients of the formation that excavated by blasting on Kadıköy-Kartal metro route, the effect of detonator delay time on amplitude. In: Proceedings of the 4th international symposium on underground excavations for transportation, organized by Turkish Tunnelling Society and Chamber of Mining Engineers of Turkey, Istanbul Branch

Barton N (2012) Reducing risk in long deep tunnels by using TBM and drill-and-blast methods in the same project–the hybrid solution. J Rock Mech Geotech Eng 4(2):115–126

Barton N (2012b) Hybrid TBM and drill-and-blast from the start. Tunnel J:22–32

Barton N, Bilgin N (2016) Fast or slow progress with TBM in ideal or faulted conditions. In: Ulusay et al (eds) Rock mechanics and rock engineering: from the past to the future. Taylor & Francis Group, London, ISBN

Bilgin N (2016) An appraisal of TBM performances in Turkey in difficult ground conditions and some recommendations. Tunn Undergr Space Technol 57:265–276

Bilgin N, Ates U (2016) Probe drilling ahead of two TBMs in difficult ground conditions in Turkey. Rock Mech Rock Eng (2016) 49:2763–2772

Bilgin N, Çopur H, Balcı C (2016) TBM excavation in difficult ground conditions. Case studies from Turkey. Ernst & Sohn GmbH & Co. 978-1-138-03265-1

Bilgin N, Copur H, Balci C, Tumac D (2017) TBM Performance prediction using laboratory cutting tests in very hard and abrasive rock formations. In: International conference on tunnel boring machines in difficult grounds (TBM DiGs), Wuhan, 16–18 November 2017

Clark J (2015) Extreme excavation in fault zones and squeezing ground at the Kargı HEPP in Turkey. In: SEE tunnel: promoting tunneling in SEE region. ITAWTC 2015 congress and 41st general assembly May 22–28, 2015, Lacroma Valamar Congress Center, Dubrovnik, Croatia

Clark J, Chorley S (2014) The greatest challenges in TBM tunneling: experiences from the field. In: North American tunneling conference, p 101–108

Gökçeoğlu C (2022) Assessment of rate of penetration of a tunnel boring machine in the longest railway tunnel of Turkey. SN Appl Sci 4:19

Home L (2015) Hard rock TBM tunnelling in challenging ground: developments and lessons learned from the past. In: International conference on tunnel boring machines in difficult grounds (TBM DiGs) Singapore, 18–20 Nov 2015

Home L (2016) The Kargı challenge, a study of TBM boring vs conventional tunneling. Tunnel Expo organized by The Turkish Tunneling Society, Istanbul

https://en.wikipedia.org/wiki/Mount_Macdonald_Tunnel, downloaded in February 2024

https://www.railwayage.com/mw/cp-plans-rogers-pass-tunnel-overhaul/?RAchannel=home, downloaded in February 2024

Jordan D, Bilgin N (2018) Turkey's hardest rock: the Bahçe-Nurdağ high speed railway tunnel. Tunnel No 4:22–28

Jordan D, Bilgin N (2018b) Excavating Turkey's hardest rock at the Bahce-Nurdag railway tunnel. WTC 2018, Dubai

Kahriman A, Özer U, Adıgüzel D, Karadoğan A, Aksoy M, Muştu R, Dincel E, Alan,E., Sefer İ (2007) The analysis of ground vibrations induced by blasting on Istanbul Kadıköy–Kartal subway tunnel route. In: The 2nd symposium on underground excavations for transportation, 15–17 November 2007, pp. 571–582, symposium organized jointly by Turkish Chamber of Mines, Istanbul Branch, and ITU Faculty of mines, Proceedings. ISBN 978-9944-89-400-5

Kahriman A (2008) Preliminary design of blasting operations for Istanbul Kadıköy-Kartal Metro Tunnels, project report submitted to the contractor, Astaldi.

Kahriman A, Karakuş Y, Kaya E, Angelini O, Özdemir A, Sefer İ, Ekmen B, Arpak K, Sak A (2010) Investigation of blasting model applied during tunneling of Kadikoy-Kartal subway construction. In: Thirty-sixth annual conference on explosives and blasting technique February 7–10, Orlando FL USA, International Society of explosive Engineers, vol 2, p 1–13

Kuesel TR, Hansmire WH (1987) Rogers Pass tunnel-bore and blast. Tunn Undergr Space Technol 2(4):385–589

Özcan AK (2011) The performance of two TBMs in Kartal-Kaynarca section of Kartal-Kadıköy Metro Project. Okan University, Civil Engineering Department, Blasting Engineering Program, İstanbul, p 70. M.Sc. thesis

Özer U (2008) Environmental impacts of ground vibration induced by blasting at different rock units on the Kadikoy-Kartal metro tunnel. Eng Geol 100(2008):82–90. https://doi.org/10.1016/j.enggeo.2008.03.006

Şenol S (2014) Interim report of M.Sc. Thesis, supervised by Bilgin, N. The comparison of TBM and drill and blast tunnelling performance in Doloyaba formation. Okan University, Civil Engineering Department, Blasting Engineering Program, İstanbul, p 84

Şenol S, Biberoglu S, Posluk E, Eitaz I (2014) The comparison of blasting excavation in the running tunnel being excavated through Doloyaba formation. In: 67th geological congress of Turkey, 14–14 April 2

Yurt Y, Özturk A, Arslan Z, Nuhoğlu C, Oystein L, Erdoğan E, Atlar B, Palakçı, Y, Bilgin N (2014) Factors affecting the performance of a double shield TBM in a very complex geology in Kargi Turkey. In: Proceedings of the world tunnel congress 2014—tunnels for a better Life. Foz do Iguaçu, Brazil

Chapter 12
Selected Extreme Cases

Abstract This chapter explains first the effect of earthquakes on tunnels excavated with conventional and mechanized tunneling methods, given examples of Bolu and Nurdağı tunnels. Seismic joints designed for both different tunneling methods are also summarized. East Anatolian and North Anatolian faults create tremendous tectonic stresses along the slipping line before the occurrence of the Earthquake. These tectonic stresses affect the performance of TBMs by squeezing the machine as did in the Doğançay tunnel, which is one of the subjects of this chapter. An amount of 7.8 km of Kargı hydropower tunnel is opened with a double shield TBM using precast segments, and 4 km is opened with drilling and blasting method without lining, using only rock bolts. After one year of the operation of the powerhouse, significant collapses were observed in the unlined section of the tunnel. The method of investigation and remedial work of the collapsed area are also given in this chapter. This chapter also details using strand jacks to ease the steering of a 13.7 m diameter double shield TBM in the T26 high-speed tunnel and also the high performance of the 13.8 m diameter XRD TBM in the Eşme-Salihli tunnel.

12.1 Introduction

This chapter deals with the extreme cases where the tunneling sector will benefit from their notable results. Selected case studies will be summarized in this chapter for different topics such as the effect of earthquakes on tunnels, the design of seismic joints in conventional and mechanized tunneling, major fault zones creating tectonic stresses and squeezing effect on tunnels, the collapse and remedial works of unlined hydropower tunnels, the use of strand jacks to solve steering difficulties of large section TBM's in soft ground, exceptional performance of a large section TBM, the effect of incorrect selection of the tunnel route and tunneling methodology.

It is usually considered that tunnels are resistant to earthquake action, as they don't experience the same high levels of shaking as surface structures. This idea was supported by the relatively good historical performance of tunnels and underground structures during big earthquakes, especially of tunnels in rock. Dowding

N. Bilgin and C. Balci, *Critical Issues in Selecting Conventional and Mechanized Tunnelling Methods*, Springer Tracts in Civil Engineering,
https://doi.org/10.1007/978-3-031-89114-4_12

and Rozen (1978) presented one of the first surveys of case studies on the effect of earthquakes on rock tunnels. They collected information on 71 tunnels and compared their behavior with estimated peak ground accelerations (PGAs) and peak ground velocities (PGVs). Their conclusions can be summarized as the collapse of tunnels from shaking occurs only under extreme conditions: no damage happened when PGAs were lower than 0.19 g, and when PGVs were lower than 0.2 m/s minor to moderate damage occurred; when PGAs were up to 0.5 g, and PGVs up to 0.9 m/s moderate to heavy damage occurred when PGAs were larger than 0.5 g. Tunnel collapse only occurred associated with movement of an intersected fault. Kontogianni and Stiros (2003) share the findings of Dowding and Rozen (1978), emphasizing that the Wrights tunnel affected by the San Francisco earthquake in 1906, the Urban Kern Country railway tunnel affected by the Kern Country railway earthquake, and the Twin Bolu tunnels affected by Düzce Tunnel in 1999. All these tunnels were constructed by conventional tunneling methods or by NATM. This chapter discusses the Bolu tunnel as an extreme case affected by an earthquake. Seismic joints used in tunnels are essential prevention measures to diminish the destructive effects of the Earthquake. They must be considered separately for conventional and mechanized tunneling methods. In this chapter, this was addressed for Bolu and Eurasia tunnels.

GPS data give a lateral sleep of 24 ± 4 mm/yr along the North Anatolian fault, and also a left-lateral slip of 18 ± 6 mm/yr on the East Anatolian fault and ~ 5 mm/yr on the Northeast Anatolian fault generating an average shear stress building at 0.15 bar/yr along most of these two major fault zones, Stein et al. (1997). This phenomenon justifies the findings of Bilgin (2016), stating that tunnels bored by TBMs near these major fault zones are exerted to squeeze TBMs, as in Doğançay, Nurdağı, Kargı, and Uluabat tunnels in Turkey. In this chapter, the Doğançay tunnel will be given as an example of the effect of tectonic stresses.

Unfortunately, the hydropower industry has experienced several tunnel failures over recent years, as explained by Brox (2019). These tunnels have been designed and constructed as unlined tunnels incorporating rock bolts and shotcrete lining only in limited areas and are subjected to dynamic operating conditions. Conventional tunneling methods are used in the majority of power tunnels. Blasting operations usually diminish the rock mass strength of weak geological zones, and collapses become inevitable. A typical example given by Brox (2019) is the 181 MW Shuakhevi Hydropower project in Georgia, which included 35 km of tunnels, two dams, and two surface powerhouses. Construction of the tunnels commenced in 2013 using drill and blast methods. Georgia is associated with complex and disturbed geology due to its location between the Eurasian and Africa-Arabian tectonic plates with continental collision. The rock types noted along the tunnels comprised highly disturbed basaltic andesites, conglomerates, and volcanoclastic sandstones. Operations commenced in mid-2017, and after about 2 months, multiple collapses and other damage occurred in the main 19 km downstream tunnel. Most of the multiple collapses were reported to have entirely blocked the tunnel. The total volume of the multiple collapses is estimated to be about 15,000 m^3. The repairs of the tunnel were planned to include the construction of a 250-long by-pass tunnel at one of the major collapse areas and shotcrete repairs at the other smaller collapse areas. The repair of the tunnel lasted around 18 months. This event is similar to the problem associated with the Kargı power tunnel in Turkey, which resulted in 12 months of tunnel repair and stopped the project's power generation, Hilara et al. (2019). This occurrence will be summarized in this chapter; inspecting and repairing such collapsed tunnels needs special care and expertise, and we believe that the readers of this book will benefit from this case study.

A significant problem emerges when using large-diameter TBMs in mixed faces with intermediate soft ground, leading to the steering or deviating/sinking of the TBM. Finamore and Bellizzi (2019) described innovative equipment (strand jack and self-retaining system) to solve the TBM steering problem in the Basci tunnel. The TBM launching in this tunnel was in very weak rock conditions, comparable with soil. 13.65 m diameter and 1200 t hard rock single shield TBM had a total length of 133 m. From the start of the tunneling works up to advance 57 m, the excavation faced several face instability events blocking the cutterhead. The cutterhead could be restarted just after performing a proper cleaning manually. The anisotropic rock behavior and face instability encountered in the tunnel produced a downward deviation of the TBM axis concerning the project level being − 135 cm with a tendency of − 5.7 cm/m. In the Basci tunnel, the strand jack system was installed on the TBM to compensate for the lack of contact force at the tunnel face. The force applied by the strand jack system generates a negative moment, which is opposed to the moments related to the friction force and weight of the TBM; hereby, it helps steer the TBM up. The system consists of four hydraulic cylinders installed on the steel thrust structure at the tunnel's inlet, appropriately reinforced steel ropes, and four hydraulic cylinders installed on the TBM bulkhead, just behind the cutterhead. In addition to contributing to the upward rotation of the TBM, it also reduces the advancing speed while keeping the same force on thrust cylinders. After the installation of the strand jack, the vertical tendency began to rise, and once the design level was reached, it stabilized around the value of 0.00 mm/m. In this chapter, the case study describes the T26 high-speed tunnel as similar to the Basci tunnels. Using a strand jack to solve the steering problems of 13.3 m diameter single shield TBM in soft ground formation is more understandable.

12.2 Problem of Active Faults in Tunneling in Turkey

Turkey is in a tectonically active region that experiences frequent destructive earthquakes. On a large scale, the region's tectonics are controlled by the collision of the Arabian Plate and the Eurasian Plate. A large piece of continental crust, almost the size of Turkey, called the Anatolian block, is being squeezed to the west. The block is bounded north by the North Anatolian Fault and southeast by the East Anatolian Fault. The East Anatolian Fault (EAF) is eastern Turkey's major strike-slip fault zone. It forms the transform-type tectonic boundary between the Anatolian and northward-moving Arabian Plates. The North Anatolian Fault (NAF) is an active right-lateral strike-slip fault in northern Anatolia that runs along the transform boundary between the Eurasian and Anatolian Plates. Seven tunneling projects affected directly by the

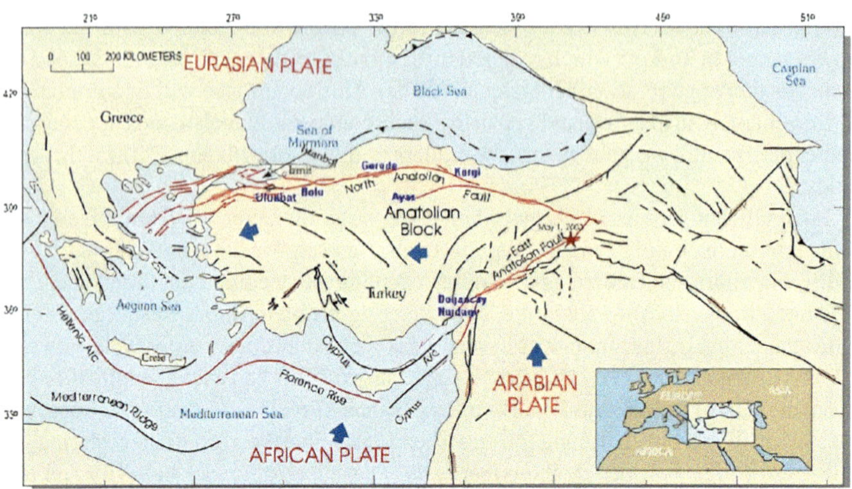

Fig. 12.1 North and East Anatolian faults in Turkey affecting tunneling activities, Bilgin (2016)

EAF and NAF faults are the Kargı, Doğançay, Gerede, Nurdağı, Ayaş, Uluabat, and Bolu tunnels, which are shown in Fig. 12.1, Bilgin (2016). The effect of active faults on tunneling emerges as face collapses, high deformations, excessive floor heave, squeezing of TBM, etc. Field observations in different tunnels enabled us to classify them as risky if they are between 0 and 5 km close to (EAF) and (NAF). The following section will be devoted to the Bolu tunnel bored by the conventional tunneling method. It was exposed to the November 12 Düzce earthquake, which caused the collapse of a part of the unfinished tunnel section. The tunnel was redesigned considering a seismic joint, Özben et al. (2005).

The Eurasia tunnel bored with a TBM crossing the Istanbul strait was 16 km close to the North Anatolian Fault. A seismic joint was also built in this tunnel. Accelerometers and displacement gauges were established in different places of the tunnel to decide whether to continue or discontinue the operation of the tunnel, Arıoğlu et al. (2023). Bolu and Eurosia tunnels will be opened to discussion as extreme cases of conventional and mechanized tunneling. Bolu Tunnel is exposed to severe damage to an earthquake and the Eurosia Tunnel may be under the influence of the expected big Istanbul Earthquake.

12.2.1 The Affect of Earthquake on Bolu Tunnel and Design of Seismic Joint

The 25.5-km-long Bolu Mountain Pass includes the 4.6-km-long four viaducts, 900-m-long three bridges, and Bolu Mountain Tunnel, built as 3 arrivals, three going lanes, and a double tube. The right tube in the direction of Ankara is 2788 m, and the left tube in the direction of Istanbul is 2954 m. The first excavation of the Bolu Mountain Tunnel started on April 16, 1990. During the project's construction, it experienced some delays due to two times flooding and two times exposed to earthquakes. The initial cost of the project was set at $ 570.5 million. Bolu Tunnel was opened on January 23, 2007, https://raillynews.com/(2024). However, as the minister of the Ministry of Transport and Infrastructures declared, the cost of the 10-year delay of the Bolu Mountain tunnel of 5.742 km in length, which was completed in 17 years, is 400 million dollars, https://www.bolugundem.com/haber/12031388/13 Mayıs 2022.

Bolu tunnels are separated by about 50 m, and emergency cross passages are every 500 m. The tunnels are being bored through seismically active faulted sequences of rocks. The geology consists of a highly tectonised and intermixed series of mudstone, siltstone, and limestone, with highly plastic pure fault gouge clay and smectite minerals changing from 35 to 50%. It should be noted that smectite is a clay mineral having a swelling characteristic. The plasticity index of clayey zones was between 25 and 55%. Overburden reaches 250 m maximum, with most of the tunnels under a 100–150 m cover. In some zones, the ground consisted of uniform fault gouges with no hard inclusions. Such zones have been encountered with a thickness of 50 m or more along the tunnel alignment. The studies carried out showed that the Bakacak fault is approximately 10 and 15 km long, and it crosses the Bolu tunnels within a 200 m-wide zone, Özben et al. (2005)

At the beginning of the excavation, every tube had two working faces. Since some good ground conditions existed, the drill and blast method was used within 400 m from the start. The remaining part of the tunnel was excavated by a conventional excavator equipped with impact hammers. The equipment used is summarized in Table 12.1.

Due to the Düzce earthquake, the Bolu tunnels experienced a wide range of damage, depending on the ground conditions, the construction method, and the construction phase. Completed tunnels performed well, but in poor ground, partially completed tunnels, where only the initial support had been installed, suffered severe damage and even collapsed. Menkiti et al. (2001) and O'Rourke et al. (2001) describe well the tunnels' performance during the Earthquake. The Bolu tunnels are part of the North Anatolian Motorway linking Ankara to Istanbul and have been under construction since 1993. Two major earthquakes occurred in Turkey in 1999. While the August earthquake left the tunnels undamaged, the November earthquake caused a collapse of part of the unfinished tunnel section. Almost 3 months after the severe Kocaeli Earthquake of August 17, 1999, another earthquake with a moment magnitude of 7.2 hit Turkey on November 12, called the Düzce earthquake, which caused nearly 1000 fatalities and 5000 injuries. Two viaducts and tunnels under construction exhibited extensive damage along Trans–European Motorway (TEM) segment near Bolu. Both Bolu tunnels were completely closed about 200 to 300 m from the Elmalı portal entrances. The collapses occurred in a tunnel section passing through a clay/weak rock zone where a temporary shotcrete lining system was in place. After the Düzce Earthquake, a seismic joint, as seen in Fig. 12.2, was designed for the Bolu tunnel. Seismic joints of 4.40 m spacing and 0.50 m spanning were constructed at the Bakacak Fault crossing.

Table 12.1 The equipment used in Bolu tunnels before redesigning the tunneling works, tunnel-builder.com›News/Bolu-Tunnel-Breaks-Through. Bolu tunnel breaks through in Turkey 01/10/2005

Job to be executed	Equipment used
Excavation	CAT 235C ME excavator with jack hammer (250 hp)
Muck loading	One CAT 966F loader (220 hp, 3.8 m^3), one CAT 9560E loader (167 hp, 3 m^3)
Mucking-out	Astra trucks BM 64.26 (260 hp, 14 m^3, 20 tons)
Forepoling	Atlas Copco MAI self-drilling anchor bolts and steel pipes
Shotcrete	Shotcrete with Bekaert's Dramix steel fibers
Forepoling	Atlas Copco MAI self-drilling anchor bolts and steel pipes, an Hydraulic truck drill, a Puntel PX 100 (120 hp) and a Puntel JET1 106 h for forepoling and boreholes
Transmixers for concrete	Astra-CIFA BM 64.31 (260 hp, 10 m^3), Iveco-CIFA 330 (2 hp, 10 m^3) and Iveco-CIFA 300 (260 hp, 9.5 m^3)
Steel rib installation	Dieci 70 mobile telescopic elevator (100 hp, 7 tons)
Installing bolts	Atlas Copco GD Anker M400 pump (600 l/h) for bolt injection, an Hydraulic truck drill

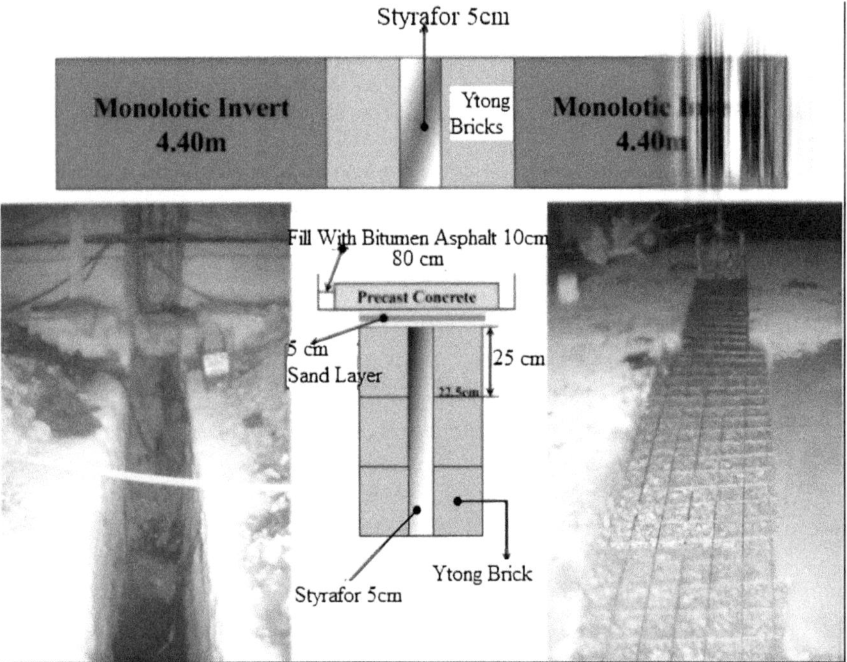

Fig. 12.2 Seismic joint designed for Bolu tunnel after Düzce earthquake, Özben et al. (2005)

What were the major drawbacks in planning the Bolu Mountain tunnels? No seismic survey was conducted before the construction of the tunnel. The Client defined a tunnel alignment corridor in 1988–1989, based mainly on geometric, topographic, and human settlement constraints. The Design and Build Contractor followed up in 1989–1990 with a limited number of surface boreholes to determine the geological characteristics of the tunnel for design purposes. Surface drilling was hampered by the high overburden (i.e., long drilling lengths and sampling depths), the rugged terrain, and the severe winters (which curtailed the annual construction season). In addition, only small volumes of ground could be sampled by drilling. As a consequence, the quality of these boreholes was poor. Refraction seismic surveys also proved imprecise in the variable ground conditions. However, due to the urgency of the Project, tunnel designs were prepared based on the standard Austrian rock classification system, Menkiti et al. (2001). It is a fact that not even tunnel alignment was picked up correctly considering the risk of an earthquake, but the place of the portals was also not selected in the right place as they should be, as the geology pointed out. On April 2, 2022, 15 years after the tunnel's opening to traffic, heavy rainfall triggered an economically significant landslide at Mount Bolu in Turkey. As seen in the Fig. 12.3, the slope above a tunnel portal is poorly designed on such an important highway, especially in a seismically active region. The debris has taken 44 h to clear as reported by in https://blogs.agu.org/landslideblog/2022/04/05/mount-bolu-1/, reported by Dave Petley.

Fig. 12.3 The landslide at Mount Bolu in Turkey. Photograph with courtesy of Turkiye Gazetesi from https://www.turkiyegazetesi.com.tr/

12.2.2 Eurasia Tunnel and Seismic Joints

The Eurasia/Avrasya Tunnel is a road tunnel crossing the Bosphorus strait. The tunnel was officially opened on December 20th 2016.The 5.4 km double-deck tunnel connects Kumkapı on the European part and Koşuyolu, Kadıköy, on the Anatolian part of Istanbul with a 14.6 km route including the tunnel approach roads. It crosses the Bosphorus beneath the seabed at a maximum depth of − 106 m. The TBM section crossing the Bosphorus is 3.4 km long, while another 2.0 km is constructed by the New Austrian Tunneling method and cut-and-cover method. The tunnel's excavation diameter is 13.7 m, which allows an inner diameter of 12.0 m with 60 cm thick precast segments.

The tunnel alignment is located in a seismically active region, about 16 km to the North Anatolian Fault Zone. To decrease the seismic stresses/strains below the permissible levels, two flexible seismic joints with displacement limits of ± 50 mm for shear and ± 75 mm for extraction/contraction were established. The design earthquake magnitude was specified as moment magnitude of 7.25 Arıoğlu et al. (2023). The 3340 m tunnel was excavated and constructed with a 13.7 m diameter Herrenknecht Mixshield Slurry TBM exclusively designed and equipped with the latest 483 mm disc cutters all atmospherically changeable and having individual wearing sensor system including 192 piece cutting knives with only 48 atmospherically changeable. Special pressurized locks for divers and material and rescue chamber for all shift members existed within the tunnel. Slurry Treatment Plant had an installed power of 3.5 MW or 4700 Hp which desilts the bentonite slurry used during excavation and the TBM had 33 kW/m^2 cutterhead power, and designed for 12.0 bar face pressure. TBM operation was completed in 479 calendar days, resulting in an average advance rate of 7.0 m/d or 9.0 m/d for working days. Seismic Joints used in Eurosia tunnel are steel-rubber composite flexible rings that allow in and out of plane movement resulting from seismic events. Main structure of Eurasia Tunnel consists of RC (Reinforced concrete) segment with "Tenon Joints" whereas seismic joint consists of steel segment with "Bolt Joints". Seismic segment system mainly includes steel bearing bars, ohm rubber and corrugated rubber. Seismic joints established in Eurasia tunnel are seen in Fig. 12.4, Arıoğlu et al. (2023). Seismic joints are placed in the tunnel where the alignment passes within the transition zone of Trakya Formation and marine sediments. As seen in Fig. 12.5, this is weakest part of the tunnel alignment where TBM torque is at minimum level. 15 accelerometer were placed on segments to measure the effects of earthquakes that will occur nearby on the tunnel, and 9 displacement gauges were placed on seismic joints to decide whether to continue or discontinue the operation of the tunnel. The places where the gauges are placed are seen in Fig. 12.6. An accelerometer placed on a segment is seen in Fig. 12.7, and a displacement gauge mounted on a seismic joint is seen in Fig. 12.8.

Fig. 12.4 Seismic joint in Eurasia tunnel, Arıoğlu et al. (2023)

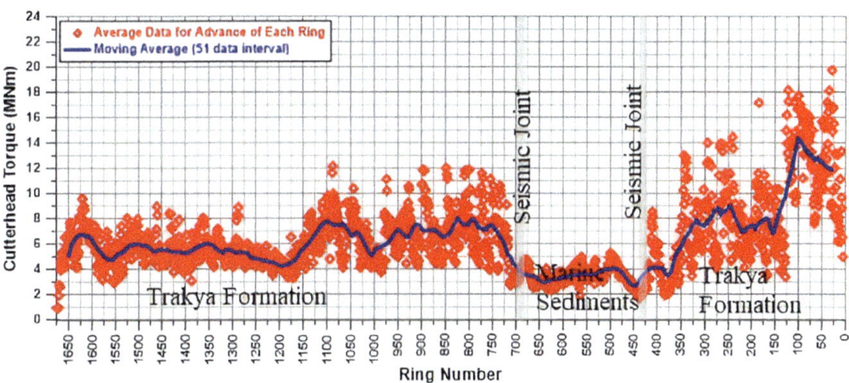

Fig. 12.5 The variation of the TBM cutterhead torque with ring numbers, Arıoğlu et al. (2023)

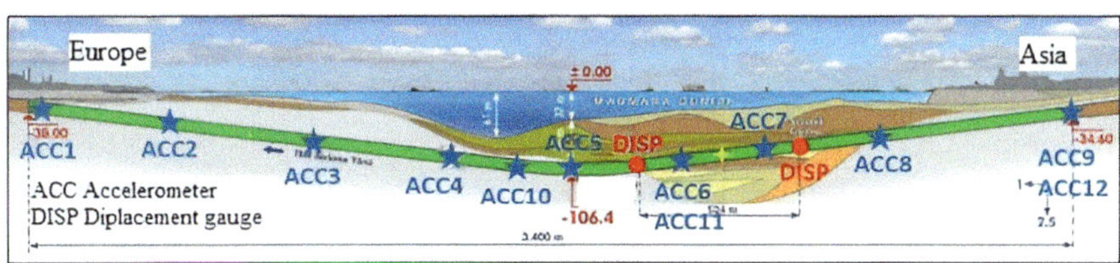

Fig. 12.6 The places of accelerometer and displacement gauges, Arıoğlu et al. (2023)

Fig. 12.7 3-axial Accelerometers on tunnel segments, Arıoğlu et al. (2023)

Fig. 12.8 Displacement gauges on seismic joints, Arıoğlu et al. (2023)

12.2.3 *Doğançay Tunnel, the Effect of Tectonic Stresses*

This section is taken from a publication of the authors Bilgin et al. (2016), and it is added to this section since we believe that it is unique in the sense that it shows the effect of the East Anatolian Fault creating tectonic stresses to affect the performance of a TBM. The Doğançay tunnel is located in the northeast of Turkey, affected by the North Anatolian Fault, and it is part of a hydroelectric project licensed by Enerji Sa. The tunnel length is 6655 m, and the excavation started in September 2012 and ended in July 2015. The tunnel was excavated by a double-shield Herrenknecht M-1665, having a diameter of 4.1 m, max. the thrust of 19,700 kN, the auxiliary thrust of 18,260 kN, the nominal torque of 1749 kNm, the gripper force at 400 bar of 25,334 kN, and the rotational speed of 0–12 rpm. The tunnel route contains Carboniferous-age limestone and shale, Upper Permian-age limestone and Lower Permian-age siltstone, claystone, sandstone, and quartzite. The frequency of encountered rock types is 54.7% limestone, 33.4% interbedded sandstone-quartzite-claystone, 7% interbedded sandy limestone and shale, and 5% shale. RMR varies between 10 and 70 and is mostly between 10 and 20 in shale. The overburden within 3 km of the tunnel route is around 1,000 m and only two boreholes could be opened in the tunnel route before starting the excavation. Tectonic stresses squeezed the TBM several times, causing considerable delays in tunnel drivage. Figure 12.9 shows the general layout of the tunnel route with the main faults (F1, F2, etc.) The squeezing of the TBM occurred at the contact areas of two faults, F1 and F2. The rock samples taken from the squeezing zones are also seen in Fig. 12.9, which clearly shows that rock samples bear traces of tectonic stresses, suggesting that the main reason for the TBM jamming is tectonic stresses. Within a site investigation carried out by the author, it was also estimated that the same type of squeezing would also occur at the contact points of the faults with the tunnel route. The TBM passed fault F3 with some small problems, but it jammed around Fault F4 on March 14, 2015. However, the TBM was rescued in 21 days by opening two rescue galleries. It is important to note that the orientation and magnitude of the tectonic stresses are important in a tunnel project, and it is highly recommended to include such measurements in tunnels that are planned to be opened around major fault zones. The ratio of torque to thrust force is a good indicator of the squeezing of a TBM. Figure 12.10 shows the variation of this ratio between 12 April, 2015 and 14 April, 2015. As can be seen from this figure the ratio started at 2, and dropped to 0.5 by the 7th ring, then later increased up to 1.5 and dropped again to 0.5, and the TBM was jammed within the 25th ring.

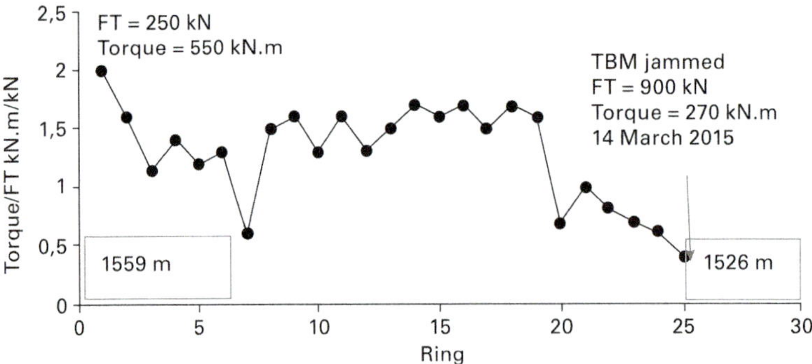

Fig. 12.9 Geological profile of Doğancay project and the evidence of tectonic stresses on the rock Samples, Bilgin et al. (2016)

Fig. 12.10 The variation of the TBM torque/thrust ratio with ring numbers within squeezing zone in Doğancay tunnel, Bilgin et al. (2016)

12.3 The Collapse of the Kargı Power Tunnel, Method of Investigation and Repair of the Tunnel

As Brox (2019) reported in one of his latest publications, the hydropower industry has experienced several tunnel failures over recent years for those opened with conventional tunneling methods since these tunnels are subjected to dynamic operating conditions and they have been designed and constructed as unlined tunnels with shotcrete lining and roof bolts only in limited areas. However, since 1995, the hydropower industry has enjoyed over 300 km of successful construction of long, low-pressure tunnels using mainly double shield TBMs in conjunction with precast concrete segmental linings with no reported problems, Brox (2021). One example of this is the Kargı hydropower project developed to utilize the energy potential of the Kızılırmak River between the towns of Osmancık and Boyabat in North Turkey. The geological map of the Kargı tunnel of the unlined part is given in Fig. 12.11. The water tunnel excavation by a double-shield TBM technology started from the lower (outfall) portal in February 2012. Initially, mechanical excavation was planned for the whole tunnel, and the segmental lining was proposed only for the initial 3 km of the tunnel, with the rest intended to be supported with shotcrete and rock bolts. After only boring 80 m, TBM became stuck in a collapsed mixed ground face of hard rock and running ground.

In the first 2 km of the tunnel, the TBM jammed 7 times due to face collapses, and bypass tunnels were required to free the TBM. Thus, it was decided to start with a conventional Drill and Blast (D&B) excavation from the upper (inlet) portal as a mitigation measure to save time. After a few collapses, the cutterhead drive torque capability was increased by more than 50%, and the thrust capacity of TBM was also increased by hydraulic power pack cylinder jacks, including overbore capabilities, shield lubrication, and belt scales. The average advance was 144.2 m/month before modifications and 407.7 m/month after modifications. A length of 7.8 km of the tunnel was excavated with a double shield Robbins TBM of 9.84 m diameter, and 4 km of the tunnel from the inlet part was opened with NATM, Home (2015). Due to problems with the mechanized excavation, a 4 km long upper tunnel section was excavated conventionally by the Drill and Blast method with single shell lining

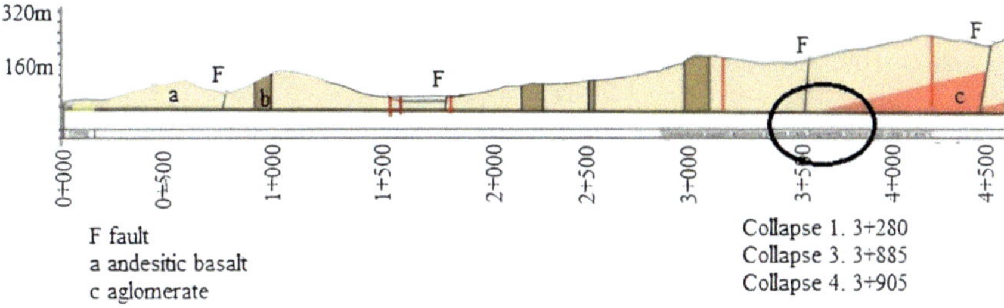

Fig. 12.11 The geological map of the Kargı tunnel opened with the conventional tunneling method from the Archive Bilgin

generated from fiber-reinforced sprayed concrete and rock bolts. Continuous probe drilling in the TBM tunnel was carried out for 5 km along the tunnel, and an umbrella arch was applied in 9 different areas to stop face collapses. The 11.8 km hydropower tunnel was finished in July 2014, and power generation started in August 2015. Significant operational problems in the powerhouse were observed in September 2016. The flooded tunnel was investigated by an underwater Remotely Operated Vehicle (ROV), allowing 3D scanning, which indicated significant collapses, Hilara et al. (2019). Operation of the power station had to be stopped, and water from a supply dam at the headrace tunnel inlet had to be discharged to enable dewatering and inspection of the tunnel. Consequent investigation revealed three major collapses and a considerable variety of other damages in the D&B section of the tunnel—Fig. 12.12. Collapse one after water pumping completion is seen in Fig. 12.12 and grouting of the collapse three is seen in Fig. 12.13. The repair and strengthening of the tunnel was completed in October 2017. It is important to note that these three collapses occurred in the transition zone between andesitic basalt and agglomerate and close to a fault. This example shows that the tunnels passing similar ground formations should be treated carefully.

Fig. 12.12 The collapse one after water pumping completion, Hilara et al. (2019)

Fig. 12.13 Grouting of the collapse area 3, Hilara et al. (2019)

12.4 T26 High-Speed Tunnel, the Problem of Steering/Sinking of TBM

The excavation of the T26 Tunnel started with the NATM method in 2010. The tunnel is in the North-East of Turkey. Due to geological problems, shear zones, fault zones, and low RQD values, the daily advance rates were very slow. The tunnel excavation method was changed to mechanical excavation using a 13.7 m diameter single shield TBM at the end of 2011. After a few hundred meters of advance, TBM started to deviate down to the tunnel. The TBM deviated vertically approximately two meters after the tunnel's portal opened. A strand jack/rope system, as defined by Finamore and Bellizzi (2019), was used to retract the deviated TBM and align it with the direction of the tunnel.

For this work, the necessary apparatus to connect the pistons to be used during pulling to the TBM is first welded onto the TBM. Then, one end of the pulling piston is connected to this apparatus. At the other end of the piston, a rope laid along the tunnel, one end of which is connected to the tensioning mechanism fixed at the tunnel entrance, is connected. Profiles for guidance are welded to prevent the rope stretched between the piston and the fixing mechanism from oscillating and to ensure that equal force falls on each rope in the rope group. The rope group is connected between the piston and the tensioning mechanism in the entrance portal, and the slack in the rope is taken and tensioned to make the system ready; the system used is seen in Figs. 12.14 and 12.15. While the rope is tensioned, the piston's stroke is fully opened. After all the preparatory work is completed, the recovery process begins. While the stroke of the piston attached to the TBM is fully open and the rope is taut, the stroke is reversed by the hydraulic pressure applied to the piston. As soon as the piston starts to close, it will begin to strain, and the system pressure will increase since one end is connected to the mechanism in the portal with ropes, and the other end is connected to the TBM. When the pressure reaches the sagging point, the TBM connected to the piston will start to move backward, as it will be easier for the TBM to move than the other end. The movement will continue until the piston is completely closed. When the piston is completely closed, the TBM will be pulled backward for its stroke length. Then, the piston will be fully opened again, the gap in the rope will be taken with the tension mechanism, the rope will be tensioned again, the piston will be closed again, and the TBM will be pulled backward. This process is repeated until the TBM reaches the desired position.

Fig. 12.14 A strand jack/rope system used for the salvage of deviated TBM from the archive of the author, Bilgin

Fig. 12.15 Robes of strand jack fixed within TBM, from the archive of the author, Bilgin

12.5 Word Record in Eşme-Salihli Tunnel with an XRE Crossover TBM

This section summarizes the success of mechanized tunneling in a well-planned and organized project with the proper choice of equipment. The Eşme-Salihli Railway Tunnel is part of the Ankara-İzmir High-Speed Railway Project of the Turkish State Railways (TCDD). The 508 km line will connect Polatlı in Ankara Province to Izmir. The Eşme-Salihli Railway Tunnel is a short section of a high-speed railway measuring 3.05 km through mixed ground conditions. A 13.77 m diameter Crossover XRE machine, a dual mode rock/EPB TBM design capable of operating in EPB mode or hard rock single shield mode, was selected for the project. The TBM was designed for hard rock and mixed ground sections in highly variable conditions. The design parameters of the TBM are given in Table 12.2, and the variation of monthly advance rates is shown in Fig. 12.16. This figure is a typical learning curve showing a steady increase in advance rates by excavation time. TBM used in this project is a complex machine. The interaction between the machine and the ground, plus increasing the experience of the tunneling crew over time, make a positive impact on the performance of the TBM, and as a consequence of this, the progress rate increase and levels off thereafter, Jordan et al. (2023), Negro et al. (2023), Kansu et al. (2022), Bilgin and Acun, (2024)

Table 12.2 Design parameters of Robbins XRE TBM, Kansu et al. (2022)

Model name	XRE Cross EPB451-39
Excavation diameter	13.77 m
Main drive system	VFD-19X220 kW
Total main drive power	4180 kW
Torque capacity	32.000 kNm @ nominal 78,000 kNm @breakout
Cutterhead rotation	Bi-directional
Cutterhead rpm range	4.6 @Hard Rock Mode 2@ EPB Mode
Thrust capacity	216,000 kN @normal 282,000 kN @exceptional
Number of disc cutters	63 × 19″ Single disc Cutters 8 × 17″ Twin Disc Cutters
Number of scrapers	277 × Face Scrapers 32 × Gauge Scraper Sets
Number of foam lines	10 × Lines for Water (100 lpm) 8 × Lines for Foam (880 lpm)

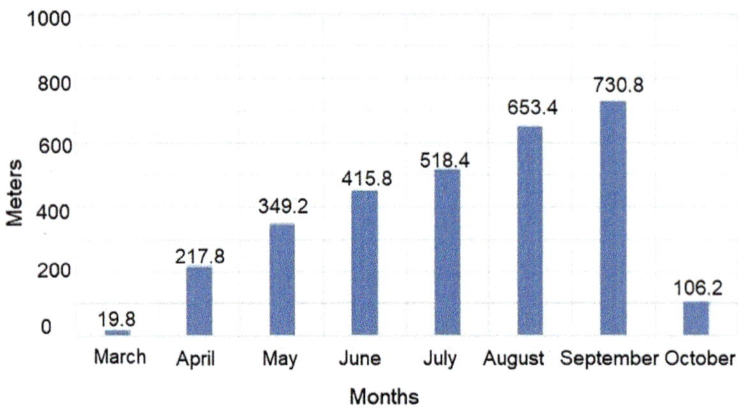

Fig. 12.16 The variation of advance rate with month, Jordan et al. (2023)

During the excavation, the machine set new world records three times for best day, week, and month for all TBMs over 13 m in diameter. Its fastest rates were set in July and August 2021, with a best day of 32.4 m, best week of 178.2 m, and best month of 721.8 m. The machine's final breakthrough occurred in October 2021. A summary of the best advance is given in Table 12.3.

What allowed the machine to perform at high rates of advance? This is the success of the tunneling crew's experience, well-organized tunnel management, and properly selected TBM with the technical support of the machine manufacturer. A detailed work analysis showed that 38% of the total time was spent on excavation, 27% on ring

Table 12.3 Best advance rates in Eşme-Salihli tunnel, Jordan et al. (2023)

Best daily advance (m)	32.4
Best weekly advance (m)	178.2
Best monthly advance (m)	721.8
Average monthly advance (m)*	486.9

*Excluding the first and last month which include launch and the breakthrough

Table 12.4 Distribution of the downtime, Kansu et al.

Cutterhead maintenance	28%
Day offs	22%
Cooling system	15%
Maintenance and hydraulics	8%
Other	8%
Conveyor belt extension	6%
Electrical	4%
Cleaning and maintenance	3%
Conveyor belt repair	3%
OG cable extension	3%
Logistics	2%
Injection system	2%

assembly, 21% on planned downtime such as maintenance, and 14% on unplanned downtime. The downtime distribution is given in Table 12.4, which will guide the success of similar tunneling projects.

In the Eşme-Salihli tunnel, 72 workers worked within the tunnel in two shifts and 36 in one shift. The distribution of workers in one shift is given as 2 Engineers; civil, mining, mechanical, geologist, or geophysicist, 2 formen, 3 pipefitters; 1 welder, 4 mechanic; 4 electricians; 7 operators; TBM, segment erector, locomotive, etc., and 13 laborers. 46 people were working outside of the tunnel. However, 120 people were employed in the segment plant: 50 iron workers, 50 mold workers, and 30 people charged with maintenance, electricians, and operators of mixers and cranes.

However, if the performance of mechanized tunneling is compared to that of conventional tunneling, it may be concluded that conventional tunneling can achieve a good performance through a highly mechanized D&B method. This method has the advantages of a more flexible excavation system, such as the easier possibility of adjustments in terms of supports, ground improvements, etc. Falanesca et al. (2023) give the following examples. For the Gothard Base Tunnel, 54.7 km of the total 153 km of excavations was carried out by conventional methods. The daily advance rate with a conventional drive for single-track tubes with overburden up to 2400 m at the mean advance rate in favorable rock conditions is 3.0 to 4.5 m/d for cross-section between 61 and 80 m^2, and the mean advance rate in unfavorable rock conditions is approximately 1.0 m/day for the cross-section between 85 and 110 m^2, and max. The advance rate recorder for the single-track tube Sedrun South East is 11.5 m/d.

12.6 Concluding Remarks

This chapter is devoted to the extreme cases where the tunneling sector will benefit from their notable results. What did we learn from them? It is usually considered that tunnels are resistant to earthquake action, as they don't experience the same high levels of shaking as surface structures. Dowding and Rozen (1978) collected data from 71 tunnels subjected to severe earthquakes. They concluded that no damage occurred when PGAs (peak ground acceleration) were lower than 0.19 g, PGVs (Peak ground velocity) were lower than 0.2 m/s, minor to moderate damage occurred when PGAs were up to 0.5 g, and PGVs up to 0.9 m/s, moderate to heavy damage occurred when PGAs were larger than 0.5 g, tunnel collapse only occurred associated with movement of an intersected fault. Dowding and Rozen's results justify the damages in Bolu Mountain tunnels that were subjected to the Düzce Earthquake in Turkey. Typical seismic joints designed for conventional and mechanized tunneling are given within this chapter as examples to diminish the damages made by Earthquakes. East Anatolian and North Anatolian faults create tremendous tectonic stresses along the slipping line before the occurrence of the Earthquake. Stein et al. (1997) computed that these two major fault zones generate an average shear stress building at 0.15 bar/yr. These tectonic stresses affect the performance of TBMs by squeezing the machine as did in Doğançay, Nurdağı, Kargı, and Uluabat tunnels in Turkey, Bilgin (2016).

A significant problem emerges when using large-diameter TBMs in mixed faces with intermediate soft ground, leading to the TBM's steering or deviating/sinking. Finamore and Bellizzi (2019) described innovative equipment (strand jack and self-retaining system) to solve the TBM steering problem in the Basci tunnel. The same deviating/sinking problem of the 17.7 m diameter TBM in the T26 high-speed tunnel in Turkey was solved using the same salvage equipment.

The hydropower industry has experienced several tunnel failures over recent years for those opened with conventional tunneling methods since these tunnels are subjected to dynamic operating conditions. They have been designed and constructed as unlined tunnels with only shotcrete lining and roof bolts in limited areas. However, since 1995, the hydropower industry has enjoyed over 300 km of successful construction of long, low-pressure tunnels using mainly double shield TBMs in conjunction with precast concrete segmental linings with no reported problems to date Brox (2021). A typical example of this is the Kargı hydropower tunnel in Turkey. The water tunnel excavation by a double-shield TBM technology started from the lower (outfall) portal in February 2012. The machine faced several problems, such as squeezing of TBM and face collapses. As a mitigation measure to save time, it was decided to start with a conventional Drill and Blast (D&B) excavation from the upper (inlet) portal. A 4 km long upper tunnel section was excavated conventionally by the drill and blast method with single shell lining generated from fiber-reinforced sprayed concrete and rock bolts. After the operation of the powerhouse, significant operational problems were observed. Operation of the power station had to be stopped, and water from a supply dam at the headrace tunnel inlet had to be discharged to enable dewatering and inspection of the tunnel. Consequent investigation revealed three major collapses and a considerable variety of other damages in the D&B section of the tunnel. The repair and strengthening of the tunnel was completed almost in one year. Notably, these three collapses occurred in the transition zone between andesitic basalt and agglomerate and close to a fault. This example points out that the tunnels passing similar ground formations should be treated with great care, especially in transition zones.

During the Eşme–Salihli high-speed railway tunnel excavation, Robbin XRE Crossover TBM set new world records three times over for best day, week, and month for all TBMs over 13 m in diameter. Its fastest rates were set in July and August 2021, with a best day of 32.4 m, best week of 178.2 m, and best month of 721.8 m. The machine's final breakthrough occurred in October 2021. It should be emphasized that the success of mechanized tunneling mainly depends on the proper selection of the TBM, on the experience of the tunneling crew, on the well-organized tunnel management, and on the technical support of the machine manufacturer.

References

Arıoğlu B, Arıoğlu E, Gökçe B (2023) Eurasia tunnel, emphasis on seismic design. In: Tunnel 2023, an international symposium organized by the Turkish Tunneling Society, was held from 23 to 24 November. The presentation is open access on the Turkish Tunneling Society web page

Bilgin N (2016) An appraisal of TBM performances in Turkey in difficult ground conditions and some recommendations. Tunn Undergr Space Technology 57:265–276

Bilgin N, Copur H, Balci C (2016) TBM excavation in difficult ground conditions case studies from Turkey, Ernst&Sohn, p 336.

Bilgin N, Acun S (2024) Practical management of tunnelling with tunnel boring machines. CRC Press, p 230

Brox D (2019) Hydropower tunnel failures: risks and causes. In: Peila, Viggiani and Celestino (eds) Tunnels and underground cities: engineering and innovation meet archaeology, architecture and art. Taylor & Francis Group, London. ISBN 978-1-138-38865-9

Brox D (2021) Practical guide to rock tunneling, 2nd edition, Vancouver, p 366.

Dowding CH, Rozen A (1978) Damage to rock tunnels for Earthquake shaking. J Geotech Eng Div Am Soc Civ Eng 104:175–191

Falanesca M, Merlini D, Schürch R, Marclay R, Neuenschwander M (2023)Benchmarking tunneling production rates: challenging case histories of mechanized and conventional tunneling in different geological conditions. In: Anagnostou, Benardos and Marinos (eds) Expanding underground. knowledge and passion to make a positive impact on the world. The Author(s), ISBN 978-1-003-34803-0

Finamore A, Bellizzi G (2019) TBM steering difficulties. Innovative equipments: strand jack and self retaining systems. In: Peila, Viggiani and Celestino (eds) Tunnels and underground cities: engineering and innovation meet archaeology, architecture and art. Taylor & Francis Group, London, ISBN 978-1-138-38865-9

Hilara M, Srb M, Nosek J (2019) A headrace tunnel reconstruction in Turkey. Acta Polytechnica CTU Proc 23:25–30

https://blogs.agu.org/landslideblog/2022/04/05/mount-bolu-1/, reported by Dave Petley
https://raillynews.com/, about Bolu mountain tunnel, uploaded in June 2024
https://tunnelbuilder.com/News/Bolu-Tunnel-Breaks-Through-in-Turkey.aspx. Bolu tunnel breaks through in Turkey 01/10/2005, uploaded in June 2024
Home L (2015) Hard rock TBM tunnelling in challenging ground: developments and lessons learnd from the past, TBM DiGs Singapore, 18-20 November 2015
https://www.bolugundem.com/haber/12031388/13 Mayıs 2022, uploaded in June 2024
https://www.turkiyegazetesi.com.tr/, uploaded in June 2024
Jordan D, Alpagut Y, Kılıç Ş (2023) Record-setting large diameter mixed ground tunneling in Turkey: the Eşme-Salihli railway tunnel. In: Anagnostou, Benardos and Marinos (eds) Expanding underground. knowledge and passion to make a positive impact on the world. The Author(s), ISBN 978-1-003-34803-0
Kansu O, Kılıç Ş, Coşkun B (2022) Ankara-İzmir the project of high peed tunnel, T-01 Eşme Salihli TBM tunnel
Kontogianni VA, Stiros SC (2003) Earthquakes and Seismic Faulting: Effects on Tunnels, Turkish Journal of Earth Sciences (Turkish J. Earth Sci.), Vol. 12, 2003, pp. 1–27.
Menkiti C, Mair RJ, Miles R (2001) Tunneling in complex ground conditions in Bolu, Turkey. In the Proceedings of the underground construction 2001 symposium, London, pp 546–558
Negro ED, Boscaro A, Barbero E, Irfan SB, Yertutan Y, Kansu, O (2023) Evaluation of the influence of ground conditioning on a large diameter EPB TBM: performance in the Esme Tunnel, Turkey. In: Anagnostou, Benardos and Marinos (eds) Expanding underground. knowledge and passion to make a positive impact on the world. ISBN 978-1-003-34803-0
O'Rourke TD, Goh SH, Menkiti CO, Mair RJ (2001) Highway tunnel performance during the 1999 Duzce earthquake. In: The proceedings of the fifteenth international conference on soil mechanics and geotechnical engineering, 27–31 August 2001, Istanbul, Turkey, vol 2, pp 1365–1368. Balkema
Özben M, Tokgözoğlu F, Işık (2005) S. Seismic assessment results and actual application in the complex ground, conditions of Bolu tunnels after the 1999 Duzce earthquake. In: Erdem and Solak (eds) Underground space use: analysis of the past and lessons for the future. Taylor & Francis Group, London, ISBN 04 1537 452 9
Stein RS, Barka AA, Dieterich JH (1997) Progressive failure on the North Anatolian fault since 1939 by earthquake stress triggering. Geophys J Int 128:594–604

Chapter 13
Emerging New Technologies/Innovations in Conventional and Mechanized Tunneling Methods

Abstract The market for tunnel construction is expected to show a significant groving rate with a remarkable CAGR (Compound Annual Growth Rate) of 9.5% during the forecast period between 2024 and 2031. The adoption of advanced tunnel boring machines (TBMs) has revolutionized tunnel construction, improving efficiency and safety and reducing construction time, making it a preferred solution for large-scale projects. With respect to those mentioned above, this chapter resumes the innovations made for the tunneling industry, summarizing advancements in safety, TBMs, robotics, new instrumentation, materials, and methodologies to decrease carbon footprint.

13.1 Introduction

The Tunnel construction market is projected to be valued at a CAGR (Compounded Annual Growth Rate) of 9.5% at US $330.28 Bn by 2031, exhibiting significant growth from the US $174.58 Bn achieved in 2024. The future of tunnel construction is ready to move with emerging advancements in robotics, automation, and materials science, which promise to transform traditional tunneling methods, enhancing safety, efficiency, and environmental sustainability. Future trends include the use of sustainable materials, advances in tunnel boring technology, and cutting-edge construction methods. https://www.fairfieldmarketresearch.com/report/tunnel-construction-market

The growing emphasis on sustainable construction practices, including low-carbon materials and energy-efficient TBMs, is a key trend as companies and governments aim to minimize environmental impact. Innovations in technology are also shaping the tunnel construction landscape. The launch of China Railway Group Limited's low-carbon tunnel boring machine (TBM), named "CREC 1237," marks a significant advancement. Designed for the overseas market, this TBM reduces emissions by 20% compared to conventional machines and reflects the industry's shift toward sustainable construction practices.

N. Bilgin and C. Balci, *Critical Issues in Selecting Conventional and Mechanized Tunnelling Methods*, Springer Tracts in Civil Engineering,
https://doi.org/10.1007/978-3-031-89114-4_13

The market for tunnel construction is expected to show a significant expansion rate with a remarkable CAGR (compound annual Growth Rate) of 9.5% during the forecast period between 2024 and 2031. The adoption of advanced tunnel boring machines (TBMs) has revolutionized tunnel construction, improving efficiency and safety and reducing construction time, making it a preferred solution for large-scale projects, Kunpeng et al. (2014).

With respect to those mentioned above, this chapter resumes the innovations made for the tunneling industry, summarizing advancements in safety, TBMs, robotics, new instrumentation, materials, and methodologies to decrease carbon footprint.

13.2 ITA Awards Given to Technical Equipment/Products Innovations in the Last Ten Years

Since 2015, ITA (International Tunneling and Underground Space Association) has awarded the most ambitious underground projects worldwide, as well as the latest innovations, techniques, new technologies, and new methods in tunneling. Well-known judges in their fields nominate awards. The entries received are analyzed by a panel of judges representing the different components of the tunneling community, owners, engineers, contractors, and academics from all parts of the world. The various members are changing regularly following a process that ITA describes. The authors of this book decided that it is best to start this chapter by describing and discussing the awards given by ITA in the field of technical innovations and later adding to this chapter new ones selected by the authors. https://awards.ita-aites.org

13.2.1 Technical Innovation of the Year in 2015

The award is for the "Innovative Vehicle-Mounted GPR (Ground Penetrating Radar) Technique for Fast and Efficient Monitoring of Tunnel Lining Structure Conditions."

The aging and deterioration of the tunnels' lining, support, and structures is a big problem for currently active tunnels. A new non-contact and non-destructive railway vehicle-mounted GPR with air-launched antennas was developed in China under the patronage of the Xi'an Railway Bureau to detect aging problems of tunnel linings and structural issues of tunnels at 0.9–2.25 m distance from the tunnel walls. Ninety-one tunnels with a total length of 122 km by railway vehicle-mounted equipment were tested with a significant speed.

13.2.2 Safety Initiative of the Year in 2015

The award is to the "MineARC Systems Compressed Air Management Solution."

Refuge chambers are inevitable in mines and tunnels where the geological structures are liable to geo-hazards, such as sudden collapses, water inrush, gas emissions, etc. When evacuation is no longer safe or practical in an emergency (such as a tunnel fire), a refuge chamber provides a safe and secure area for personnel to gather and await evacuation. However, they should provide breathable air, which should be durable for a long time. The Compressed Air Management System (CAMS), developed by MineARC's engineering team, improves operational safety during an emergency. It provides clean, breathable air through a four-phase filtration process. Some of the significant benefits of CAMS include optimization of mine air services, guarantee against over-pressurization of the refuge chamber, gas toxicity monitoring, and flood protection. A MineARC to be used with a refuge chamber is seen in Fig. 13.1

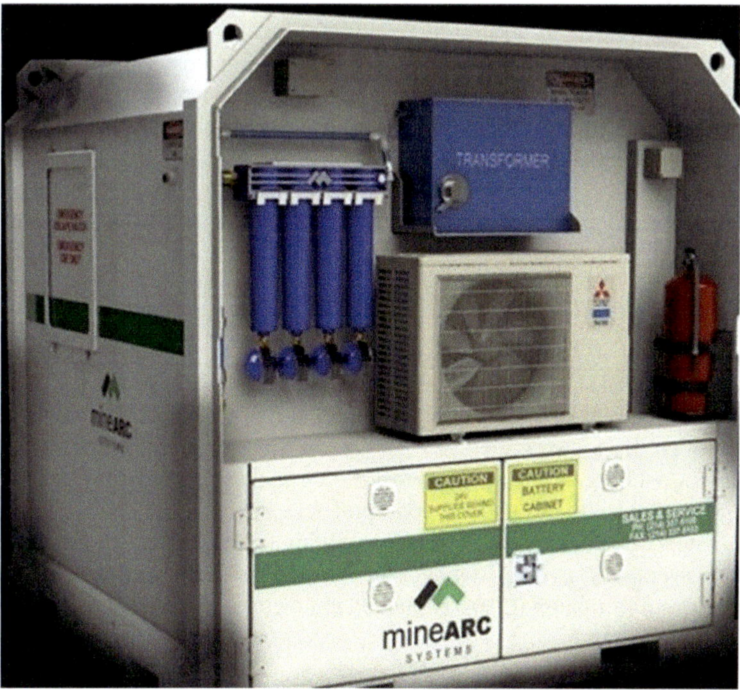

Fig. 13.1 Mine ARC Systems to be mounted into a refuge chamber, with courtesy of Mine ARC Systems

13.2.3 Technical Innovation of the Year in 2016

The award is to the "Large Diameter Shield Tunneling in Pure Sands with Hybrid EPB Shield Technology."

One of the main application areas of slurry TBMs is sands due to their high permeability and grain size. However, the Earth Pressure Balanced EPB TBM technology has advanced in recent years with improved conditioning agents and equipped with slurry injection facilities and alternative mucking methods. The fact that the excavation chamber is filled with the muck of higher density in hybrid EPB shields, compared to slurry shields, provides a higher level of safety such that in case of short-term wrong operation, the risk of sinkholes as a consequence of a face collapse is reduced since less material can enter the excavation chamber. These new so-called hybrid EPB machines have steadily increased the range of applications of the EPB technology.

The shield-driven Tunnel of metro line 4 in Rio de Janeiro (Brazil) has an approximate length of 5.2 km. The excavation was performed using a hybrid earth pressure balance shield with an excavation diameter of 11.51 m in a complex geology that included a long band of pure sand bounded by two bands of hard, highly abrasive rock. This hybrid EPB technology improved tunneling efficiency considerably, reduced materials for conditioning consumables, and reduced energy and power consumption, which is well explained by Maidl and Pierri (2014) and Maidl et al. (2015).

13.2.4 Safety Initiative of the Year in 2016

The award is to ABSIS (Activity Based Safety Improvement System).

It is a platform developed by SP Powergrid Limited that captures work activities in a video. By viewing the video footage of themselves carrying out the work, the workers, without any language barriers, can see and witness the lack of safety they subject themselves to and the good practices they achieve and can adapt. It is used efficiently in a cable tunnel project involving the construction of a 35 km long tunnel, on average 60 m deep, to house the 400 kV and 230 kV transmission cables. Crutchfield and Roughton (2014) explained the methodology.

13.2.5 Technical Product/Equipment Innovation of the Year 2017

The award is to "Strength Monitoring Using Thermal Imaging (SMUTI)."

SMUTI is an entirely novel method of monitoring sprayed concrete strength gain. It allows the strength of the whole shotcrete lining to be monitored continuously in real-time from a secure position, bringing considerable benefits to safety, quality control, and productivity. SMUTI is easier to use, safer, and quicker than existing methods, and the data is accessible, audible and traceable. This is documented in Jones and Li (2013).

13.2.6 Safety Initiative of the Year in 2017

The award is for the "Telemach Cutterhead Disc Robotic Changing System."

Changing disc cutters in hyperbaric conditions requires trained drivers and their health care. It is also costly and time-consuming. So, robotic interventions were the aim of several companies. One of them, developed by Bouyges under the trade name Telemach, has found a successful application in a Tunnel of 14.1 m diameter TBM (Supplier NHI/NFM Technologies) with 77 discs in a project in Hong Kong.

Telemach is a semi-automatic multi-purpose robotic arm installed inside of the front shield of TBM aiming to safely replace won disc cutters with the operator remaining inside the TBM cabin. During the stoppage for TBM maintenance, the arm can maneuver into the cutterhead chamber, clean, and replace worn discs with a new unit. This 10 years development project is part of the demonstration that the Robotic industry, adapted to the Tunneling environment, can contribute to a better and safer tunneling environment, Depoorter and Jeanne (2017). One of the most ambitious projects realized in Hong Kong was full with risk. The ground is poor, the Tunnel is deep up to 55 m and the water pressures are enormous up to 5.5 bar, Hansford (2017). The contractor is Bouygues subsidiary Dragages Hong Kong, which has rolled out two key robotic innovations to significantly reduce the human intervention needed in the cutter heads of EPB-TBM. A robot arm was built to manipulate the new disc-cutter assembly in hyperbaric conditions without human help. Time taken per disc exchange was 4 h with human intervention and 1.5 h with Telemach. Telemac Robot changing a disc cutter is seen in Fig. 13.2.

13.2.7 Technical Project Innovation of the Year in 2018

The award is to a "Mechanized Method with Large Section Horseshoe Shape EPB-TBM First Applied in Loess Mountain Tunnel."

Constructing mountain tunnels in Loess usually requires an advanced support system, sequential excavation method (SEM), and double lining, which is slow and lacks safety. Loess is a clastic, predominantly silt-sized sediment that is formed by the accumulation of wind-blown dust formation with density changing from 16 to 18 kN/m^3, internal friction angle changing from 27.10 to 27.50, cohesion from 20.7 to 20 kPa^{-2}. CREG has constructed the world's first-ever horseshoe shape EPB-TBM with multiple cutter heads for the mountain tunnel project in Loess, Fig. 13.3.

Fig. 13.2 Telemac Robot disc changer, with courtesy of Telemac, Bouygues

The section of the machine is 10.95 m high and 11.9 m wide and has a main drive power of 1990 kW. The construction method is successfully used in the Baicheng tunnel of Menghua Railway, which is 3345 m long and lies in the sandy new Loess. The range of the overburden is from 7 to 81 m, as Qiu et al. (2022) reported. The machine started boring on December 29, 2016. After the 1064th Ring, hard and cohesive old Loess was encountered, which prevented excavation. So, a conic soil breaker was added at the front of the TBM to ease the excavation. As of June 2017, Baicheng Tunnel has successfully advanced by 1704 m, with an average daily advance of 12 m Li (2017). Compared with single circular cutterhead EPB-TBM, the manufacturing cost of the TBM is decreased, the utilization ratio of the section is increased, and the quantity of excavation and construction material is decreased.

Fig. 13.3 Horseshoe shape EPB-TBM, with courtesy of CREG

13.2.8 Technical Product/Equipment Innovation of the Year in 2018

The award is given to "Multifunctional Energy-Storage and Luminescent Material for Sustainable and Energy-Saving Lighting (LUMA)."

Entrances and exits of road tunnels are high-risk areas for traffic accidents. The average energy cost for the lighting in road tunnels is estimated to be around 150,000 US dollars per kilometer annually. Safety and energy saving are always hot issues of road tunnel lighting. Recent studies show that the design of operation lighting for road tunnels should not only focus on the brightness and uniformity of the road but also minimize the "black hole" and "white light" phenomenon at entrances and exits, respectively. When the primary peak wavelength of the illumination light falls in the range of 490 to 570 nm, the visibility of small objects to human eyes is significantly increased. Tunnels are coated with a unique, rare earth material. Such a visual reference frame suitable for human vision helps to mitigate visual fatigue. It is Presented by Zhuang et al (2018) (Fig. 13.4).

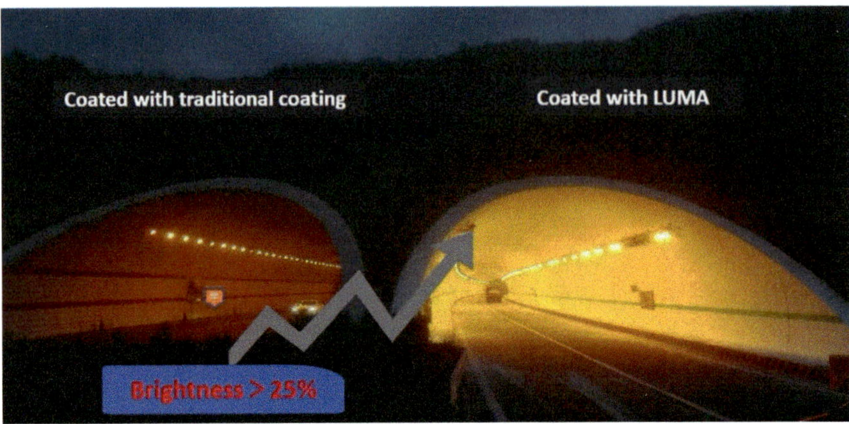

Fig. 13.4 Comparison of brightness in a tunnel coated with traditional coating material and coated with LUMA, Zhuang et al. (2018)

13.2.9 Technical Product/Equipment Innovation of the Year 2019

The award is to "Autonomous TBM."

The world's first Autonomous TBM (A-TBM) system comprises custom Artificial Intelligence control algorithms. It was developed by the in-house team, which allowed for autonomous control of the TBM. This state-of-the-art system analyses machine data in real time and assumes control of steering, advance, excavation, and slurry sub-systems with minimal human input. Deployed and proven across four distinct geological formations of the Kuala Lumpur MRT2 alignment, it has completed over 3 km of tunneling with Herrenknecht—Multimode Variable Density TBMs utilizing both EPB and slurry processes. The A-TBM has demonstrated faster response times, improved accuracy, and tangible productivity benefits, including prolonged cutting tool life and shorter cycle times, Xiong et al. (2000).

13.2.10 Technical Innovation of the Year in 2020

The award is to "An Innovative Automated Geological Forward prospecting Technique Mounted on Hard Rock TBM."

Without the identification of unknown geology in advance, tunneling with hard rock TBM may suffer from water in a rush, faults, fractured zones, face collapse, and significant deformation. One method of forward prospecting in tunnels is probe drilling, Bilgin and Ates (2016). This method is time-consuming, and there is a probability of missing geological hazards that might be encountered in front of the

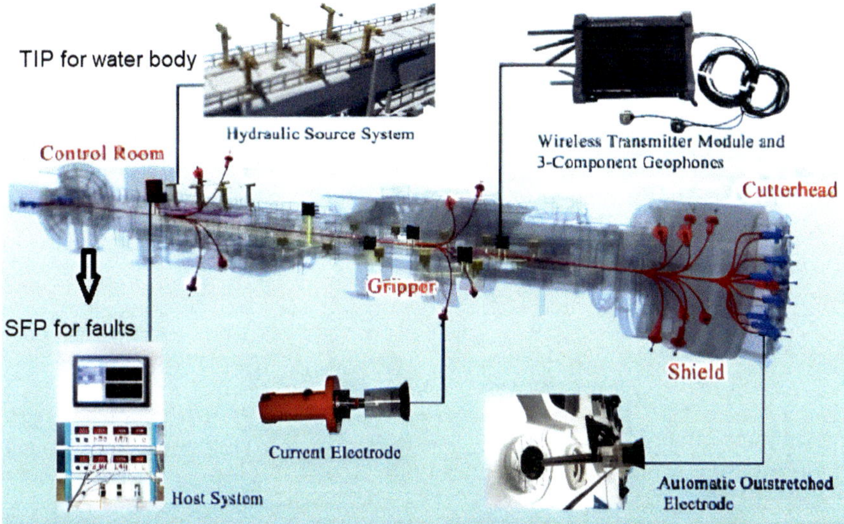

Fig. 13.5 TBM-mounted forward-prospecting instruments, after Li and Li (2018). This system was developed at Shandong University; full credit should be given to the staff who worked on this project

tunnel face. However, the geophysical prospecting developed at Shandong University proved to be a quick and efficient forward prospecting method.

For adverse geology like faults, this system was developed as an active-source seismic prospecting system mounted on TBM coupled with a real-time forward-prospecting technique using TBM drilling noise to realize accurate 3D imaging of faults and fractured zones within 120 m ahead,

Precisely, the detection system of the proposed IP (Induced Polarization technique) is for water bodies, and seismic methods using TBM disc cutter noises are all mounted on TBM. Through remote control, automatic data acquisition can be realized within 10 min. This prospecting technique has been applied in 21 engineering projects 1365 times (67.42 km in total) without missing any primary disaster sources, which played an essential role in TBM tunneling efficiency optimization and safety assurance, Li and Li (2018). TBM-mounted forward-prospecting instruments are seen in Fig. 13.5.

13.2.11 Technical Innovation of the Year in 2022

The award is "Unprecedented In-Tunnel Diameter Conversion of the Largest Hard Rock TBM in the U.S."

Changing the diameter of a TBM from 11.6 to 9.9 m inside the Tunnel requires innovative technology. That is why the award for technical Innovation of the year is

Fig. 13.6 Robbins gripper TBM of 11.6 m diameter, credit. Quell: Robbins, Willis (2021)

given to Robbins Company. The Mill Creek Drainage Relief Tunnel is 8 km long and is planned to provide flood protection for east Dallas. The project required a drainage tunnel starting at a diameter of 11.6 m and ending with a diameter of 9.9 m. The upstream 5.2 km is designed with a circular cross-section for a peak flow of 425 m^3/s. The downstream 2.8 km running between the outfall shaft to the East Peaks Branch intake was initially designed with a horseshoe cross-section to allow a higher peak flow of 566 m^3/s. The horseshoe section was to have been excavated by TBM initially and expanded by roadheader to create a flat invert. One TBM was selected instead because of the cost and time to excavate by roadheader. Robbins took on the challenge of designing a TBM with a larger diameter of 11.6 m, designed with skins and spacer segments that would work for both diameters. It could accommodate the constraints of the conversion process to the original diameter of 9.9 m. TBM used in this project is given in Fig. 13.6. The entire conversion process has taken 3–4 months, with some delays within the project duration, Willis (2021), Steve and Evan (2023).

13.2.12 Product/Equipment Innovation of the Year 2022

The award is "Integration of Robotics into the construction works of the Chuquicamata underground mining site."

In the modern world, human life has the highest value, and therefore, workers are replaced by robots at hazardous production facilities. Using robots for underground mining allows for improved employee safety, performance of exploration (e.g., mapping), and monitoring work in potentially dangerous areas, access to which is restricted to human safety regulations, and carrying out rescue operations in an emergency event. Using robotic systems for open mining is already well developed; there are many mass-produced devices, such as a loading–unloading machine Sandvik LH621, Plotnikov et al. (2020).

The second award given to Robots by ITA is SPOT. Since November 2021, Acciona has been testing the SPOT quadruped robot (Boston Dynamics' Spot) in the underground mining site in Chile, aiming to reduce risk for human beings while increasing process control and productivity. The Robot is currently performing the following tasks: high precision scanning for section control, high precision scanning after applying shotcrete to perform quality control by measuring the thickness, thermal monitoring of the shotcrete to estimate setting level and early mechanical resistance, image acquisition of the tunnel face that allows generating a geologic report and robotic exploration of the Tunnel after blasting, and to identify misfired explosives. The Robot has demonstrated the ability to move through the aggressive environment of a mine with unstable rocks, water, and mud without many problems, Crespo and Rodríguez (2023). The robot developed by Boston Dynamics is seen in Fig. 13.7

13.2.13 Technical Innovation of the Year 2023

The award is to "Building Blocks in a Foundation Pit-Prefabrication and Assembly Construction Technology for Metro Stations."

"Prefabrication and Assembly Construction Technology for Metro Stations" is a new construction method for underground metro stations. It uses precast concrete structural components and assembles them like "building blocks" on-site in a foundation pit to quickly assemble into tunnel structures, minimizing on-site construction activities.

This innovative technology was initially developed to solve the problem of the impossibility of constructing open-excavated metro stations in the severely cold winter of Changchun by using industrialized construction methods. Still, actual project experience using this technology has demonstrated that it also has outstanding advantages in various other environmental conditions. This technology was first researched and developed in Changchun, China. Thus far, six prefabricated metro stations on Changchun Metro in China have been constructed using this new technology, and it is expected to be applied and gain popularity in other cities. Yang and Lin (2021).

Fig. 13.7 A robot developed by Boston dynamics, Courtesy of Boston Dynamics, Acciona

13.2.14 Product/Equipment Innovation of the Year in 2023

The award is "Hard Rock Shield + Earth Pressure + Slurry three-mode TBM three-mode-TBM."

The section of Luogang station, Guangzhou Rail Transit, is 1086.5 m long, with 19.73 m overburden and 28‰ maximum gradient, with the Tunnel's outer diameter of 6 m. The geology is complex and variable. The TBM excavates mainly through granite, residual soil, boulders, upper-soft and lower-hard rocks, and full-section hard rocks, with a high strength of up to 140 MPa. During the construction, the shield continuously bored under important buildings and structures such as the airport high-pressure oil pipeline, tram line 1, and existing operating subway tunnels.

It is predicted that using dual-mode TBM might cause surface settlement, damage buildings, and structures, and make it incapable of effectively controlling the overall risk during construction. The newly developed Hard Rock Shield + Slurry + Earth Pressure Three-Mode TBM could effectively reduce risk and prevent the settlement of the earth's surface and surrounding Structure during tunneling. The Structure of the three-mode shield machine, courtesy of CREG, is seen in Fig. 13.8

CREC820 tri-mode shield machine was launched from Luogang Station on May 18, 2021; after 235 days, it safely and smoothly passed through the soft soil layer,

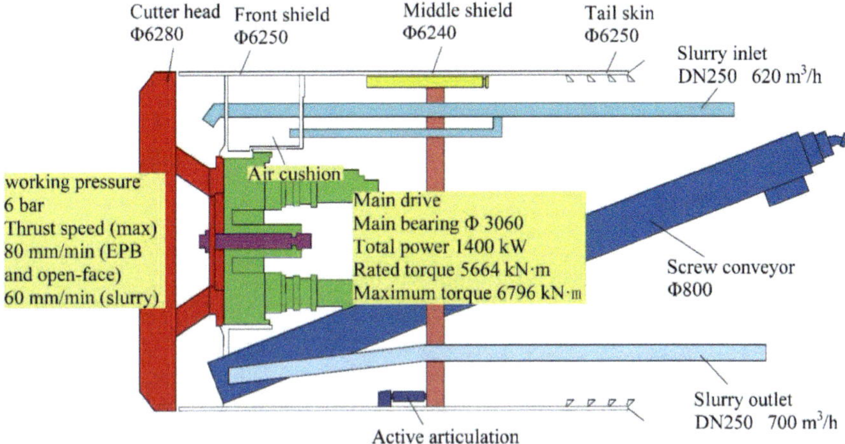

Fig. 13.8 The structure of the three-mode shield machine, with courtesy of CREG

upper soft and lower hard formation, full-section hard rock formation, district government office hall, University Building, and other sensitive structures. It was successfully arrived at Shuixi Station on January 08, 2022. The mean advance rate in the soil was 4.2 rings per day, the mean daily advance rate in hard rock was 6.2 rings, and the highest advance rate reached 11 rings per day, Zhong and Zhu (2023).

13.2.15 Technical Innovation of the Year in 2024, Gold Award

The Gold award was to the "Disc Cutter Wear Sensor Package and TBM Monitoring System."

Disc wear monitoring systems is a topic that is most investigated and published. Some of them are Mosavat (2017), Lan et al. (2019), Han et al. (2023), and Akhlaghi et al. (2024). We could not differentiate the award content from the ITA web page and previously published works. However, it is mentioned as: "*Newly developed technology that maximizes excavation efficiency by remotely measuring disc cutter wear using a package that consists of magnetic sensors, hall sensors, a power supply battery, a wireless communication module, an Arduino board, and an external casing. As the disc cutter ring wears, the relative distance between the Ring and the magnetic sensor increases. This affects the magnetic flux density, which can be detected by the sensor, thereby obtaining voltage measurements. Using these voltage values from the sensor, the relative distance of the disc cutter ring can be calculated. A deep learning-based algorithm for assessing wear conditions and predicting the remaining cutter life has been developed, which can be provided to users through the monitoring system. The Innovation lies in measuring the wear of disc cutters through sensors instead of relying on man-entry interventions, allowing for data-based decisions on when to replace the disc cutter.*"

13.2.16 Technical Innovation of the Year in 2024, Silver Award

The silver award was to "MIRET-Tunnel AI" in 2024.

MIRET, Management, and Identification of the Risk for Existing Tunnels is a methodology and a series of tools for planning and managing infrastructures and tunnels, aiming to digitally transform the whole asset management and life cycle, from element inspection to decision-making. MIRET is the approach to analyzing the elements focusing on integrated workflow to connect survey inspection data to geology, digitization, diagnostic, and design. This approach can be defined through the following milestone: Survey and Inspection (SI), Infrastructure and Slope Digitalization (DI), Priorities Analysis (PA), Planning and Design (PD), Works and Maintenance (WM), Monitoring (MO). Knowing and predicting the conditions of the infrastructure allows the adoption of new planning and intervention strategies. Digitization and data engineering become information to support the decision maker; this is the purpose of MIRET, the innovative methodology for diagnostics of existing tunnels through multidimensional mobile mapping systems and a new approach for risk management and identification supported by artificial intelligence allows the adoption of a non-invasive, efficient, environmentally friendly approach. Minet is a product of ETS Engineering Company.

13.2.17 Technical Innovation of the Year in 2024, Bronze Award

Bronze Award was to the Advanced Tunneling Assistance System in 2024.

The Advanced Tunneling Assistance System (ATAS), developed in Italy, monitors and implements real-time TBM data analysis using Al/machine learning and estimates the variation of geological conditions at the excavation face. The system has been applied for the first time on one of four 8.78 m diameter TBM drives excavated for a rail node development project in the center of the very congested city of Łódź in Poland. The ATAS uses existing data to optimize TBM excavation and reduce excavation risks.

13.2.18 Product/Equipment Innovation of the Year in 2024, Gold Award

Gold Award was to “Ultra-small Turning Radius Hard Rock TBM”.

The horizontal turning radius of a conventional TBM is about 300–500 m. However, the 4 km route of the 3.53 m diameter drainage corridors from the first to the fourth layer for the Jinyun Pumped Storage Power Station in Zhejiang Province required a TBM with a turning radius of less than 30 m, about 10 times the tunnel diameter. A TBM was designed and developed by CREG, China Railway Engineering Equipment Group, and was launched in November 2021. After boring 22 sections with a turning radius of 30 m, it achieved maximum monthly advance rates of 660.5 m and a maximum daily advance rate of 38.38 m. In December 2022, the TBM successfully broke through, completing the whole route six months ahead of schedule.

13.2.19 Product/Equipment Innovation of the Year in 2024, Silver Award

The silver award is to “SOGUN: Geometric Control System for shotcreting and other tunneling works.”

The R&D Department of Spanish contractor Dragados developed it. The SOGUN system represents a significant advancement in the execution cycle of the shotcrete in tunnels. SOGUN is a compact unit whose functionality is automated in a process that scans the Tunnel in 3D, generating a surface point cloud. This data is then compared to the pre-loaded theoretical tunnel surface, and an image is projected on the tunnel wall, indicating areas and amounts of excess or missing concrete thickness. This process is completed in under a minute, far quicker than the traditional spray painting method, which typically takes one and a half to three hours and requires at least two people, a man lift, and a dedicated total station. Shotcrete robot operators clearly understand the work to be performed. SOGUN enhances productivity, reduces costs, and minimizes workers’ risks, introducing superior quality finishes and safety. It also promotes sustainability by optimizing material usage and reducing waste, leading to a lower environmental footprint.

13.2.20 Product/Equipment Innovation of the Year in 2024, Bronze Award

The bronze award is to “Advanced Hybrid Wireless Sensor Network (H-WSN) System for Safety Risk Sensing in Tunnel and Underground Engineering.”

The advanced Hybrid Wireless Sensor Network (H-WSN) System developed in China has systematically and structurally defined the next generation of WSN monitoring for safety risk sensing in underground construction. It is reported that since March 2021, the H-WSN has served 1200 commercial projects in 20 countries and is credited with avoiding 300 major accidents by timely warnings in the real world. Extensive collaboration with international university researchers and industrial colleagues has generated significant economic, social, and environmental benefits for the Tunnel and underground engineering community.

13.3 Rectangular TBMs, for Soft Ground

The face cut by a rectangular TBM enables the best use of shallow overburden and reduced construction costs. It is mainly used for pedestrian passages, traffic tunnels, underground utilities, underground parking lots, and underground reservoirs. A rectangular TBM is an efficient and convenient solution to urban traffic congestion.

Currently conventional circular TBMs cannot completely meet the requirements of underground space exploitation regarding the cross-section and space-utilization ratio. However, non-circular TBMs, the tunneling equipment for an ideal cross-section, have become the new market growth point. Relevant engineering practice shows that non-circular TBMs with customized design and manufacture have significant advantages regarding construction schedule, settlement control, and space utilization. Li (2017), Sun et al. (2023). The results indicate that rectangular tunnels are suitable for shallow depths in weak ground as they have lesser settlement than circular tunnels. This is crucial for tunneling beneath structures such as railway lines and existing roads, Vinod and Khabbaz (2019).

Traditional rectangular tunnel boring machines have low tunneling efficiency and poor soil mixing effects, hindering the construction of long-distance and large-section box jacking projects. A new type of full-face excavation boring machine was developed to overcome these limitations based on a planetary transmission mechanism. This machine achieves full-face excavation by utilizing three eccentric cutter heads that revolve around the cutter plate's central axis and their shafts. Furthermore, a case study evaluated the feasibility of the RTBM in a challenging, shallow-buried project beneath National Highway No. 6 in Hitachinaka City, Japan. In conclusion, the planetary transmission mechanism-based RTBM offers a promising advancement for box-jacking projects, particularly in urban environments with strict settlement requirements. However, further research and development are needed to refine the machine's scalability, cost-effectiveness, and adaptability to a broader range of geotechnical conditions Ma et al. (2024). Table 13.1 summarizes a. detailed survey on the manufactured fifteen rectangular TBMs from 2001 to 2023, with size changing between 882 × 882 mm and 7400 × 4600 mm, with overburden ratio from 0.25 D to 21.d mm in sandy and silty soil, gravel sand/clay, excavation rate changed from 15 to 35 mm/min.

Table 13.1 A summary of rectangular/square shaped TBMs manufactured beyween 2002 and 2023, with kind permission of Dr. Yahong Zhao, the corresponding author of the refernce Ma et al. (2024)

Year	RTBMs	Size/mm	Jacking distance/m	Buried depth ratio/m	Soil condition (*N*-value)	Ground average settlement/mm	Excavation speed/mm/min
2002		882 × 882	972.4	18.71 m 21.21 D	Sand and silty N: 0–18	0.6	20–25
2008		2800 × 2520	36	1.92 m 0.76 D	Sandy soil N: 0–8	2	15
2010		3300 × 2500	220.97	0.97 m 0.39 D	Sandy soil N: 0–13	16	25
2012		3600 × 3600	57.00	2.68 m 0.74 D	Silty soil N: 0–10	8	15
2013		3000 × 3180	50.08	1.37 m 0.43 D	Clay soil N: 2–13	3	20
2014		2390 × 3310	29.57	2.78 m 0.84 D	Sand soil N: 1–7	5	25
2017		3000 × 3000	224.78	2.50 m 0.83 D	Silty sand N: 2–8	7	30
2017		5000 × 6300	32.00	1.60 m 0.25 D	Gravel sand N: 5–20	13	10
2017		4900 × 6150	40.50	1.78 m 0.29 D	Gravel clay N: 10	10	35
2019	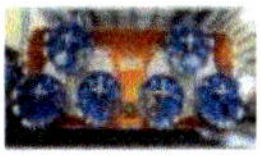	4700 × 2200	146.16	1.56 m 0.71 D	Sand soil N: 2	17	16

(continued)

Table 13.1 (continued)

Year	RTBMs	Size/mm	Jacking distance/m	Buried depth ratio/m	Soil condition (*N*-value)	Ground average settlement/mm	Excavation speed/mm/min
2021		2360 × 1600	281.14	4.00 m 2.50 D	Clay and gravel N: 3–68	10	20
2022		7400 × 4600	122.20	7.50 m 1.63 D	Sand clay N: 0–2	8	15
2021		5300 × 2900	19.00	0.30 m 0.10 D	Gravel sand N: 5–28	3	35
2023		4000 × 3900	121.33	2.88 m 0.74 D	Gravel sand N: 5–28	13	20
2023		3500 × 2000	10.80	3.17 m 1.59 D	Rhyolite/ Soft rock N: 50/ 27–50/15	5	25

Rectangular TBMs are mainly built in China by CREG and in Japan by Ugitec. In 2021, Kawasaki Heavy Industries Ltd and Hitachi Zozen Corporation integrated the TBM business and newly established it as "Underground Infrastructure Technologies Corporation" or Ugitec. Figure 13.9a–c show some of the machines built by this corporation.

Rectangular TBM has already been successfully used in China. On December 25th, 2015, the world's largest-face rectangular TBM by CREG (10.42 m × 7.55 m) was manufactured in the assembly line in Dalian, China, and bored the underpass tunnel across Heiniucheng Road in Tianjin Metro Line 11 Project. Its size breaks the record set by a previous CREG-made 10.12 × 7.27 m rectangular TBM, which bored the underpass across Zhengzhou Zhongzhou Avenue in China.

13.4 Rectangular TBM for Hard Rock

Although several rectangular TBMs were built in the market, we came across only one hard rock rectangular. TBM is manufactured and tested in the field. The rectangular TBM, known as the Robbins MDM5000, with 5.0 m × 4.5 m dimensions, can

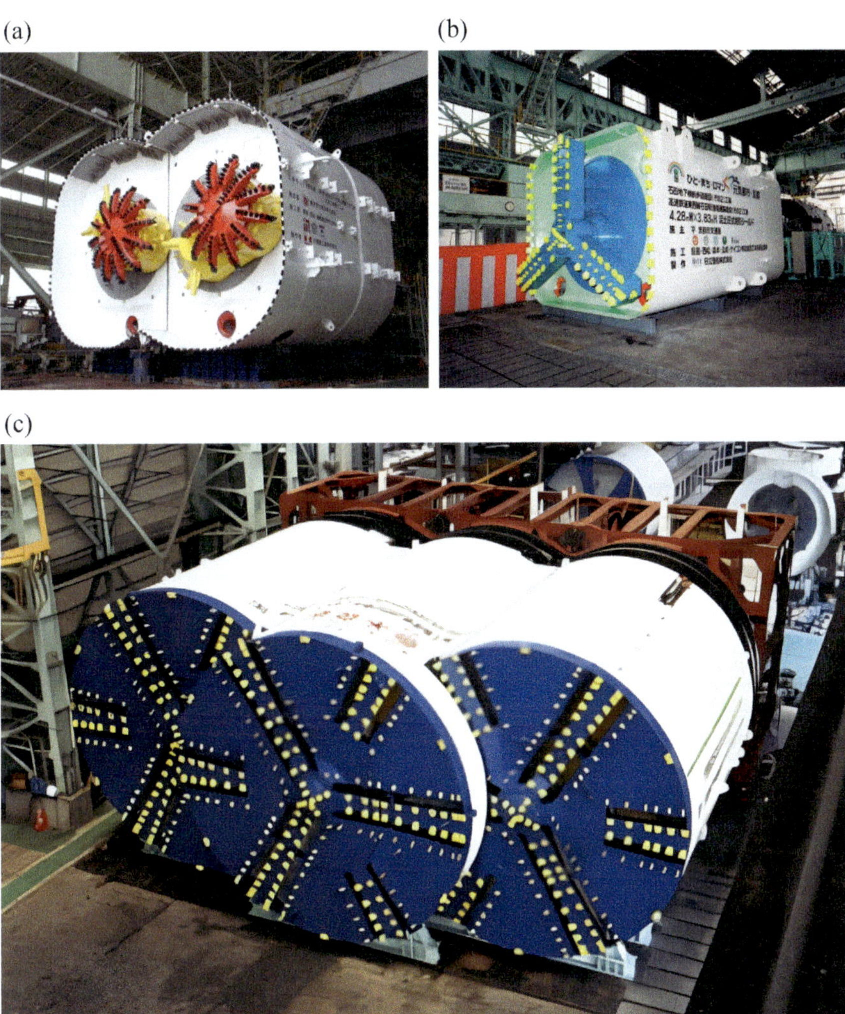

Fig. 13.9 **a** A rectangular TBM built by Ugitec, photograph by the kind permission of Ugitec. **b** A rectangular TBM built by Ugitec, photograph by the kind permission of Ugitec. **c** A TBM with three cutting head built by Ugitec, photograph by kind permission of Ugitec

excavate a flat tunnel invert for immediate use by rubber-tired vehicles. Employed by Mexico's oldest silver mine, Fresnillo Plc, TBM bored a mine access tunnel in andesite and shale with quartz intrusions. The technology is not only beneficial for the mining industry but also for many applications in civil tunneling. While other machines have been developed for soft ground, this machine represents the first successful advance into rectangular hard rock tunneling. In a traditional circular tunnel, the invert is filled, but with the MDM, 30% less rock is required to be removed from the profile. With more than 1700 m of advance until September 2021, at rates

Fig. 13.10 Rectangular hard rock TBM developed by Robbins, Credit/Quelle: Robbins

up to 52 m in one week and 191 m in one month, the MDM is significantly faster than drill & blast excavation. However, the machine has lower productivity due to the cyclic, swinging cutting action compared to a traditional, rotary TBM that cuts a circular profile. With the early success came some overall observations due to the cyclic swinging cutting action. While there are still many benefits, mainly a flat roadbed, the comparisons must be carefully evaluated for any significant mine excavation considering this method, Ofiara and Lewis (2022), Robbins MDM5000 is illustrated in Fig. 13.10.

13.5 TBMs Developed for Cross-Passages (CPs)

Cross passages are required to connect twin tunnels for the tunnel fire safety strategy to facilitate self-rescue and intervention. The structure is usually constructed by excavating through an opening created by dismantling segments through opening segments of the completed bored tunnel section. Construction can be technically challenging and expensive. However, it is legally required, i.e., according to NFPA (US Agency for National Fire Protection), the distance between CPs should be less than 244 m. For example, in the Gayrettepe-Istanbul Airport Metro Project, 104 cross passages were opened with a total of 47,925 m^3 excavated material, Bilgin (2024) (Figs. 13.11, 13.12 and 13.13).

In most cases, the construction of cross passages is one of the most critical steps in the overall tunnel construction program. For mechanized cross-passage execution, a clear design is the basis for precise project preparation to achieve a fast and successful

Fig. 13.11 Typical cross-passage, after Bilgin (2024)

Fig. 13.12 Breaking segments for a cross passage, with impact hammers, if the rock is hard controlled blasting is applied, after Bilgin (2024)

Fig. 13.13 The final cross-section of a cross passage in Gayrettepe Istanbul New Airport Metro Project, after Bilgin (2024)

execution. Similar to TBMs used in boring tunnels, various types of cross-passage TBMs are available, which need to be selected to suit the hydrogeology, geology, and project-specific requirements, Schmaeh (2017). Cross-passage TBMs are shield machine types or pipe jacking machine types. They can excavate cross passages with different lengths and curve radii in main tunnels with different diameters. They are mainly used for soft ground, such as clay, mud, silty clay, and silt. Some field results are as follows: (a) It took 18 days to break through the 17 m long cross passage Ningbo Rail Transit Line 3 in China with shield type TBM on 29th December 2017. (b) It took 16 days to break through the 14 m long cross passage of Wuxi Metro Line 3 in China with a pipe jacking type TBM on 19th October 2019. (c) On 19th October 2019, with 10.8 m clear spacing, the cross passage in Qingdao Metro Line 8 TBM was launched and took 17 days to break through. This project is the first submarine cross-passage constructed mechanized in China. (d) On 11th February 2022, the cross passage TBM was launched, taking 12 days to break through. This project is the first cross passage excavated by the first dual-mode cross passage TBM (pipe jacking machine type) in Hangzhou Airport High-Speed Railway China, which overcame the construction difficulties of tunneling in large overburden and high-pressure underground water. A typical schematic diagram of a cross-passage TBM is seen in Fig. 13.14. https://tunnelingonline.com/cross-passage-construction-with-creg-tbms/.

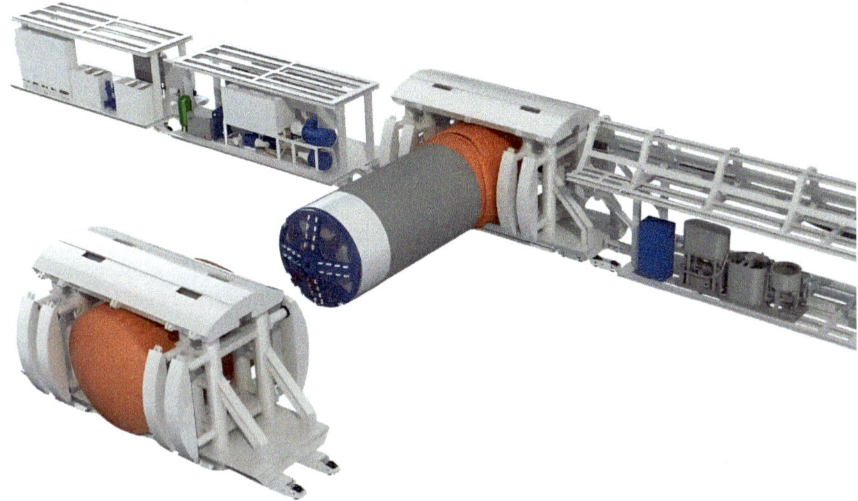

Fig. 13.14 A schematic diagram of a cross passage TBM is seen, Courtesy of CREG

13.6 Innovation on Low Carbon Concrete and Grout for Tunneling

Climate change is affecting the globe, causing extreme weather events such as flooding, extreme heat waves, heavy downpours, and rapidly changing climate conditions. The Intergovernmental Panel on Climate Change (IPCC) and the 2015 Paris Agreement limited global warming to 1.5 °C to achieve carbon neutrality. It is essential to understand where carbon dioxide and other greenhouse gas emissions come from so that innovative actions can be taken to reduce emissions. To limit global warming to 1.5 °C, The European Climate Law established a target to attain climate neutrality for the region by 2050 and a corresponding goal to decrease detrimental emissions by a minimum of 50% by 2030.

A carbon footprint is defined as the total amount of greenhouse gases emitted into the atmosphere, such as carbon dioxide (CO_2), methane (CH_4), nitro oxide (N_2O), etc. expressed in equivalent tons of CO_2. The carbon footprint is a valuable tool for measuring the contribution to climate change by individuals, organizations, products and services, and more. Much research has been done on lowering the construction industry's carbon footprint, focusing on alternative binders and supplementary cementation materials (SCMs). The utilization of machinery and equipment, along with emissions from construction vehicles, adds to the overall carbon footprint of concrete construction. Between 1990 and 2020, the global production of concrete and mortar reached an estimated annual volume of around 26 gigatons in 2020. Approximately the yearly demand for concrete is four gigatons of cement utilized in 2020. Traditional concrete production significantly contributes to CO_2 emissions due to the high energy required to produce cement, the primary binding agent in concrete etc. The literature has reported that approximately 0.9 tons of CO_2 are emitted while producing 1 ton of cement. Nearly 3.4 gigatons of CO_2 were emitted from the manufacturing of cement and concrete in the year 2020, Althoey et al. (2023).

The primary CO_2 generator found in the tunnel construction process is the lining element; the percentage varies between 50% in tunnels with smaller diameters (4–5 m) and 75% for tunnels with larger diameters (9–10 m), followed by the auxiliary elements, 16%, and the operation of the tunnel boring machine itself, 11.2%, while the other parts remain in a range between 1.3 and 5.7%, Rodriguez et al. (2024).

Sposetti (2021) reported that in a standard grout in the tunnel, the average quantity of CO_2 emissions stands at around 300 kg of CO_2 for each m^3 of grout injected into the ground, with the large majority of CO_2 emissions coming from cement. Master Builder Solutions plans to reduce its carbon footprint by around 67% at zero cost while increasing compressive strength to 3.0 MPa.

Balancing the growing demands for concrete production with environmental sustainability is essential. It calls for adopting innovative approaches to reduce the carbon footprint and enhance the efficiency of concrete manufacturing processes using mix design optimizations, nanotechnology, and carbon capture techniques, from innovative materials like clinker replacements, supplementary materials, and cement-less innovative binders to the latest. Limestone calcined clay cement (LC3) is a type of cement that incorporates calcined clay and limestone as its primary constituents. It is an environmentally friendly alternative to traditional Portland cement, as it reduces the carbon footprint associated with cement production. Utilizing SCMs (Supplementary CementitiousMaterials) instead of a portion of OPC (Ordinary Portland Cement) stands out as the most efficient method to simultaneously reduce the carbon footprint of concrete construction while maintaining elevated levels of strength and durability, Althoey et al. (2023).

A more comprehensive innovation to reduce carbon footprint when producing cement was realized for The Fehmarnbelt project, consisting of a 27 km long immersed tube tunnel linking Denmark and Germany. Aalborg Portland has developed a type of cement known as Futurecem, which can reduce the CO_2 footprint from concrete production by 25 percent compared to traditional concrete. In this cement, approximately 50 percent of cement clinker has been replaced by a unique combination of clay and limestone. This process significantly reduces the CO_2 footprint compared to the types of cement typically used in Denmark. This trial is essential to demonstrate how the concrete performs in a real traffic environment with frost and salting. If this project achieves the expected results, it will open up the possibility for the new concrete, with a reduced CO_2 footprint, to be used in future construction projects, Thomas (2024).

13.7 Concluding Remarks

Selection and summarizing the latest innovative technological developments of the last 10 years in tunneling is not easy. However, thanks to ITA, we formulated a solution to the problem. Since 2015, ITA has awarded the most ambitious underground projects worldwide, as well as the latest innovations, techniques, new technologies, and new methods in tunneling. First, we did a literature survey on the 19 awards given in the field of innovation on the latest technologies or products. We tried to understand the philosophy behind the selections of the awards. Apart from the prefabrication of metro stations in difficult weathering conditions, we grouped them according to the field of application, and later, we selected new ones that would be future nominations on the latest new technologies.

The first group, comprising seven awards, went to the innovation on safety. These are, MineARC systems compressed air management solution, which is used in refuge chambers; ABSIS activity-based safety improvement system; Telemach cutterhead disc robotic changing system especially in hyperbaric conditions; LUMA multifunctional energy-storage and luminescent material for sustainable and energy-saving lighting" to increase safety and increase brightness and uniformity of the road minimizing the "black hole" and "white light" phenomenon at entrances and exits; integration of robotics into the construction works of the Chuquicamata underground mining site; MIRET identification of the risk for existing tunnels; a new approach for risk management and identification supported by artificial intelligence; an advanced hybrid wireless sensor network (H-WSN) system for safety risk sensing in tunnels.

The second group, comprising six awards, went to the innovation on TBMS. They are, large diameter shield tunneling in pure sands with hybrid EPB Shield Technology, mechanized method with large section horseshoe shape EPB-TBM First Applied in Loess Mountain Tunnel; "Autonomous TBM," which comprises custom Artificial Intelligence control algorithms; unprecedented in-tunnel diameter conversion of the largest hard rock TBM in the U.S; Hard rock shield + earth pressure + slurry three-mode TBM; Ultra-small Turning Radius Hard Rock TBM.

A third group, comprising five awards, went to the innovation on in situ testing of aging of lining, strength of shotcrete, geological forward prospecting, and monitoring disc wear. Such as: the innovative vehicle-mounted GPR (ground penetrating radar) technique for fast and efficient monitoring of tunnel lining structure conditions; strength monitoring of shotcrete using thermal imaging (SMUTI); an innovative automated geological forward prospecting technique mounted on hard rock TBM; disc cutter wear sensor package and TBM monitoring system; SOGUN: geometric control system for shotcreting and other tunneling works.

Plus to awards given by ITA, we found worth mentioning that innovation on the planetary transmission mechanism-based rectangular TBMs offers a promising advancement for box-jacking projects, particularly in urban environments with strict settlement requirements; the technological development on rectangular hard rock TBM is not only beneficial for the mining industry but also for many applications in civil tunneling; Similar to TBMs used in boring tunnels, various types of innovative cross-passage TBMs are available which increase safety and decrease the time of construction. Finally, it is worth mentioning that innovative works to decrease the carbon footprint in the manufacturing process of cement. We cordially support the academic and technological efforts that target the decrease in the carbon footprint in the Fehmarnbelt tunnel project using innovative cement technology.

References

Akhlaghi MA, Bagherpour R, Hoseinie, SH (2024) Real-time monitoring of disc cutter wear in tunnel boring machines: a sound and vibration sensor-based approach with machine learning technique. J Rock Mech Geotech Eng. Available online August 24

Althoey F, Ansari WS, Sufian M, Deifalla AF (2023) Advancements in low-carbon concrete as a construction material for the sustainable built environment. Dev Built Environ 16(2023):100284, p 20

Bilgin N (2024) Critical Issues when opening cross passages in complex geologies, the need of consolidation grouting, and some examples from Turkey and International projects, planning, execution and quality control of geotechnical injection during TBM excavation, 3rd International workshop, VersuchsStollen Hagerbach AG, Switzerland

Bilgin N, Ates U (2016) Probe drilling ahead of two TBMs in difficult ground conditions in Turkey. Rock Mech Rock Eng. https://doi.org/10.1007/s00603-016-0937-9. First online: March 02, 2016

Crespo C, Rodríguez F (2023) Integration of robotics in underground mining construction works. In: Anagnostou, Benardos and Marinos (eds) Expanding underground—knowledge and passion to make a positive impact on the world. The Author(s), ISBN 978-1-003-34803-0. Open Access: www.taylorfrancis.com, CC BY-NC-ND 4.0 license

Crutchfield N, Roughton J (2014) Developing an activity-based safety system (Chapter 11). In: Safety culture an innovative leadership approach, pp 213–233. ISBN 978-0-12-396496-0. https://doi.org/10.1016/C2011-0-07188-0

Depoorter L, Jeanne L (2017) Cutterhead disc robotic changing system "Telemac". Presentation for ITA Awards 2017, ITA-AITES, Paris– 15 November 2017

Han J, Xiang H, Feng Q, He J, Li R, Zhao W (2023) Detection for disc cutter wear of TBM using magnetic force. Machines 11(3):388. https://doi.org/10.3390/machines11030388

Hansford M (2017) Rise of the machines. The cutting edge, New Civil Engineer, May 09

https://awards.ita-aites.org

https://tunnelingonline.com/cross-passage-construction-with-creg-tbms/

Jones BD, Li S (2013) Strength monitoring using thermal imaging. Tunnell J (Dec 2013/Jan 2014):40–43

Kunpeng C, Lianhu J, Yuefeng Y, Xiaobo W, Yali F (2014) Exploration and application of green TBM. In: Tunneling 2024, symposium organized by Turkish Tunneling Society, 12–13 December, Istanbul, p 10

Lan H, Xia Y, Ji Z, Fu J, Miao B (2019) Online monitoring device of disc cutter wear, design and field test. Tunnell Undergr Space Technol 89:284–294

Li J (2017) Key technologies and applications of the design and manufacturing of non-circular TBMs. Engineering 3(2017):905–914

Li S, Li Y (2018) An ınnovative hard-rock TBM-mounted system for geological forward-prospecting. Shandong University, China, Chuzhou-Nanjing 7th November 2018, ITA tunneling awards 2018

Ma P, Zhao Y, Shimada S, Zhou H (2024) Effectiveness analysis of a novel rectangular tunnel boring machine with planetary transmission for box jacking. Sci Rep 14:27059. https://doi.org/10.1038/s41598-024-78206-8

Maidl U, Pierri J (2014) Innovative hybrid EPB tunneling in Rio de Janeiro. In: Tunnels for a better life, WTC 2014, Brazil

Maidl U, Henrique C, Comulada M, Mahfuz A, Coutinho A (2015) First experiences gained with the hybrid EPB technology in Rio de Janeiro sands. In: SEE Tunnel: promoting tunnelling in SEE region, May 22–8, 2015, Dubrovnik, Croatia: ITA WTC 2015

Mosavat K (2017) A smart disc cutter monitoring system using cutter ınstrumentation technology. In: RETC (Rapid excavation Tunneling Conference) 2017, San Diago, California, USA

Ofiara D, Lewis M (2022) A novel, non-circular tunnel boring machine for underground mine development. In: ITA-AITES world tunnel congress, WTC2022 and 48th General Assembly Bella Center, Copenhagen 2–8 September 2022

Plotnikov NS, Kolokoltseva EU, Volkova YV (2020) Technical review of robotic complexes for underground mining. In: International science and technology conference, Earth Science, IOP Conference Series: Earth and Environmental Science 459(2020):042025. IOP Publishing. https://doi.org/10.1088/1755-1315/459/4/042025.

Qiu W, Ai X, Zheng YC (2022) First application of mechanized method using earth pressure balance TBM with large horseshoe-shaped cross section to loess mountain tunnel: a case study of Baicheng tunnel. Tunnell Undergr Space Technol 126:104547

Rodriguez R, Bascompta M, Garcia H (2024) Carbon footprint evaluation in tunnels excavated in rock using tunnel boring machine (TBM). Int J Civ Eng 22:995–1009. https://doi.org/10.1007/s40999-023-00935-0

Schmaeh P (2017) Mechanized solutions for cross passage construction. In: Proceedings of the world tunnel congress 2017—surface challenges—underground solutions. Bergen, Norway

Sposetti M (2021) Environmentally friendly solutions for the TBM tunneling industry. In: Tunnel Turkey 2021 ınternational tunneling symposium organized by Turkish tunneling society, Istanbul, 21–22 December 2021

Steve C, Evan B (2023) Unprecedented in-tunnel diameter conversion of the largest hard rock TBM in the United States. Min Eng 75(3):13

Sun W, Han F, Liu H, Zhang W, Zhang Y, Su W, Liu S (2023) Determination of minimum overburden depth for underwater shield tunnel in sands: comparison between circular and rectangular tunnels. J Rock Mech Geotechn Eng 15(2023):1671–1686

Thomas T (2024) Fehmarnbelt tests CO_2-reduced concrete of the future. Tunnel J, 10 December 2024

Vinod M, Khabbaz H (2019) Comparison of rectangular and circular bored twin tunnels in weak ground. Underground Space 4(2019):328–339

Willis D (2021) Diameter change—USA's largest hard rock TBM undergoes planned in-tunnel conversion, Tunnel 03/2021

Xiong J, Shen LKS, Batty RJ, Ho JCJ (2000) The pursuit of an autonomous tunnel boring machine, WTC2000, Kuala Lumpur, Malaysia

Yang X, Lin F (2021) Prefabrication technology for underground metro station structure. Tunn Undergr Space Technol 108(6):103717. https://doi.org/10.1016/j.tust.2020.103717

Zhong CP, Zhu WB (2023) Innovation and application of triple-mode shield machine/TBM. In: Anagnostou, Benardos and Marinos (eds) Expanding underground—knowledge and passion to make a positive impact on the world. The Author(s), ISBN 978-1-003-34803-0 Open Access: www.taylorfrancis.com, CC BY-NC-ND 4.0 license

Zhuang X, Zhongyi A, Feng S (2018) Multifunctional energy-storage and luminescent material for sustainable and energy-saving lighting for tunnels (LUMA), Chuzhou-Nanjing 7th November 2018, presented for ITA tunnelling awards for 2018